WITHDRAWN

MANAGEMENT OF ADVANCED MANUFACTURING TECHNOLOGY

ETM WILEY SERIES IN ENGINEERING & TECHNOLOGY MANAGEMENT

Series Editor: Dundar F. Kocaoglu

PROJECT MANAGEMENT IN MANUFACTURING AND HIGH TECHNOLOGY OPERATIONS
Adedeji B. Badiru, University of Oklahoma

MANAGERIAL DECISIONS UNDER UNCERTAINTY: AN INTRODUCTION TO THE ANALYSIS OF DECISION MAKING
Bruce F. Baird, University of Utah

INTEGRATING INNOVATION AND TECHNOLOGY MANAGEMENT
Johnson A. Edosomwan, IBM Corporation

CASES IN ENGINEERING ECONOMY
Ted Eschenbach, University of Missouri—Rolla

MANAGEMENT OF ADVANCED MANUFACTURING TECHNOLOGY: STRATEGY, ORGANIZATION, AND INNOVATION
Donald Gerwin, Carleton University, Ottawa
Harvey Kolodny, University of Toronto, Ontario

ENGINEERING ECONOMY FOR ENGINEERING MANAGERS
Turan Gönen, California State University at Sacramento

MANAGEMENT OF RESEARCH AND DEVELOPMENT ORGANIZATIONS: MANAGING THE UNMANAGEABLE
Ravinder K. Jain, U.S. Army Corps of Engineers
Harry C. Triandis, University of Illinois

STATISTICAL QUALITY CONTROL FOR MANUFACTURING MANAGERS
William S. Messina, IBM Corporation

KNOWLEDGE BASED RISK MANAGEMENT IN ENGINEERING: A CASE STUDY IN HUMAN-COMPUTER COOPERATIVE SYSTEMS
Kiyoshi Niwa, Hitachi, Ltd.

FORECASTING AND MANAGEMENT OF TECHNOLOGY
Alan L. Porter, Georgia Institute of Technology
A. Thomas Roper, Rose-Hulman Institute of Technology, Indiana
Thomas W. Mason, Rose-Hulman Institute of Technology, Indiana
Frederick A. Rossini, George Mason University, Virginia
Jerry Banks, Georgia Institute of Technology
Bradley J. Wiederholt, Institute for Software Innovation, Georgia

MANAGING TECHNOLOGY IN THE DECENTRALIZED FIRM
Albert H. Rubenstein, Northwestern University

MANAGEMENT OF INNOVATION AND CHANGE
Yassin Sankar, Dalhousie University

PROFESSIONAL LIABILITY OF ARCHITECTS AND ENGINEERS
Harrison Streeter, University of Illinois at Urbana—Champaign

MANAGEMENT OF ADVANCED MANUFACTURING TECHNOLOGY

Strategy, Organization, and Innovation

DONALD GERWIN AND HARVEY KOLODNY

A WILEY-INTERSCIENCE PUBLICATION
JOHN WILEY & SONS, INC.
NEW YORK   CHICHESTER   BRISBANE   TORONTO   SINGAPORE

In recognition of the importance of preserving what has been
written, it is a policy of John Wiley & Sons, Inc., to have books
of enduring value published in the United States printed on
acid-free paper, and we exert our best efforts to that end.

Copyright © 1992 by John Wiley & Sons, Inc.

All rights reserved. Published simultaneously in Canada.

Reproduction or translation of any part of this work
beyond that permitted by Section 107 or 108 of the
1976 United States Copyright Act without the permission
of the copyright owner is unlawful. Requests for
permission or further information should be addressed to
the Permissions Department, John Wiley & Sons, Inc.

Library of Congress Cataloging in Publication Data:
Gerwin, Donald, 1937–
    Management of advanced manufacturing technology : strategy,
organization, and innovation / Donald Gerwin and Harvey Kolodny.
        p.   cm. —(Wiley series in engineering management)
    "A Wiley-Interscience publication."
    Includes bibliographical references and index.
    1. Production management. 2. Computer integrated manufacturing
systems. 3. Flexible manufacturing systems.   I. Kolodny, Harvey.
II. Title.   III. Series.

TS155.G38   1992
670.42—dc20   91-16040

    ISBN 0-471-63574-X   CIP

Printed in the United States of America

10 9 8 7 6 5 4 3 2 1

To Marie-Christine
You made the difference

To Joyce
It could have been a lovely year

CONTENTS

Preface   xiii

Acknowledgments and Permissions   xv

1. Advanced Manufacturing Technology   1

   Description of AMT, 4
      Computer-aided Manufacturing, 8
      Utilization of AMT, 11
   The Emerging Role of AMT, 13

PART I STRATEGY

2. Strategic Implications of Flexibility   21

   The Conceptual Model, 22
      Implications of the Model, 24
   Uncertainty, Strategy, and Flexibility, 27
      Mix Flexibility, 28
      Changeover Flexibility, 33
      Other Flexibility Dimensions, 36
      Implications of the Flexibility Classification, 40

3. Design and Justification of AMT   42

   Designing CAM for Flexibility, 42
      CIM Facilities, 47

Capital Justification and the Value of Flexibility, 48
  Intangible Benefits and Costs, 50
  Modified Discounted Cash Flow, 54
  Multiattribute Approaches, 58
  The Strategic Approach, 61
  Evaluation of Current Approaches, 62

4. Performance Measurement 65

Discrepancy Analyses, 66
  Required Greater than Potential, 66
  Potential Greater than Actual, 69
  Other Discrepancies, 71
Measuring Flexibility, 73
  Conceptual and Methodological Issues, 73
  Flexibility Scales, 77
Flexibility in Practice, 81
  The Pilot Study, 82
  The Mail Survey, 87
  Comparison of the Two Studies, 90

5. Evolutionary Processes 92

Relationships Among the Feedback Loops, 92
Creating the Flexible Factory, 96
Avoiding Maturity, 99
  The Abernathy-Utterback Model, 99
  Evolutionary Paths, 104

PART II ORGANIZATION

6. Organizational Structure and the Management of Uncertainty 111

The Effects of Uncertainty and Interdependence, 111
Structural Adaptation to Uncertainty, 114
  Product and Process Uncertainty, 116
  Level of Analysis, 120
Organizational Structure, 121
  Technology and Structure, 122
  Other Influences on Organizational Structure, 124
  Interdependence and Coordination, 124
  The Collapse of Organizational Boundaries, 128

Organizational Forms, 132
    Functional Forms, 133
    Divisional Forms, 134
    Project/Matrix Forms, 134
    Network Organization, 137
    Integrative Organizational Forms, 139

7. Sociotechnical Systems and Product-focused Forms   141

Sociotechnical Systems Theory, 142
The Design of Sociotechnical Systems, 145
    The Primary Work System, 147
    The Work Group, 149
    Self-Regulation of Work Groups, 150
    STS Design Principles and Processes, 152
    STS Critiques, 159
Alternative Factory Organization Designs, 161
    Engineered Organizational Design, 162
    The International Motor Vehicle Program, 164
Product-focused Forms, 168
    Forsaking the Functional Hierarchy, 168
    An Hierarchy of Product-focused Forms, 169
    Work Groups, 171
    Group Technology/Manufacturing Cells, 173
    Assembly Cells, 175
    Flexible Manufacturing Systems, 180
    Flexible Assembly Systems, 180
    Flexible Focused Factories, 182
Summary and Conclusions, 184

8. AMT at the Level of the Individual   186

The Changing Relationship of Work and Technology, 187
The Individual and the Characteristics of AMT, 188
    Integration and Coordination, 190
    Deskilling or Reskilling, 192
The Nature of Skills, 194
    Visualization, 195
    Conceptual Thinking (or Abstract Reasoning), 196
    Understanding of Process Phenomena, 197
    Statistical Inference, 198
    Verbal Communication, 198

Attentiveness, 199
Individual Responsibility, 199
Role Changes in AMT Environments, 200
   Operators, 200
   Maintenance Workers / Skilled Trades / Craftsworkers, 201
   Supervisors, 202
   Technicians and Computer Specialists, 204
   Professional Engineers and Managers, 205
Work Design, 206
   Complementarities between Technology and Workers, 207
   Choice in the Control of Work, 208
   Job Design and New Skills, 209
Human Resource Support for AMT, 215
   Selection, 216
   Training and AMT, 216
   Reward Systems, 222
   Union-Management Relationships and AMT, 224
Summary and Conclusions, 224

PART III INNOVATION

9. The Innovation Process: Adoption   229

Conceptual Framework, 231
   Basic Assumptions, 232
   Participants in the Process: The Champion, 233
The Adoption Process, 235
   Awareness of the Innovation, 235
   Selection of the Innovation, 238
Summary and Conclusions, 245

10. The Preparation Process   248

Knowledge of the Required Infrastructure, 249
   Transferring Knowledge, 250
Opportunity to Design the Required Infrastructure, 252
   Short-Run Issues, 252
   Long-Run Issues, 255
Summary and Conclusions, 257

11. The Implementation Process   261

Expectations for the Innovation, 262
The Innovation's Performance, 263

Conflict and Resistance to Change, 267
   Analyzing the Reasons for Conflict, 268
   Methods for Reducing Conflict, 271
Revising the Technology and Infrastructure, 276
Summary and Conclusions, 281

12. Implications of the Innovation Theory — 285

Understanding the Innovation Process, 285
Designing the Innovation Process, 287
Implications for Innovation Research, 290
Radical Manufacturing Innovation as an Interorganizational Exchange, 294
   Nature of the Relationship, 296
   Determinants of the Relationship, 297
   Activity Analyses, 299
   Future Research Possibilities, 301

13. Lessons for the Future — 303

Uncertainty and Its Implications, 303
Strategic Considerations, 306
Organizational Considerations, 312
Issues in Innovation, 316
   Suggestions for Redesign, 320
Integration of the Key Factors, 321
   The Flexibility Cycle, 322

Appendix 1. Characteristics of Some Empirical Studies — 324

Appendix 2. American Manufacturing Activities in the Pilot Study — 327

Appendix 3. Questionnaire — 330

Appendix 4. Reasons for the Changes in Flexibility in the American Activities of the Pilot Study — 351

References — 357

Author Index — 383

Subject Index — 389

PREFACE

This is a book on the management of new types of computerized manufacturing technologies with emphasis on management rather than technology. Managing is defined here as devising strategies, designing organizational arrangements, and introducing new technologies. Operational issues such as requirements planning and production scheduling are not included. The technologies include computer-controlled processes used in product and process design, the production process, and manufacturing planning and control. However, we emphasize computer-aided manufacturing in the production process of firms that manufacture discrete products, and we focus on what occurs within individual companies. There is some discussion of interorganizational issues and virtually no attention is paid to societal level ramifications.

The book was written as a reference work. It is intended for faculty and graduate students who do research and teaching in the management of process technology. Such work is now carried out in business schools, industrial relations schools, social science departments, and industrial engineering departments among others. For these people we evaluate current work and suggest future research directions. Individuals in industry who hold or aspire to specialized managerial positions in strategic planning, technology, human resources, manufacturing engineering, product design, or accounting, for example, should also benefit from the book. For them it discusses problems that companies have and proposes alternative solutions. Although the book does not provide technical information, it is also useful to engineers, operations researchers, accountants, and others engaged in designing process technologies and accompanying systems and procedures. By demonstrating the implications of their choices, the book may lead to more "organization-friendly" designs. Due to its foundation in managerial research, the book will have less appeal to general managers, but if they are patient they too will find much relevant information.

This book was written for at least three reasons. First, so much material has now appeared in the general area that a need exists and a challenge arises to put it all in some logical order. The frameworks we have devised should help tomorrow's research build on the past, and increase the research's acceptance by industry. Any effort, however, to impose some order on such a diverse body of literature implies decisions on the selection and location of material that others would not have made. We make no claim to have compiled an objective distillation of all that is significant in the field. In some respects, this remains a highly personal view, but we are encouraged by the large amount of work that seems to fit into our categories.

Second, we highlight for managers the significance of (1) strategy, (2) organization, and (3) innovation in the selection and use of computerized manufacturing processes. These issues are often not given much attention, whereas considerable time and resources are lavished on the technology. However, a new technology's potential is almost never realized if the strategy is inappropriate, if the organization and people are not supportive, and if innovation processes do not exist to facilitate adoption and implementation. Once more, the three topics are usually not considered susceptible to research because such work does not typically result in neatly packaged "deliverables" with measurable benefits and costs. Eschewing these unstructured managerial issues is one symptom of the short-term orientation that continues to plague North American industry. We believe a tremendous payoff exists for companies willing to experiment with new managerial approaches even though the gains are difficult to determine precisely. The technology itself, which seems to open up new competitive opportunities, is partially responsible for the large returns from creative management. The book helps managers realize that these opportunities exist and offers reasoned guidelines for taking advantage of them. Of course, by placing the area on a more scientific footing it also helps to allay concerns about fuzziness.

Third, our purpose is to put the technology's role in the modern factory in perspective. There seems to be a typical psychological process that accompanies the introduction of any new technology. To get the technology accepted, advocates create high expectations among potential users. Inevitably, firms willing to take a risk are disappointed. As the news spreads, a backlash sets in which makes it even harder to gain acceptance. Since computerized manufacturing processes have fallen victim to this cycle, one needs a better idea of their strengths and limitations. Are these technologies an industrialized country's main hope for competitive survival, and are they creating a revolution in manufacturing comparable to that set in motion by the assembly line? Or do they have little to offer except in a few factories in a limited number of industries, and should most managers instead spend their time and resources getting back to the basics? We want to help the reader make up his or her own mind.

DONALD GERWIN
HARVEY KOLODNY

ACKNOWLEDGMENTS AND PERMISSIONS

Several individuals and organizations have helped bring this book to fruition. Financial support for various aspects comes from the U.S. Department of Transportation (Contract DTRS 5682-C-00026), the Cost Management Systems Program of Computer Aided Manufacturing-International (CAM-I), and the French National Foundation for the Teaching of Management (FNEGE). Jean-Claude Tarondeau deserves special mention for his collaboration in parts of the strategy section. Lakshmanan Prasad helped with the work on customer-vendor relations, and Jane Sorci contributed information on Texas Instruments through a class project. Steve Warnick and Luk Pellerin assisted with the case material from Cummins Engine Company's Jamestown plant and General Electric Canada's Bromont plant, respectively. Individuals who offered useful suggestions for improving the manuscript include Jerry Dermer, Vijay Jog, Sunder Kekre, and Jack Meredith. Else Brock and Edith Kosow deserve thanks for typing the manuscript. Finally, we want to thank the scores of individuals in American, Canadian, and European companies who were so generous with their time.

Permission of the publishers to use material from the following books and articles is gratefully acknowledged: S. Aguren and J. Edgren, *New Factories*, Stockholm, Sweden Employers' Confederation, © 1980; J. Bessant and P. Senker, "Societal Implications of Advanced Manufacturing Technology," in T. D. Wall, C. W. Clegg, and N. J. Kemp, Eds., *The Human Side of Advanced Manufacturing Technology*, pp. 153–171, Chichester, England, Wiley, © 1987; G. Betcherman, K. Newton, and J. Godin, *Two Steps Forward: Human Resource Management in a High-Tech World*, Ottawa, Canadian Government Publishing Centre, © 1990, Ministry of Supply and Services; F. E. Emery, *The Emergence of a New Paradigm of Work*, Canberra, Australia, The Centre for Continuing Education, Australian National

University, © 1978; Donald Gerwin, "Manufacturing Flexibility in the CAM Era," *Business Horizons*, 32(1), 78–84, © 1989, School of Business, Indiana University; Donald Gerwin, "Strategies for Manufacturing Flexibility," 1989 CAM-I; Donald Gerwin, "A Theory of Innovation Processes for Computer Aided Manufacturing Technology," *IEEE Transactions on Engineering Management*, 35(2), 90–100, © 1988 IEEE; Donald Gerwin, "Innovation, Microelectronics, and Manufacturing Technology," in Malcolm Warner, Ed., *Microprocessors, Manpower, and Society*," Aldershot, England, Gower Press, © 1984, St. Martin's Press; Donald Gerwin, "Control and Evaluation in the Innovation Process: The Case of Flexible Manufacturing Systems," *IEEE Transactions on Engineering Management*, EM-28(3), 62–70, © 1981 IEEE; Donald Gerwin and Jean-Claude Tarondeau, "International Comparisons of Manufacturing Flexibility," in Kasra Ferdows, Ed., *Managing International Manufacturing*, Amsterdam, Elsevier-North Holland, © 1989; Donald Gerwin and Jean-Claude Tarondeau, "Consequences of Programmable Automation for French and American Auto Factories: An International Case Study," in Benjamin Lev, Ed., *Production Management: Methods and Studies*, Amsterdam, Elsevier-North Holland, © 1986; P. G. Gyllenhammer, *People at Work*, Reading, MA, Addison-Wesley, © 1977 P. G. Gyllenhammer; D. Jenkins, *The Age of Job Design: National Patterns of QWL Activities in Western Europe*, Stockholm, Sweden, Eurojobs, © 1982; T. A. Kochan, "On the Human Side of Technology," *ICL Technical Journal* (November), 391–400, © 1988 Oxford University Press; M. Liu, H. Denis, H. Kolodny, and B. Stymne, "Organization Design for Technological Change," *Human Relations*, 43(1), 7–22, © 1990; R. Lund and J. Hansen, *Keeping America at Work*, New York, Wiley, © 1986; P. W. Swenson, "Fast Track to CIM: Automated Machining," *Manufacturing Engineering* (December), 31–33, © 1990, Dearborn, MI: Society of Manufacturing Engineers; and E. L. Trist, *The Evolution of Sociotechnical Systems*, Ontario Quality of Working Life Centre, Ontario Ministry of Labour, 1981.

D. G.
H. K.

MANAGEMENT OF ADVANCED MANUFACTURING TECHNOLOGY

1
ADVANCED MANUFACTURING TECHNOLOGY

The role of production technology in the modern corporation has been an important theme of managerial research for some time. For continuous-process industries such as oil or chemicals, in which there is a flow of liquids or gases, attention has focused on computerized controls. In discrete parts manufacturing, where metals, other materials, and electronic components are fabricated and/or assembled, the production or assembly line was "king" until recently. Recommendations flowed to companies to reap the benefits of this technology if they were large enough or to grow to be able to enjoy them. Now, due to changing market conditions, the predictable world of the assembly line has collapsed and research is concentrating on computer-controlled or advanced manufacturing technology (AMT).

Some of the factors accounting for the technological revolution in discrete parts manufacturing are as follows (Bolwijn et al. 1986; Reich 1983):

- Delivery problems arising from rigidity and a susceptibility to disturbances
- The switch to quality as opposed to high output as a competitive weapon
- The increasing diversity of and temporal changes in products
- Concerns over the quality of worklife and the related issues of employee motivation, absenteeism, and turnover
- The evolution of alternative forms of work organization
- The ability of less developed countries to accomplish more cheaply high volume production with unsophisticated workers

Why is it necessary to explore the managerial aspects of the new technology? A number of interconnected arguments exist. First, its diffusion has been disappointing by almost any standard and managerial reasons are more to blame than technical ones (Rosenthal 1984). An often-cited example is the difficulty in conducting financial evaluations of prospective investments in AMT due to a preponderance of intangible factors. Second, companies that have bought the technology are not using it effectively, and again managerial factors are cited more frequently than technical ones (Jaikumar 1986). For example, one cannot realize AMTs potential by using it as a substitute for mass production technology. Third, since the technology is readily available to firms in advanced countries, competitive advantages go to those enterprises that are most effective in selecting and using it (Boddy and Buchanan 1986). Fourth, the stunning ascendancy of Japanese economic power is based on the effective management of technology as opposed to just acquiring large amounts of leading edge equipment (Hayes and Wheelwright 1984). Comparisons of the Japanese and American auto industries have made this point (Abernathy, Clark, and Kantrow 1981; Womack, Jones, and Roos 1990).

Three interrelated managerial issues form the core of this book. First, manufacturing strategy is covered because many firms do not recognize the implications of AMT for seizing competitive initiatives and therefore do not take advantage of some of its most significant benefits. Other firms, however, may expect much more from the technology than it can deliver. To help unravel these issues it is necessary to recognize that AMTs strategic benefits derive from its property of flexibility. Our emphasis on flexibility leads to consideration of these principal questions:

1. What is the meaning of flexibility and what are its strategic implications?
2. Can a manager verify that the design of proposed AMT will meet a company's needs for flexibility?
3. Is it possible to consider the strategic benefits of flexibility when financially evaluating a proposed investment in AMT?
4. Can a manager determine whether a discrepancy exists between the flexibility an existing manufacturing process should have and actually has? Can he or she adopt procedures that will help eliminate any difference?
5. What has been the flexibility of AMT in practice?
6. How does the use of AMTs flexibility affect the way in which a manufacturing process matures in the long run?

Second, individual, work-group and higher-level organizational issues are treated due to the following anomaly. Although there will undoubtedly be fewer employees in the factory of the future, the remaining individuals, by virtue of their responsibility for expensive and critical resources, will have considerable influence over the production process. They must be highly motivated and work together as effectively as possible. The following questions are therefore investigated:

1. Are some structural arrangements better than others for organizing, and do these arrangements differ for different kinds of AMT?

2. As organizational boundaries collapse, both vertically and horizontally, how is management and coordination being accomplished?
3. How do managers come to understand the alternatives they have in linking social and technical subsystems, particularly when technology is flexible and social system alternatives exist?
4. What organizational arrangements allow companies to reduce cycle time and respond more quickly?
5. How has AMT affected the skill requirements of workers and professionals? How do their roles change and how should their work be designed to accommodate the skill changes? What kind of training support is required?

Third, innovation, the process by which AMT new to an organization is introduced, is also discussed. New AMT has significant impacts on all hierarchical levels and functions of a plant. Strategic executives, supervisors, workers, and technical specialists are all affected. Managers need to know the nature of the resulting problems and how to deal with them. The principal questions investigated include the following:

1. How does a company become aware of the existence of new types of AMT?
2. What are the rational and nonrational factors that influence a decision to purchase new AMT?
3. How can a firm develop a support system that is required to operate, maintain, and control new AMT?
4. Is it possible to conduct a successful initial trial of new AMT, or is it doomed to a rocky start?
5. What are the sources of conflict and resistance to change with respect to AMT, and what methods are available to handle these problems?

We believe that a unifying theme underlies a good deal of the research in all three areas. AMT is both a response to and a generator of uncertainty for a firm. Here, uncertainty is defined as a lack of information needed for decision making. In one sense AMT is viewed as a way to manage certain long-term uncertainties that continually plague a firm. They may arise in the marketplace or lurk in the production process. In another sense AMT, due to its complexity, introduces short-run uncertainties into a firm that may never have been dealt with before. Issues arise in the technical domain (e.g., what are the technology's capabilities and limitations?), in the financial arena (e.g., what are future net cash flows from an investment in the technology?), and in the social area (e.g., what is the technology's impact on organizational conflict?).

Each of the book's three major parts puts a different emphasis on AMTs role in the response to and generation of uncertainty. Part I, "Strategy," deals with how AMT can help in managing market and production process uncertainties. Part II, "Organization," inquires into the kinds of structures needed to support AMT as it copes with these long-run factors. It also deals with the impact of uncertainties generated by AMT on structural requirements. Part III, "Innovation," stresses how introducing AMT into a firm creates short-run uncertainties. These technical, financial, and social factors lead to problems and solutions that are discussed in detail. Chapter 13 reviews what has been learned from this book. We begin, however, with a description of advanced manufacturing technologies and a discussion of their emerging role in the modern factory.

One can draw on an increasing number of empirical studies to discuss strategy, organization, and innovation. Many of these studies are qualitative case studies, which reflects a need to understand at a detailed level the managerial revolution associated with AMT. During the second half of the 1980s, however, studies began to appear that conducted statistical analyses on data collected from large samples. Appendix 1 provides information such as type of study, sample characteristics, and nature of the technologies investigated for some of the research which is frequently mentioned in this book. Due to the lack of standard definitions for the various types of AMT, one must be careful in assuming that what is called a certain technology in two or more studies is actually the same.

DESCRIPTION OF AMT

Technology represents practical knowledge by which some organizational activity's inputs are converted into outputs. In a business firm, operating activities such as purchasing, manufacturing, and distribution; support activities such as research and development; maintenance activities such as personnel; and control activities such as accounting all have their own technologies. Technical knowledge exists in at least three forms: machines, codified procedures including computer programs, and informal routines stored in individuals' memories (Collins, Hage, and Hull 1988, Gerwin 1981b). In sum, technology is more than the machines used in production—it is a system of hardware, software (codified procedures), and humans, which exist to perform any organizational activity.

Advanced manufacturing technology is used mainly in the activities of product and process design, manufacturing planning and control, the production process, and in their integration (Gunn 1987). Integration creates links among elements lying within or between the other activities. Although the technology is highly automated it can be adapted to a variety of uses through computer programming or related means. Although it contains hardware, software, and human components, as do traditional manufacturing technologies, the relative proportions of each differ. One feature that distinguishes AMT from traditional technologies is the relative emphasis on software, specifically computer programs, for control purposes. A less obvious feature is the need for the human element to "control the computer controls" when unexpected developments arise (Hirschhorn 1984). To the extent that an experienced person's internal routines can be made explicit, the human component also acts more and more as a source of computerized procedures.

Due to their generic nature and the continual appearance of new examples, one cannot readily establish a domain for, categorize, or define the specific technologies that comprise AMT. Table 1.1 indicates some of the more popular technologies, most of which are discussed here because they have been the subject of managerial research. Several clearly overlap the categories within which they have been placed. Expert systems, for example, are applied in product and process design as well as in manufacturing planning and control. The following definitions have been adopted from a number of different sources primarily Gunn (1987), Noori (1990), U.S. Congress (1984), and U.S. Department of Commerce (1989).

TABLE 1.1 Glossary of Advanced Manufacturing Technology

Product and Process Design

Computer-aided design (CAD)
Computer-aided engineering (CAE)
Computer-aided process planning (CAPP)
Group technology and cellular manufacturing

Manufacturing Planning and Control

Material requirements planning (MRP)
Computerized preventive maintenance (CPM)
Expert systems

The Production Process—Computer-aided Manufacturing (CAM)

CONVERSION

Numerical control (NC)
Computer numerical control (CNC)
Programmable robots
Flexible manufacturing systems and cells (FMS, FMC)

AUTOMATED MATERIAL HANDLING (AMH)

Automated storage and retrieval systems (AS/RS)
Automated guided vehicle systems (AGVS)

PROCESS CONTROL

Computer-aided inspection (CAI)
Programmable controllers (PC)

Integration

Local area networks (LAN)
Intercompany or wide-area networks
CAD/CAM
Computer-integrated manufacturing (CIM)

Product and Process Design

Computer-aided design (CAD). Helps create or modify engineering designs and stores the information in a computerized database.

Computer-aided engineering (CAE). Uses simulation and other techniques to analyze the properties of a design such as the stress in a mechanical part.

Computer-aided process planning (CAPP). Helps in planning the sequence of production operations a part will go through on the shop floor.

Group technology. Takes advantage of similarities in design or manufacturing requirements to cluster parts into families with similar attributes.

Cellular manufacturing. Involves the rearrangement of production equipment on the shop floor into groups, each of which is responsible for one or more part families.

Manufacturing Planning and Control

Material requirements planning. A computer-based information system for production scheduling, ordering, and inventory control of all the items that go into a finished product (materials, components, subassemblies).

Computerized preventive maintenance (CPM). Aids in planning, scheduling, record keeping, and report preparation for routine maintenance aimed at reducing equipment failure.

An expert system. A computer program that applies rules of thumb termed heuristics to find acceptable solutions to a given complex problem; the rules are often obtained from individuals with experience in solving the problem.

The many production process technologies that collectively are referred to as computer-aided manufacturing (CAM) are often divided into the three categories of conversion, automated material handling, and process control.

Conversion

Numerical control (NC). Uses instructions coded on tape to control the operation of a general purpose machine tool or machining center.

Computer numerical control (CNC). Uses a program stored in a mini- or microcomputer attached to a machine tool or machining center to control its operation.

A programmable robot. A multifunctional manipulator that moves a part, tool, or specialized device through variable motions to perform a variety of tasks such as assembly, material handling, or inspection.

A flexible manufacturing system or cell (FMS or FMC). Consists of two or more machines for cutting or forming metal with automated material handling controlled by computers; as compared to a cell an FMS usually has more machines, more sophisticated components, and multiple paths which the parts can follow.

Automated Material Handling (AMH)

Automated storage and retrieval systems (AS/RS). Computer-controlled stocking systems in which parts are stored on racks and received and retrieved using computerized robots, cranes, and/or similar devices.

Automated guided vehicle systems (AGVS). Provide unmanned transportation of materials in the factory using vehicles equipped with automatic guidance devices programmed to follow certain paths.

Process Control

Computer-aided inspection (CAI). Helps check the conformance to specifications of parts typically using an automated programmable device termed a coordinate measuring machine or robots equipped with sensors.

Programmable controllers (PC). Electronic process control devices with programmable memories that execute the steps taken in a certain logical sequence and collect information.

Integration Technologies

Local area networks (LAN). Telecommunications linkages for exchanging information between different points on the factory floor or within design and engineering departments.

Intercompany or wide-area networks. Telecommunications links for exchanging information between the factory, subcontractors, suppliers, and/or customers.

CAD/CAM. Design information is used to automatically generate parts programs that are electronically downloaded to control a machine's production of a part.

Computer-integrated manufacturing (CIM). Bringing together the individual advanced manufacturing technologies used in product and process design, manufacturing planning and control, and the production process under unified computer control using information technology.

The essence of CIM is shared data bases and communications among the programmable technologies in the three manufacturing activities (U.S. Congress 1984). A CAD system would directly interface with CAM technology (CAD/CAM) and would access data from inventory records on the costs of raw materials being considered for a new product. A CAPP system would automatically interpret design information from CAD in order to develop efficient process plans. Manufacturing planning and control systems would be automatically updated using performance data from CAD and CAM databases. Some of the many CIM linkages are already a reality to a degree in a few corporations. Interfaces exist in practice between CAD and CAM, CAD and MRP, CAPP and MRP, and MRP and AS/RS (Gunn 1987). Other connections are far from being realized.

The terms "computer-integrated enterprise" (CIE) or "computer-integrated business" (CIB) have been given to proposals that extend the CIM concept beyond the walls of a single plant (Browne, Harhen, and Shivnan 1988). Integration may be with other factories in a multiplant system, distribution outlets, suppliers, and/or customers at a regional, national, or international level. Consider, for example, two alternative scenarios developed by the sandpaper division of a multinational company manufacturing abrasives (Orne and Hanifin 1986). The division has an international network of product design, manufacturing, and distribution centers. Scenario 1 centralizes most manufacturing in one location to take advantage of economies of scale and learning curve effects. CIB technology provides enough flexibility to handle a variety of products. To avoid long lead times a telecommunications network transfers orders and other information between the central manufacturing site and the regional distribution centers. Scenario 2 decentralizes most manufacturing to the regions where distribution occurs. This alternative profits from local comparative advantages in factor costs and exchange rates. CIB facilitates the sharing of manufacturing experience. Product and process definitions, for example, are held in common to avoid each facility engaging in needless duplication.

Computer-aided Manufacturing

Managerial research has concentrated on computer-aided manufacturing, that portion of AMT applied in the production process. Conversion technologies, NC, CNC, robotics, and flexible systems and cells have received the most attention. In succeeding chapters on strategy, organization, and innovation, reference to AMT will frequently imply some aspect of computer-aided manufacturing.

To understand more about the different kinds of CAM technology, it is useful to divide the world of discrete parts manufacturing into four overlapping categories: (1) fabrication, or the cutting and forming of parts; (2) assembly, putting parts together; (3) batch production, where a number of different parts or products are made in distinct lots, the sizes of which run from low to medium volume; and (4) mass production, where a factory makes one or very few similar parts or products in high volumes.

CAM technology has had a great impact in batch fabrication especially in the aerospace, defense, and industrial and farm equipment industries. Before its arrival, batch producers had two kinds of equipment from which to choose, neither which completely met their needs. The first, dedicated machinery such as transfer lines, best suited for mass production, permits low unit costs but also inhibits flexibility. The second, general purpose machine tools such as conventional turning and milling machines, is best suited for one of a kind or very small batch production. Costs per unit tend to be high, but so is flexibility.

As the result of CAM, two distinct subcategories of batch fabrication seem to be emerging (Hull and Collins 1987). A "technical" category based on CAM is highly automated, integrated, and sophisticated. A "traditional" category based on conventional technologies has less of each of these characteristics. Evidence for the existence of two classes was compiled on the assumption that the degree of technical expertise required for production operations is a distinguishing variable. The research subdivided batch fabricators based on the median percent of professional and technical workers they had. Some statistically significant organizational differences were found. Firms in the technical category had narrower spans of control, more occupational specialization, and more decentralization of strategic decisions. The technical class also had a greater capacity for innovation in terms of more research and development employees, and patent applications per employee.

The technology has also had an impact on mass assembly, but it has been concentrated in the auto and electronics industries. Specific applications have appeared in batch assembly for small- and medium-sized motors, printed circuit boards, contactors, and other items. AMT is beginning to appear in mass fabrication as computer-controlled transfer lines are developed for the production of auto engine components.

Batch and Mass Fabrication. CAM technology in fabrication consists of integrated systems and stand-alone equipment. The former category, used primarily in batch and mass production, has four main components. First, there are a number of work stations where metal cutting, inspection, or other operations are carried out. The work stations may consist of single pieces of NC or CNC equipment or cells comprising groups of machines.

Second, there is an automated material-handling system for parts and tooling, which loads these items into the system (often with human help), transports them between the work stations, temporarily stores them, and unloads them from the system (often with human help). A number of different principles are available. An automated guided vehicle system uses battery-powered transporters controlled by wires in the floor. A robot system uses movable robots hanging from overhead supports. A towline system has tow chains in the floor that provide power to four-wheel carts by means of a tow pin attached to each cart. In a power roller conveyor system the transporter moves along an elevated track, and chains mounted on the track turn rollers that move the transporter.

Third, centralized computer controls send operating instructions to the work stations when parts arrive, move the transporters through the system, and compile operating information from the work stations. In complex systems involving hierarchical levels of work stations there is a corresponding hierarchy of computer control. Programmable controllers are used in some applications, especially in mass production.

The fourth element of an integrated system is its workforce, which may monitor the computer controls and machines, load and unload parts, perform tool-setting activities, maintain the equipment, and perform some inspection. Workers may be organized using traditional person-to-person approaches involving a supervisor with overall authority and subordinates with specialized tasks and little planning responsibility. Alternatively, the work organization may involve a group approach in which the supervisor is more of a consultant, and group members, who are multiskilled, rotate tasks and have some planning responsibility.

Integrated systems come in three general varieties. The flexible transfer line (FTL) used in mass- and large-batch production, is somewhat more flexible but has higher unit costs than a conventional transfer line. It produces a few similar parts at a relatively high production rate starting at around 2000 units of the same part per year but often much more. Each different part moves through the same fixed sequence of work stations except that a given part may skip one or more stations.

A flexible manufacturing system, designed for midbatch production, has moderate degrees of flexibility and unit costs. It usually has from about five to twenty work stations to handle anywhere from about 2 to 800 different parts in volumes of about 15 to 15,000 units each per year. The parts are typically grouped into one or more families based on similar sizes, shapes, compositions or other characteristics. An FMS is a multipath system; not all parts follow the same operational sequence. When there is no automated material handling, the system is referred to as direct numerical control (DNC).

A flexible manufacturing cell (FMC) with high flexibility and high unit costs is designed for small-batch production. It usually has from two to four machines to produce at least 300 different parts grouped into families at a volume of less than about fifty units per year each. Some typical applications include making service parts and prototypes of new parts or serving as a work station in an FMS. Besides being smaller in scale than an FMS, an FMC has less sophisticated computer controls and material handling, and the parts are more likely to follow a single path. A machining center and a lathe serviced by an overhead robot provide one example.

Stand-alone NC and CNC equipment, used primarily in batch production, consists of individual machines that perform specific functions such as milling, drilling, or turning, and machining centers, which combine different functions. It is the most flexible but highest unit cost CAM alternative in fabrication. Stand-alone machines are best suited for small-batch production of many different parts each in annual volumes of say 200 units or less.

NC equipment is controlled by coded instructions on a tape that is run into the machine each time a unit of a part is worked on. The operator, who does not cut the metal, performs ancillary activities such as tool selection, loading and unloading of parts, and monitoring. A parts programmer usually prepares and debugs the tapes at least in the United States. NC is used extensively in aerospace and military industries where the need for precision and the use of exotic materials are the rule.

A CNC machine is controlled by a mini- or microcomputer attached to it. Before a batch of a certain part is worked on, a tape is fed in and the instructions are stored in the computer's memory. The instructions are called on each time a unit is run. Rather than preparing a new tape each time modifications are needed due to a debugging problem or a change in a customer's requirements, the internally stored instructions are revised. The computer also compiles operating information for management such as tool wear or utilization time.

Batch and Mass Assembly. Assembly is poorly understood as compared to fabrication mainly because humans have routinely been able to do it (Nevins, Whitney, and DeFazio 1989). CAM technology for the former manufacturing task is therefore less well developed and less often utilized than for the latter. Programmable assembly is most frequently encountered in putting electrical components into printed circuit boards. Since this is essentially a two-dimensional problem, it is relatively easy to automate. Programmable assembly of mechanical components, a three-dimensional task, is usually less flexible since the technical problems are more manageable if there is a single product or a small product range.

Integrated systems in assembly roughly parallel those in fabrication with the most popular alternatives being at the extremes (Groover, Weiss, Nagel, and Odrey 1986, Skoog and Holmquist 1983, Williams 1988). A flexible assembly line (FAL), which approximately corresponds to a FTL, has some flexibility and relatively low unit costs. The assembly is transported on a track or conveyor usually running in a line and stops at each work station where an operation such as welding, fastening, or inserting is performed. The work stations use some combination of people, robots, special-purpose programmable machines, and dedicated equipment. Components are supplied manually or through an automated material-handling system. Computers and programmable controllers are used in the hierarchical control of the system. One important application is in auto body framing, an example of which is given in Appendix 2.

A flexible assembly cell (FAC), roughly analogous to an FMC, has relatively high flexibility but high unit costs. It has a robot that identifies, selects, and puts together parts. In one configuration the robot is located in the middle of the cell and equipment is arranged around it in a partial circle. In another the robot moves to the equipment on a rail system or an automated guided vehicle. A feeder automatically provides, separates, orients, and presents components to the robot. The cell may be used as a work station in a FAL.

Utilization of AMT

Two studies, one in the United States and the other in Germany, help indicate the extent to which AMT is being used. A 1988 survey by the U.S. Bureau of the Census covered 10,526 business establishments with twenty or more employees in the following two-digit SIC groups: fabricated metal products, industrial machinery and equipment, electronic and other electrical equipment, transportation equipment, and instruments and related products (U.S. Department of Commerce 1989). These industries account for 43 percent of all employees and value added recorded in the 1987 Census of Manufacturers. The other survey covered 1100 firms in the West German capital goods industry during 1986/87 (Schultz-Wild 1989). Comparisons among the two studies must be made with a great deal of caution. They undoubtedly define specific technologies in different ways. It is not clear, for example, whether the German robots are just programmable or whether they include pick-and-place devices. Once more, the industries surveyed are not completely the same.

Advanced manufacturing technologies are not widely deployed. According to Table 1.2, no technology in either country, except for NC and CNC in West Germany, appears in a majority of the firms surveyed. The most extensively used are NC/CNC (41.4 percent), CAD/CAE (39.0 percent), and PCs (32.1 percent) in America, and NC/CNC (50 percent) in West Germany. The least employed are AGVS (1.7 percent), AS/RS (3.2 percent), and, surprisingly, programmable robots

TABLE 1.2 Percent Use of Advanced Manufacturing Technology

	Certain U.S. Industries	West German Capital Goods Industries
Product and Process Design		
CAD/CAE	39.0	17.0
CAPP	—	18.0
Manufacturing Planning, Control		
MRP	—	15.0
The Production Process		
NC/CNC	41.4	50.0
Programmable robots	5.7	8.6
FMS/FMC	} 10.7	4.0/3.0
FAS/FAC		5.1
AS/RS	3.2	4.8
AGVS	1.5	0.7
CAI		
Incoming, in-process	10.0	} 8.0
Final product	12.5	
PC	32.1	—
Integration		
LAN for engineering data	18.9	—
LAN for factory use	16.2	—
Intercompany links	14.8	—
CAD/CAM	16.9	2.7

Source: U.S. Department of Commerce (1989), Schultz-Wild (1989).

(5.7 percent) in the United States; and AGVS (0.7 percent), CAD/CAM (2.7 percent), flexible systems and cells (3.0 percent to 5.1 percent), and AS/RS (4.8 percent) in West Germany.

Another study compiled information on how sixty-four American FTLs, FMSs, and FMCs were used (Darrow 1986). Thirty-nine percent of the systems made parts for heavy equipment such as tractors, trucks, construction vehicles, and tanks. Thirty-two percent manufactured parts for aircraft and missiles. Machine tool parts (7 percent) and automative parts (2 percent) accounted for the rest of the known uses. Six percent of the systems were unclassified. Most systems manufactured prismatic parts whose shapes fit into a given-sized cubic box. The cube sizes usually ranged between 1 foot and 6½ feet. Some systems accounted for rotational parts that revolve around a central axis in order to be cut. Very few systems made both kinds of items together. Sixty-one percent of the systems fabricated iron and steel, 25 percent aluminum, and the rest handled other materials.

None of these studies inquired into how widely AMT is used within a factory. Any establishment using at least one robot, for example, was included as employing this particular technology. Other research suggests that AMTs penetration of factories is very thin. In the overwhelming majority of the small and medium U.S. metal working establishments studied by Hicks (1986) less than 50 percent of the machines were controlled by NC and CNC, the most popular versions of AMT.

THE EMERGING ROLE OF AMT

The symptoms and causes of America's competitive decline have been classified into societal level economic and social factors such as monetary and fiscal policy and the nation's educational system; and firm level economic and social issues such as investment in plant and equipment and the way in which manufacturing is organized and managed (Abernathy, Clark, and Kantrow 1983). In exploring the managerial aspects of AMT the emphasis here is clearly on the latter; these are issues over which managers exercise some degree of control. However, broad societal factors cannot be completely disregarded as they often severely constrain the options available to executives.

The significance of specific macro and micro issues has been debated long enough to already be familiar to most of the readers of this book. It will not be elaborated here, except to stress one point. Most of the factors initially identified well over a decade ago have persisted, some of them have even intensified. Let us consider those which are especially relevant for the management of AMT. The United States still leads the world in labor productivity, and the rate of growth of this measure during the 1980s was somewhat better than during the 1970s. However, productivity growth remains below the long-term average for the United States, and below that of some of our ablest competitors notably Japan (Dertouzos, Lester, and Solow 1989).

A persistent gap has remained throughout the 1980s in the cost of capital for Japan and the United States (U.S. Congress 1990). Not surprisingly, therefore, the short-term orientation of American businessmen, initially brought to our attention about ten years ago (Hayes and Abernathy 1980), is still regarded as a significant factor in limiting capital investment in AMT (Dertouzos, Lester, and Solow 1989). It is little wonder that from 1976 through 1987 Japanese investment in machinery and equipment ran from 14.9 percent to 20.6 percent of the Gross National Product versus from 7.5 percent to 9.0 percent in the United States (U.S. Congress 1990).

American firms' inability to find well-educated, trained employees from the executive suite to the shop floor who can make the best use of sophisticated AMT is another recurring problem. In the 1960s American public school students performed as well as students anywhere on tests of mathematical competence. In the mid-

1980s they ranked in the lower half of those countries among which comparisons were made (U.S. Congress 1990). The functional literacy rate of American workers is falling and is now among the lowest in the developed world (Harris 1990). There is a continued lack of attention to on-the-job training, one of the most critical ways of learning about AMT (Dertouzos, Lester, and Solow 1990).

The debate over competitiveness shows some signs of changing in at least one respect. As more and more jobs are lost to efficient producers overseas the ability to function in a global economy may become a central issue in American politics (Harris 1990). Polls indicate that a large majority of Americans believe that the United States is declining relative to Germany and Japan. Nine out of ten Americans are disturbed by these developments, and six out of ten view them with serious alarm.

What role can AMT play in helping to restore our competitiveness? The importance to an industry's productivity growth of new manufacturing processes in general and AMT in particular is illustrated by case studies conducted by Baily and Chakrabarti (1988). They collected data on product and process innovations available in each of six U.S. industries by surveying technical trade journals for articles and advertisements and then checking with industry experts. They calculated the average number of innovations in each of three time periods between the late 1960s and early 1980s. Information was also compiled on an industry's rate of productivity growth during the same periods. Examination of the values of the two quantities indicated that, in general, they tended to move together.

In the textile industry, for example, from the late 1960s to the early 1980s there was no slowdown in new manufacturing processes, the principal type of innovation, and, correspondingly, no slackening in productivity growth. The industry took advantage of a constant flow of new technologies from equipment suppliers. In the chemical industry a sharp downturn in the innovation of new products and processes was associated with a slowdown in productivity growth.

The pattern in the metal-cutting segment of the machine tool industry, where product and process innovation are synonymous, is of particular relevance here. In general the pace of innovation fell off after 1970 and picked up again after 1977. Productivity growth fell from 1973 to 1979. According to the theory it should also have picked up subsequently, but, in fact, it fell even more. The researchers attributed this to the overvalued dollar, product innovation, which did not take advantage of the technical advances in computer control, and weak profitability, which impeded investment.

Baily and Chakrabarti believed that the slowdown in innovation in industries such as chemicals and machine tools resulted from a failure to exploit available technological opportunities as opposed to an exhaustion of new technological ideas. The substantial U.S. lead in computer technology afforded our machine tool industry a chance to surge to the forefront in computer-controlled equipment. Due to the industry's lack of commitment, Japanese machine tool makers were able to seize the initiative in computer controls and robotics. For American manufacturers of machine tools this failure led directly to slow productivity growth and created serious structural weaknesses in demand as other countries took over the U.S. market.

The claims that have been made about AMTs strategic implications are also significant in the quest to improve competitiveness. By virtue of its flexibility AMT provides options in markets that are continually changing in uncertain ways. A CAM-based production process, for example, has an ability to:

- Handle a mix of different parts
- Cut manufacturing lead times by reducing time-consuming setups
- Substitute new parts for old ones in the mix
- Reduce manufacturing's share of new product introduction time by facilitating small experimental runs of new parts
- Handle customers' requests for design modifications in an existing part
- Meet a need for sudden increases or decreases in production volumes

Although there has been a great deal of exaggeration of these benefits, studies indicate that they do exist to a limited degree, but rarely all in the same machine or system (Gerwin and Tarondeau 1989). Still, imaginative exploitation of AMTs flexibility will help a company realize an existing competitive strategy or provide an opportunity to alter strategy in new directions.

AMTs strategic properties have consequences for national as well as firms' competitiveness. It seems that an industrialized nation's economic future lies in so-called flexible systems of production, that is, technically advanced and skill-intensive industries that make customized products (Ayres 1984, Choate and Linger 1986, Cohen and Zysman 1987, Reich 1983). Industrially developed countries can retain their competitive advantage by offering technologically superior products aimed at market niches as a response to standardized items from abroad. Accomplishing this reorientation means a thorough transformation of traditional smokestack industries as well as the growth of completely new industries. Either way, the success of a flexible strategy hinges on the effective application of advanced manufacturing technologies.

In sum, the nature of global competitiveness requires manufacturers to become more productive and flexible at the same time. The auto industry is faced with a need to lower prices and to develop more innovative cars. Traditional technological solutions tend to trade off one of these characteristics against the other. Conventional production processes, based on automation alone, offer improved productivity at the expense of flexibility (Abernathy 1978). AMT has some potential to finesse this dilemma. By virtue of its automation aspect it creates higher productivity, and through its programming feature it contributes greater flexibility simultaneously, at least under certain conditions.

It is no wonder that some firms have seized on AMT as the panacea for their competitive dilemmas. Yet in their haste to bring the technology on line they have overlooked significant managerial issues. Reports from the firing line talk about a "tricky technology" that has created just as many problems as solutions (Nag 1986). Since 1980 General Motors has spent about $40 billion for modernization and new facilities, a good deal of it involving AMT (Keller 1989). Yet as we enter the 1990s GM's once preeminent position in the auto industry has still not been restored.

GM's bittersweet experiences with the new technology are reflected in the startup of its Hamtramck, Michigan's assembly plant constructed in the mid-1980s to handle new models of Cadillacs. Undoubtedly the most technically advanced factory of its kind ever built to that time, it featured the latest in computer-controlled manufacturing processes. From the outset it was beset with numerous problems, such as equipment failure, that severely affected its efficiency. By late 1988 studies concluded that it was about as efficient as a comparable Ford plant with much less computerized automation. The company's troubles had particular significance for many other U.S. manufacturing concerns that tend to adopt technologies pioneered by the auto companies.

What accounts for the discrepancies that firms have found to exist between expectations and performance? A *technological* approach to AMT, which contains some rather dubious assumptions left over from the halcyon days of the assembly line, shares in the blame (Jaikumar 1986, Keller 1989). There is a belief, for example, that new technology in itself is sufficient to solve competitive problems. This translates into a willingness to obtain the highest level of technological sophistication available. The result is deployment of the "bleeding edge" of technology, that is, the next generation of equipment even if it is not fully tested. When new technology is attempted in conjunction with a new product and new employees, as at Hamtramck, tremendous uncertainties are created that are difficult to manage.

A belief exists that comprehensive technical solutions are needed all at once in order to remain ahead of competitors. There is no chance to tackle one thing at a time, and pilot projects do not have sufficient impact. The result is an underestimation of the complexities of integrating separate islands of automation with consequent breakdown, quality, debugging, and stress problems. Ultimately, a company may fall behind competitors who have learned from its mistakes.

There is a belief that high production costs are due to excessive compensation to hourly workers and that union control over the shop floor must be broken. Therefore, AMT is used to eliminate workers rather than to improve their productivity. Remaining workers are not properly trained to handle the technology even though they are responsible for very expensive equipment. No attempts are made to benefit from their knowledge of the production process.

Other more prosaic reasons have been offered for the disruptive consequences of AMT. The human, hardware, and software problems could simply be due to the inevitable transient effects of introducing a revolutionary new technology and they will disappear in due time as learning occurs (Baily and Chakrabarti 1988). Unavoidable and non-recurring crises may, however, account for a lower limit on the problems which arise. Poor technology management, undoubtedly, greatly intensifies their frequency, magnitude, and duration.

If AMT is not a panacea it does not have to be a disappointment. Proper management of the technology is the key to getting the most out of its potential, even if that is sometimes more limited than we would like to believe. Although there are few definitive guidelines that can be currently employed the broad outlines of an *integrative* approach to handling AMT are emerging. Its implications for AMTs role in the modern factory are very different than those envisioned by early enthusiasts who saw the technology as *the* answer to our competitive problems.

Primarily, AMT should not be considered an end in itself as indicated by guidelines to always purchase the most sophisticated technologies on the market. Company and manufacturing strategy should be the basis for the selection of production processes so that AMT becomes a tool for realizing the firm's competitive objectives. When a company allows AMT to be an end in itself, purchased technology influences the kinds and amounts of factory flexibility, and that constrains strategic objectives in ways not foreseen by management. When strategic needs are paramount they dictate the need for certain types of manufacturing flexibility, which influences the selection of compatible technology. With this shift in orientation the critical questions are no longer "How large do we want our AMT system to be?" or "How fast do we want to implement it?" but "What is the most appropriate system for our needs?" or "Are there less expensive ways than AMT of meeting our objectives?"

Second, a company wants to eliminate unnecessary operations from the manufacturing process so that they are not eventually automated (R. W. Hall 1987, Schonberger 1987). Converting from process layouts to cells reduces travel distances, thus obviating the need for sophisticated material-handling equipment. Reducing inventories through the Just-in-Time (JIT) philosophy lessens the need for automated storage and retrieval systems. Redesigning a product with fewer, simpler, and more standardized parts eliminates fabrication and assembly operations. In IBM's Proprinter project a multidisciplinary team assigned to design a new computer printer reduced the number of components from 160 to 63. An assembly worker could put the new printer together in three and one-half minutes largely precluding the need for the highly automated and expensive assembly plant that had already been built (Dertouzos, Lester, and Solow 1989).

Third, an enterprise wants to investigate whether utilizing its existing technology more effectively may create the same results as purchasing the latest equipment (R. W. Hall 1987, Schonberger 1987). Rearranging existing machines in cells represents a less expensive and risky solution than buying a flexible manufacturing system. Diminishing machine breakdowns and quality defects helps cut manufacturing lead times without resorting to advanced technology. Ironing out these problems also helps pave the way for eventual deployment of AMT. To work effectively computerized equipment demands harsh control over its immediate environment. Robots, for example, cannot be successful assemblers if parts are not designed and fabricated correctly. Once more, the integration usually associated with AMT magnifies the effects of a breakdown or defect. A downed machine that is part of a CAD/CAM system can have an immediate impact on the product design department as well as the production process.

To illustrate, General Motors' CPC Division purchased the most technologically advanced and expensive stamping presses in order to reduce setup times. The division was saddled with tremendous debugging problems and then overcapacity when GM's market share fell. To accomplish the same objective the BOC Division decided to make incremental changes in its existing presses. By maintaining dies better, achieving faster throughput times, and standardizing die bolts so that workers would need only one tool to make changes, setup times were reduced from twelve hours to eighteen minutes (Keller 1989).

Once a firm has a more efficient manufacturing process, it is time to seriously explore the gradual implementation of AMT. In choosing which technology to purchase a firm must consider strategically related variables such as a product's characteristics, the degree of change in a product's design, the predictability of demand, and production volumes (Gunn 1987, Venkatesan 1990); develop a sophisticated human support system to operate, maintain, and control the new technology; and take advantage of employees' training, skills, and experience for ideas on how to realize the technology's potential. In other words, once operations have been made more efficient it is time to consider the issues that this book addresses.

PART I
STRATEGY

2
STRATEGIC IMPLICATIONS OF FLEXIBILITY

Until recently, process technology was not seen as having much import for a company's competitive strategy. Now most authorities would agree on at least the need for compatibility between the two (Hayes and Wheelwright 1984, Porter 1983, Skinner 1985). This change in thinking has arisen in part because of the development of AMT. AMTs strategic impacts distinguish it from previous generations of manufacturing technology.

Porter (1983, 1985), who identified three general competitive strategies for a company, illustrates the new thinking. First, overall cost leadership means having the lowest cost in each of a broad range of industry segments, that is, geographical markets, customer groups, distribution channels, or product lines. Second, overall product differentiation emphasizes competing on product attributes in each of a broad range of industry segments. Third, niche development focuses on a specific segment and tries to achieve cost leadership or product differentiation.

No matter which of the three strategies is selected, a firm must have subsystem policies that are consistent with it. Consider, for example, technological policy. It has product and process characteristics as its focus. In order for overall cost leadership to work, process technology must exhibit economies of scale and facilitate learning curve improvements. Overall product differentiation requires process technology that is flexible, turns out reliable products, and contributes to fast delivery times. A successful niche strategy demands that manufacturing processes are oriented toward the targeted segment's needs.

Porter's ideas were not specifically influenced by the strategic impacts of flexible automation. Susman and Dean (1989) mentioned ways that the new technology could modify his thinking. First, they suggested a fourth general strategic alternative based on AMTs flexibility: switching from one industry segment to another. Second, they questioned Porter's belief that competing on the basis of cost leadership and product differentiation results in an unsatisfactory compromise as compared to pursuing either one. Recent trends, they argued, are forcing some American firms to compete on the basis of cost, quality, and lead times. Furthermore, Susman and Dean saw AMT as instrumental in being successful with these multiple objectives. Consider General Electric's major appliance group, which uses robot assembly and automated inspection to produce dishwashers with low cost and high quality.

THE CONCEPTUAL MODEL

To explore the acknowledged links between strategy and AMT it is useful to work within a broad context. The conceptual framework of Figure 2.1 is a modified contingency model that grew out of the organization theory and manufacturing strategy literatures (Child 1972, Skinner 1985). Company and manufacturing strategy and investments in production technology including AMT are related by virtue of their interactions with environmental uncertainties, manufacturing flexibility, and performance measurement. An organization's task environment is the driving force in that management must learn to cope with uncertainties arising in the product market or in the production process and its inputs. Management's learning is reflected in a company and a manufacturing strategy that tries to defensively adapt to the uncertainties and/or attempts to proactively control them (Swamidass 1988).

Manufacturing flexibility is needed to adapt to uncertainties; it allows a corporation to respond effectively to changing circumstances (Zelenovic 1982). Theoretical support comes from research conducted by Tombak and DeMeyer (1988). They developed an economic model to maximize utility for a firm with a single output and two inputs in which the costs of flexibility in the manufacturing process are considered. One of the model's implications is that the need for flexibility goes up as process or market uncertainties increase. Specifically, more random variation in the production function coefficients or in consumers' preferences as manifested in price leads to more investment in being able to vary input proportions. Empirical support for flexibility's adaptive characteristic is found in a mail survey of 169 manufacturing firms in Minnesota conducted by Wharton and White (1988). They

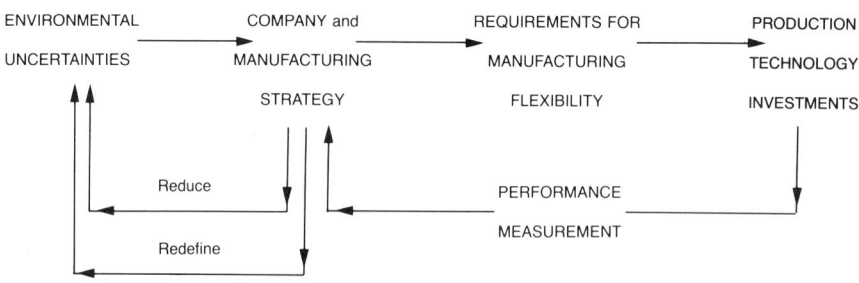

FIGURE 2.1. The conceptual model.

found that as market unpredictability and competitiveness increased the flexibility of the production process went up.

Any need for flexibility in the conceptual model is met through capital investment in production technology, thus making the design and justification of AMT a salient issue. Flexibility, however, is created through a variety of other means as well. Sources include product design, subcontracting, work organization, materials management, or a "generic" product such as baking soda for which new uses are readily found. Too little is known about how to achieve flexibility effectively by properly balancing these methods (Whitney 1986). Providing answers is of great concern to operations managers because they rely less and less on inventories to solve their flexibility problems.

An enterprise may also try to proactively redefine market uncertainties as indicated by the relevant feedback loop in Figure 2.1. Market needs are based in part on what customers have come to expect from a particular industry (Hayes and Abernathy 1980, von Hippel 1988). A firm in that industry may be able to encourage customers to think in terms of shorter lead times or more frequent new product introductions than they currently expect and then provide these new levels of service through its superior manufacturing flexibility. By causing more uncertainties for competitors the firm has established a powerful competitive advantage. Management must be creative to devise the redefinitions and risk taking to carry them out.

Presently this brand of thinking is more characteristic of Japanese than American companies. When Yamaha challenged Honda's commanding position in the world motorcycle market Honda chose a proactive approach instead of adapting to the new situation (Stalk 1988, Stalk and Hout 1990). By significantly augmenting its rate of change of new products Honda introduced or replaced over 100 models in a year and a half. As customers began to expect more frequent product changes, Yamaha's cycles began to look out of date to them, spelling the end of its challenge to Honda's preeminence.

A proactive firm may also try to reduce contingencies as indicated by the remaining feedback loop from strategy to uncertainty. Management attempts to stabilize product market uncertainties through long-term contracts with customers, design for manufacturability, and leveling of demand. Uncertainty reduction in the manufacturing process and its inputs occurs through preventive maintenance, long-term contracts with suppliers, and total quality control. To the extent that management is successful the need for flexibility is reduced and production technology can be chosen mainly for its efficiency. Power in the marketplace is needed for this approach to be successful, which usually translates into large company size relative to customers and competitors. Another requirement is a manufacturing process that is understood enough to be controlled.

Instead of discarding its excess flexibility a company may decide to "bank" it, that is, hold it in reserve to meet a future need. It may eventually be used defensively to adapt to a sudden upheaval in the market. If American auto companies had been able to draw on stores of flexibility, their adjustment to Japanese competition would have been much smoother. Alternatively, it may be employed proactively in the manner of Honda. One existing example is the "surge capacity," which the U.S.

Department of Defense requires that military contractors maintain in order to quickly increase production volume during a national emergency. Banking flexibility requires that investment alternatives with long-run intangible returns are not unduly penalized.

The framework's evolutionary implications indicate the relationship between defensive and proactive approaches. As environmental uncertainties change over time, existing strategy becomes ineffective resulting ultimately in low performance. Failure to meet performance targets is the motivation in the model to revise strategy as indicated by the feedback loop between these two concepts. Changes in manufacturing flexibility and production technology follow. Abernathy (1978) uncovered a developmental process in certain industries in which temporal diminutions in market uncertainties correspond with shifts in firm strategy from adaptation to uncertainty reduction. Simultaneously, product differentiation gives way to cost leadership, manufacturing flexibility to production efficiency, and general purpose equipment to dedicated automation.

Recent events have indicated that even firms in an advanced stage of development are threatened by pronounced increases in market uncertainty due to international competition and new superior products. Abernathy, Clark, and Kantrow (1983) discussed whether these companies could handle their new situations. Since their dedicated production processes severely constrain manufacturing flexibility Susman and Dean (1988) argued that they should rely on AMT. Although the new technology is not as efficient as dedicated automation it is more effective at minimizing risk. In other words, AMT is a depository of flexibility for companies in mature industries worried about the possibility of competitive upheavals.

Implications of the Model

To review the major features of the conceptual model, strategy and production technology are embedded in a broad context embracing several variables. Within this context the two factors are directly linked by manufacturing flexibility. Strategy and technology are compatible to the extent that the flexibility demanded by the former matches that delivered by the latter.

Flexibility's pivotal role in the model corresponds to its growing salience to manufacturing managers. The 1986 Manufacturing Futures Survey queried manufacturing managers or technical officers in 214 Japanese, 186 American, and 174 European corporations (DeMeyer, Nakane, Miller, and Ferdows 1987). The European and American companies both ranked quality and dependable delivery as first and third on their competitive priority lists. Flexibility to introduce new products was only sixth and flexibility to adjust production volume was only eighth on both lists. The Japanese, however, ranked these two flexibilities as second and fourth due to increasing market uncertainties. Quality was third and dependable delivery was fourth (tied with volume flexibility). To introduce more flexibility into their operations the Japanese are using flexible manufacturing systems and broader ranges of tasks for workers.

DeMeyer et al. believed that American and European firms are still concentrating on deficiencies in quality and delivery. The Japanese who are successfully overcoming these problems are turning their attention to flexibility. Over the next few years American and European companies will have to give this issue more priority if they are to compete.

The model highlights the following interrelated strategic processes.

1. Flexibility is normally considered as an adaptive means of handling environmental uncertainties. American managers view it in this way when they talk about adjusting to changing customer needs. AMT is one way to provide the required flexibility. Overemphasis on adaptation, however, breeds an inability to be proactive.
2. By proactively encouraging customers to rethink their market needs a company redefines an industry's competitive uncertainties to its own advantage. Through superior manufacturing flexibility, due perhaps to AMT, it can meet the new needs better than its competitors. American managers need to think more along these lines.
3. Various marketing and manufacturing policies exist for proactively reducing uncertainties. The need for flexibility and correspondingly AMT is thereby diminished. American firms do take advantage of this alternative, but many are not using it judiciously. They need to ensure that flexibility is not being added at the same time the need for it is being removed.
4. A firm may bank flexibility to use proactively or defensively in the future. AMT acts as a depository when its potential flexibility is intentionally not tapped completely until a need or opportunity arises. There is little indication of how many firms consciously make this type of investment.

The model has some normative implications. It asserts, for example, that strategy should influence manufacturing flexibility requirements and hence the choice of production technology. This reflects a top-down planning process in which CAM is used as a tool for realizing company objectives (Gold 1980, Gunn 1987, Slack 1988). A process of this type does not preclude lower level involvement to ensure commitment.

In practice this preferred direction of flow is sometimes reversed. Low level engineers want to imitate competitors or to increase the manufacturing process' degree of automation. The technology they purchase influences the type and degree of flexibility as a by-product. In turn flexibility constrains strategy in ways not compatible with where top management wants the firm to be moving. In this bottom-up planning process the tools are selected first and strategic objectives may never be addressed (Gunn 1987).

Senker and Beesley (1985), who studied the introduction of computer-aided production and inventory control systems in British industry, provided some evidence. Nine of their companies started from clear goals such as the need to increase volume. Six rated their new systems as successful, three rated them in a middle ground of being operational, and none believed them to have failed. Four companies had vague goals such as "We need to computerize," or "Everyone else has one." None of them rated their systems as successful, one rated theirs as operational, and three believed them to be failures.

Lim (1987) studied twelve British FMS users with a mail survey and a structured questionnaire. He found that strategic objectives were inconsistent with the types of flexibilities respondents said they needed. Although three companies at most pursued new product development as a strategy, eleven said they needed quick changeovers to new products in manufacturing. Lim's conclusion was that if some of the companies had done more careful planning they would have introduced FMSs with different capabilities.

Since it is based on just five variables the model is parsimonious. At the same time it subsumes other frameworks in the literature. Swamidass and Newell (1987) also start from environmental uncertainty influencing strategy. In their view, strategy has a content and a process aspect. The content component reflects required flexibility, which includes goals for the frequencies of new product introductions, product variety, and product features relative to the industry. The process component is the degree to which manufacturing managers participate in strategy making. Both aspects are hypothesized to influence performance. Their model has no feedback processes from performance to strategy and from strategy to uncertainty.

Swamidass and Newell tested their ideas by administering questionnaires to executives in thirty-five machinery and machine tool firms in the Seattle, Washington, area. As predicted they found a significant positive relationship between uncertainty and strategy content, but only 10 percent of the variance in the latter variable was explained. Uncertainty was also significantly related to strategy process but, contrary to predictions, in a negative direction. The two strategy variables were significantly and positively related to performance and explained about 36 percent of its variance.

Slack's (1988) flexibility hierarchy based on a study of ten British manufacturing companies is similar to that part of Figure 2.1 involving strategy, flexibility, and production technology. Strategy includes a priority ranking of the need for productivity, product availability, and product dependability. Next, a firm must determine the importance of various production system flexibilities by understanding how each contributes to the three strategic objectives. Finally, the information on flexibility guides the selection of appropriate production system resources such as production technology.

In succeeding pages the conceptual model will function as a road map for more detailed investigations of strategy and AMT. The next part of this chapter travels along the path from uncertainty to flexibility via strategy. It provides a deeper understanding of manufacturing flexibility and its interrelationships with the other two variables. Chapter 3 discusses how to provide manufacturing flexibility through investment in AMT. It analyzes choices on design and justification. Chapter 4 on performance measurement shows how to investigate whether AMT is delivering enough flexibility. It also presents evidence on the flexibility of CAM in practice. Chapter 5 completes the tour by traversing the model's feedback loops and concentrating on the framework's evolutionary implications.

UNCERTAINTY, STRATEGY, AND FLEXIBILITY

The model's branch, which represents adaptation, links uncertainty, strategy, and flexibility. Their interrelationships are explored by unbundling each abstract concept into a number of more specific dimensions. Since no rigorous methods exist for indicating the domains of the concepts the approach adopted here utilizes the limited amount of relevant theory as well as interviews with manufacturing managers. Table 2.1 presents six different sets of uncertainties, strategies, and flexibilities. The top four sets are market oriented because the relevant uncertainties exist in the demand for products. The bottom two are process oriented since their uncertainties exist in the manufacturing technology or its inputs.

Discussion will be centered on flexibility because of the critical role it plays in the conceptual model. Each type of flexibility is considered to have two aspects—range and time (Slack 1983). One manufacturing process is more flexible than a second if it can handle a wider range of possibilities. For example, it may be able to vary production through a greater range of volumes. The second process however may be more flexible in the sense that it can attain a new production level within its range faster than the first.

Table 2.2 indicates the range and time aspects of each type of flexibility. It helps us see how important concepts in the manufacturing strategy literature mesh with the framework developed here. The range aspect of mix flexibility, product variety, is tied up with the concept of the focused factory. The time aspects of mix and changeover flexibility, lead time, and startup time, respectively, are the bases for time-based competition. JIT increases mix flexibility through product variety and lead times. CIM has the potential to improve the range and time aspects of mix and changeover flexibility. There is, of course, some overlap; time-based competition uses elements of the other concepts.

TABLE 2.1 Uncertainty, Strategy, and Flexibility

Nature of Uncertainty	Strategic Objective	Type of Flexibility
Product Markets		
Demand for the kind of products offered	Diverse produce line	Mix
Product life cycles	Product innovation	Changeover
Appropriate product characteristics	Customer responsiveness	Modification
Amount of aggregate product demand	Market share	Volume
Process Operations		
Machine downtime	Meet customer due dates	Rerouting
Meeting raw material standards	Product quality	Material

TABLE 2.2 Range and Time Aspects of Flexibility

Flexibility Type	Range Aspect	Time Aspect
Mix	Variety of parts handled	Lead time
Changeover	Variety of major design changes	Startup time
Modification	Variety of minor design changes	Time to manufacture a minor design change
Volume	Amount of change in the production level	Time to accomplish the change in production level
Rerouting	Degree to which machine sequence can be changed	Time to reroute and do the work
Material	Range of compositional and dimensional variations	Time to make the adjustments

Slack (1988) proposed a method for investigating whether the range and time aspects of a flexibility dimension are related. A range-response curve shows the extent of change possible from the existing point in the range as a function of the time needed to accomplish the change. For example, Slack used a curve for changeover flexibility developed for a British electronics company. The range variable involved categories running from customized variants of existing modules, the smallest change, through novel modules using new technologies, the largest change. Time was represented by the budgeted duration to achieve a given category.

What are the implications of range-response curves? They are useful in comparing the flexibility of two or more manufacturing systems. If one system's curve dominates another's, it is higher at any given abscissa value, then it is unequivocally more flexible on that dimension. Where domination does not exist, it is possible to learn over what sets of values for range and time one system is more flexible than another. Presumably the concept can also be extended to illuminate how the various flexibility dimensions interact with each other. One could compare how the range or time aspect of one dimension changes as a function of the range or time aspect of another dimension.

Slack (1983) included cost as a third aspect of flexibility, but it seems more appropriate to consider it as an economic impact. First, suppose two manufacturing systems have similar range and time values for a given flexibility dimension, but one delivers them more cheaply. It is not more flexible, but it is more efficient. Second, by keeping the cost and flexibility concepts separate one can determine relationships between cost and range or time, which are useful for comparative purposes. The innovation cost-time trade-off function discussed by Mansfield (1988) is an example.

Mix Flexibility

It is impossible to predict which type of color television sets made by Philips will be sold in what numbers for more than three months ahead (Bolwijn et al. 1986).

This uncertainty as to which products are acceptable to customers leads to the strategic objective of product diversity. A firm desires a broad product line or a number of related lines. From 1980 to 1982 the number of different types of Philips' color TVs went from 400 to 800. Mix flexibility is a necessary ingredient of the manufacturing process in this situation. With it the capability for switching readily among currently produced products or variants is built into the system. The range aspect signifies the breadth of the product line or the extent of product variety, while the temporal aspect reflects setup time, an important component of manufacturing lead time. A flexible manufacturing system with a random processing capability provides good performance on both features.

Mix flexibility's significance lies in its challenge to the more established principle of the focused factory. Japanese companies are moving to compete on the basis of product variety, whereas many Western firms are still trimming product lines to gain focus (Abegglen and Stalk 1985). What are the advantages and disadvantages of each approach?

The Focused Factory. Focused manufacturing relies on the benefits of task specialization as opposed to task variety (Hayes and Wheelwright 1984, Skinner 1985). Narrowing the range of demands placed on a manufacturing facility allows its human and technical resources to concentrate on a few key priorities instead of a large number of conflicting ones. It is also a way of reducing uncertainties by cutting the number of problems encountered.

Focusing is compatible with an overall leadership strategy as well as a niche strategy as indicated in Table 2.3. To promote overall leadership many American and European enterprises maintain a broad product line at the business unit level, whereas individual factories have limited tasks. A good example cited by Hayes and Wheelwright (1984) is Reliance Electric Corporation's motor division, which decided to make small-, medium-, and large-size motors using seven factories, each one concentrating on a narrow size range.

Allocating a broad business unit task so that each plant has a well-defined specialization is done according to any of the following bases: markets served (e.g., original equipment manufacturer, consumer, industrial), production volume (e.g., high, medium, low), product families, or degree of product customization. For example, low volume production of many diverse items is incompatible with high volume production of a few standardized items. They belong in different facilities.

TABLE 2.3 Implications of Mix Flexibility

	Focused (Overall)		Focused (Niche)		Flexible		Traditional	
	Range	Time	Range	Time	Range	Time	Range	Time
Business unit	Broad	Depends	Limited	Depends	Broad	Short	Broad	Long
Factory	Limited	Depends	Limited	Depends	Broad	Short	Broad	Long

Since the business unit level must contend with a broad mission, the primary benefit of focus, a reduction in conflicting demands on shared resources, is mainly enjoyed at the factory level. People, technologies, systems, and structures no longer have to reconcile incompatible priorities. It is unnecessary to disregard some demands or to make unsatisfactory compromises. The consequent reduction in complexity and confusion lowers overhead costs. There is less need for technical, staff, and supervisory people to ensure smooth operations. Setups, inventory, and material-handling costs are reduced. Stalk (1988) estimated that each time that variety doubles, these costs increase at a rate of 20 percent to 35 percent per unit.

Although some evidence exists on the cost advantages of focus, it is not based on statistical tests of significance or on controls for size, vertical integration, or other factors. In one company, overhead costs, including data processing, quality control, production scheduling, and supervision were compared among a facility manufacturing components for 125 models and a facility producing components for only 30 models. The latter facility's overhead costs per equivalent component (presumably an average taken over components made in both plants) were much lower (Hayes and Wheelwright 1984). A forklift truck manufacturer had six factories making from three to twenty-eight product families each. The average overhead cost per unit per plant went up as the number of product families increased (Abegglen and Stalk 1985). In a third company with seven regional plants, gross operating margin was compared to various measures of focus. The profitability measure increased as the number of products in a factory decreased, as the total number of company divisions served by a factory decreased, and as the percent of a factory's output going to its own division (versus other divisions) increased (Hayes and Wheelwright 1984).

If variety-related unit costs tend to decrease, scale-related unit costs may rise. Instead of a few large-scale plants there are a greater number of small-size factories. Each factory has a full complement of its own resources to avoid the penalties of resource sharing. Small factories, however, may not have enough production volume to justify the efficient use of these duplicated people and technologies.

One way of utilizing duplicated resources more efficiently is to share them when the opportunity arises. A transfer is arranged when one facility has an excess of what is in short supply at another. Plants focused on different market groupings may manufacture some of the same products. If one needs inventory that another is holding, a transfer may benefit both units. There is, however, little incentive to share due to the premium placed on keeping factories independent, the cost of establishing equitable transfer mechanisms, and the built-in competition among units.

To compensate, some resources must be less specialized than before, an ironic result of focusing. A manufacturing engineer who previously specialized in equipment justification in a large factory will have to perform a variety of tasks in a smaller one. Consequently, specialized expertise is no longer available, specialists lose their skills over time, and the viewpoints of various specialties are underrepresented in decision making. Alleviating these problems by also maintaining centralized staff units results in even higher overhead costs.

The plant-within-a-plant concept is a compromise that allows sharing of some expensive resources while keeping the rest separate (Skinner 1985). Focused manufacturing processes within a single plant are separated spatially and organizationally. Although these processes share the same facility, each has its own workforce, technical people, and systems.

Table 2.3 indicates that the niche version of focus involves limited product variety at the business unit and factory levels. Concentration is usually placed on high volume production of a few standardized items as was the situation for many Japanese concerns when they broke into international competition. Data on operating margins versus number of product lines for eleven companies in the same industry testify to the strategy's effectiveness (Hayes and Wheelwright 1984). The five firms making only one or two lines had the highest margins, whereas the six firms producing from four to six lines had the lowest.

Concentration on high volume production avoids complexity and confusion penalties at the business unit and factory levels. It fosters scale economies due to relatively little duplication of resources. Broad-line competitors, with higher costs at similar volumes, are forced to add new products in order to compete. As a result, their costs continue to increase allowing the focused enterprise to gradually expand its line. Gradual expansion is necessary due to a limited growth potential in the narrow range especially as product maturity sets in. If, however, focusing does not result in a cost advantage in mass markets the strategy is doomed (Abegglen and Stalk 1985).

Although a focused business unit, due to its limited task, provides better customer service, it may exhibit poor market responsiveness if the existing situation changes radically. If its facilities emphasize high volume production of a few items on dedicated machinery it is especially liable to this problem. Technological developments in the industry may switch customer allegiances to other versions of a product. The enterprise has simply placed its bet on the wrong alternative and is too specialized to adjust to the new reality in time. This is the classic case of the "Model T Syndrome." In the late 1920s when General Motors introduced more comfortable closed-body vehicles it took Ford over a year to reorganize its operations. Its lack of responsiveness cost Ford market leadership in the industry.

Product Variety and Short Lead Times. Having mix flexibility in contrast to focus implies a relatively broad product line at the business unit level by virtue of each individual factory making an expanded range of items. The firm exhibits market responsiveness when uncertainties exist as to which products will succeed. The chances of something being accepted by the market increase as a firm's portfolio widens. The time dimension of mix flexibility provides another kind of market responsiveness. Manufacturing lead times are cut since it is possible to make small quantities of numerous items to satisfy immediate market needs. As indicated in Table 2.3 traditional factories may make a range of products but usually at the expense of long manufacturing lead times. With focused manufacturing the lead time implications depend upon the basis used to create a focus. In a flexible factory both the range and time aspects are available.

Wernerfelt and Karnani (1987) indicated two qualifications on the portfolio effect of an expanded range of products. First, it works best for large companies with abundant resources. A small firm may spread its resources too thinly by trying to hedge its bets. Second, assume that one firm in an industry hedges its bets through mix flexibility while each of the others places its bet on a single product through focusing. No matter which product eventually proves to be successful, the flexible company will be dominated by one of its competitors.

Mix flexibility supports overall cost leadership by facilitating market segmentation, but the first kind of focusing achieves the same thing. Suppose a heterogeneous market can be segmented into subpopulations each consisting of homogeneous sets of consumer preferences. If a diverse product variation is tailored to each subpopulation larger sales will result than if a single product is offered to the entire population. There is a better matching of product attributes and customer needs (Talaysum, Hassan, and Goldhar 1987). When segmentation also divides a previously coordinated or concentrated population into a diverse group of uncoordinated sectors, a company's environmental dependencies are broken up (Pfeffer 1987). In other words, environmental uncertainties are reduced—an example of how mix flexibility can support proactive uncertainty reduction.

Mix flexibility offers a capacity advantage versus focusing when it is based on common equipment. The capacity of a facility handling several products on the same equipment is less than the sum of the capacities of the plants that each manufacture a different product. With the flexible alternative a shortfall in demand for one product opens up capacity that can be used by another whose demand is greater than expected. Under the same conditions one focused plant will have excess capacity and another will be unable to meet all of its demand.

Consider one of our Big Three automakers that went into the minivan business a few years ago. It based its new factory's capacity on imperfect forecasts, not unusual for a new product. When demand turned out to be much greater than anticipated there was no way to take advantage of it immediately. The new vehicle could not be produced by the other assembly plants where mix flexibility was too low.

A French auto company intends to deal with the same problem in the following way. Each assembly plant will be the major producer of one car model and the reserve producer of another when demand warrants. The company will use AMT to have activities such as body framing, painting, and inspection handle two different models. Its only manually operated plant, flexible enough to handle almost any model on short notice but at a cost and quality penalty, is being phased out.

In a flexible factory ways have been found to enjoy the benefits of product variety and short lead times without incurring substantial increases in variety related costs (Stalk 1988; Stalk and Hout 1990). At least two alternatives currently exist, Just-in-Time and computer-integrated manufacturing, although there is a growing belief that the former should pave the way for the latter. Both have implications for manufacturing technology.

A factory employing JIT principles affects variety-related costs in several ways. Organizing according to product flow reduces material handling. There are multiple lines operating in parallel for different product families. Engineering improvements cut setup times, allowing more frequent production of the complete product mix. Producing to meet the immediate needs of downstream activities reduces the need for inventory. Total quality control pares the costs of defects.

Computer-integrated manufacturing employs a single manufacturing process flexible enough to handle a variety of products. Variety-related costs diminish since engineering design, process planning, production control, and other overhead activities are embodied more in software than in people. Such a plant may exhibit economies of scope. It can make a range of items at lower unit cost than separate plants each manufacturing one item (Goldhar and Jelinek 1983).

Empirical research concerning a flexible strategy is almost totally lacking. Using the Profit Impact of Marketing Strategies (PIMS) database, Kekre and Srinivasan (1990) found that a broader product line at the business unit level leads to higher market share and profitability. Once more, there was no strong positive relationship with inventory and manufacturing costs. In fact, for companies selling to industrial markets, product line breadth had a small but significant *negative* association with manufacturing costs. The authors speculated that perhaps the use of JIT, CIM, and related activities accounted for the impact on costs. Note, however, that no distinction was made between business units that adhered to a flexible strategy versus an overall focus strategy.

Changeover Flexibility

Uncertainty in the length of product life cycles is especially troublesome when life cycles are shortening on average. Half of Philips' products have a commercial life of three years or less (Bolwijn et al. 1986). Qualls, Olshavsky, and Michaels (1981) analyzed changes in the length of the introductory and growth stages of the life cycle for thirty-seven household appliances over a fifty-year period. The products were divided into three distinct groups based on when they were introduced; twelve prior to World War II, sixteen after the war, and nine after 1964. The average length of time for the first two stages of the life cycle was 46.3 years, 26.5 years, and 8.8 years, respectively, a finding that was statistically significant.

To adjust to this type of uncertainty an enterprise should foster the strategic objective of product innovation. Then the manufacturing process requires changeover flexibility, the ability to quickly substitute a wide range of new products for those currently being produced. The range aspect indicates the variety of major design changes that can be accommodated, whereas the temporal aspect refers to manufacturing startup time—the portion of new product introduction time that occurs in manufacturing. Changeover flexibility reflects an ability to alter the product mix.

Range Aspects. Firms most convinced about the benefits from changeover flexibility's range aspect have had some disastrous experience with dedicated automation. A Midwest agricultural equipment manufacturer decided to purchase a transfer line to make a part already fabricated on an FMS when demand appeared to be increasing. Unfortunately, the heavy market response did not materialize. The transfer line, which could not be adapted to any other use, had to produce the part inefficiently. There was little difficulty in finding other parts to fill the FMSs unused capacity. A Texas-based defense contractor was told by the government to build a new plant for the next generation of a missile it had been producing. The firm lost the contract, however, and then had problems in getting rid of the machines, 50 percent of which were dedicated. It is now determined to build future plants around flexible automation.

An opposing viewpoint is that investing in the added costs of a changeover capability is wasteful if new technological developments will probably render a manufacturing process obsolete by the time new products are introduced. A preferable alternative is to have a process without changeover capability, scrap it when existing products are phased out, and build a new one with the latest technological advances for the next generation of products. This view was expressed in Sakurai's (1990) investigations of Japanese users of AMT.

First of all, this argument does not take into consideration the possibility of demand faltering for existing products before the expected end of their life cycles. Then their manufacturing process must be able to accommodate other products to recoup the capital investment made in it. Second, changeover flexibility in the current process acts as an incentive to incorporate technological advances as they occur rather than when new products are introduced. When much of a process will outlive existing products, management is more willing to invest in incremental improvements. Since a process without a changeover capability must be scrapped when the associated products are phased out, there is little interest in incorporating new advances into it especially if phasing out is in the near future. A third point is that being able to make only relatively minor adjustments in the existing manufacturing process to accommodate a new product reduces the time to introduce it.

Another argument makes the benefits of a changeover capability contingent on the nature of market uncertainties. When these uncertainties are very erratic, it is hard to predict what product designs will be like in the future. When it invests in a given range of changeover flexibility a company takes a gamble that the manufacturing process will be able to make the next generation of products. If, however, the amount of uncertainty is less extreme, process designers can identify a dimensional envelope and be reasonably sure the next generation of products will fit into it. Dimensional needs are based on already established plans for future products. Corporations most likely to be in the former situation manufacture end products in hi-tech environments. Firms that make basic components for a large number of stable uses fit into the latter situation.

Time Aspects. An important aspect of time-based competition is the reduction of new product introduction times as illustrated by the previously discussed H-Y War between Honda and Yamaha. Manufacturing startup time—the time dimension of changeover flexibility—is a significant component. Some methods for reducing manufacturing startup time, which have implications for AMT, include the following:

- Frequently introducing small product changes, which means increased but smaller-scale revisions in the manufacturing process

- Simultaneous design of new products and production processes using cross-functional teams
- Shifting pilot production from a central facility to the plants where flexible automation allows running small pilot lot sizes without disturbing normal operations
- Tying customers into a firm's computer system so they can communicate directly with the factory

Bolwijn et al. (1986) believed that North American and European companies could regain supremacy over Far Eastern firms in Western markets by competing on the basis of shortening the time to introduce new products. They argued that the wide-ranging distribution networks and production facilities of Western companies provide an inherent advantage on which to build. Data, however, from the Manufacturing Futures Survey (Miller and Roth 1987) suggest that the Japanese may have already taken the lead. Of the top eight competitive priorities of Japanese enterprises, rapid design changes was number two from 1983 through 1986. For North American and European companies, rapid design changes ranged from number five through number seven over these years.

The most compelling evidence, however, would be international comparisons of the actual cost and time of new product introductions. Mansfield's (1988) investigation, one of the very few which have been conducted, indicates the issue of which country is in the lead is more complex than is generally portrayed. He differentiated between internal innovations, those for which a major portion originates within the innovating firm, and external innovations, those for which a major portion is based on imitation or licensing. For each type of innovation he also considered development (mainly R&D activities) separately from commercialization (manufacturing and marketing startup).

Most of Mansfield's results are based on a random sample of thirty pairs of Japanese and U.S. firms matched for size and industry. Contrary to popular belief he discovered that the average cost and time to innovate internal products was comparable in Japan and the United States. The cost and time to innovate external technology, however, was much lower in Japan than in America, principally due to the expense and length of commercialization activities in our country.

The Japanese are also more willing to pay the increasingly heavy cost of constantly reducing new product introduction times, reflecting their commitment to time-based competition. Mansfield computed elasticities of innovation cost with respect to time, that is, the percent cost increase required to reduce innovation time by 1 percent. The Japanese elasticities were roughly double the American elasticities for external development and commercialization and internal development and commercialization.

Mansfield interpreted his results using the innovation cost-time trade-off function, a curve with a negative and increasing slope. For external innovations the Japanese curve is much lower than the American. They operate at a point on their curve representing lower cost and time. Since the point is further up on its curve a given reduction in innovation time can only be obtained at a greater incremental cost.

However, this is of little import for the Japanese because of the considerable cost and time advantage they already have.

For internal innovation the Japanese and American curves are close together and may, in fact, intersect. Both countries operate at points on their curves representing similar costs and times. The Japanese, however, are at a point on their curve at which further reductions in time are relatively more expensive. On average, American firms engaging in time-based competition can spend less to achieve the same time reduction or can spend the same amount to achieve a greater time reduction. This represents a significant competitive advantage if utilized.

Other Flexibility Dimensions

Although most companies currently give top priority to mix and changeover flexibility, attention is also being paid to the other dimensions.

Modification Flexibility. Uncertainty as to which product attributes customers desire arises when a new product is introduced or when a standard design is customized on an *ad hoc* basis. It calls for modification flexibility in the manufacturing process, the ability to implement minor design changes in a given product. The range aspect refers to how many different kinds of minor changes are possible, whereas the time aspect reflects the speed with which a given change is accomplished. CNC machine tools and robots, which typically require a modest amount of reprogramming and refixturing to handle minor design changes, provide this capability. Modification flexibility is associated with the strategic objective of responsiveness to customer specifications.

Modification flexibility allows product engineers to react to market feedback by improving a design's functioning and/or appearance. They have more freedom to explore the consequences of alternative designs since typically it is only necessary to modify parts programs. In Gerwin (1981a) an American company had to choose between a modified transfer line and an FMS to machine a new product line. The transfer line was more cost efficient, but the FMS could more easily handle the inevitable design changes. The latter system was selected in part for this reason.

There is a danger, however, in an overdependence on modification flexibility. Engineers may feel less pressure to get a design right the first time if they can count on catching problems when prototypes are machined. New product introduction times are consequently lengthened. Ideally, this type of flexibility should be used to handle unavoidable adjustments in new products and planned changes in existing products.

To enhance modification flexibility some companies are experimenting with computer-integrated enterprise (CIE) approaches in which electronic links exist between customers, factories, and vendors. One American firm is developing a software package that customers will use to input specifications for a made-to-order product. The output is a product design and a signal to the factory to order components through computer links with vendors. Interventions by the product engineering or applications engineering department will be no longer required.

Volume Flexibility. This type of flexibility permits increases or decreases in the aggregate production level of a manufacturing process. Its range aspect refers to the amount of change that is accomplished, whereas its time aspect deals with how long it takes to make the change. Being able to adjust aggregate production usually requires either very high capacity limits, limits that are not very rigid, or subcontracting.

A need for volume flexibility occurs in response to uncertain swings in the amount of customer demand. Firms in the computer industry are subject to the "hockey stick" effect* in which as much as 50 percent of quarterly sales may occur in the last two weeks due to the offering of discounts during this period. Sometimes there is an unanticipated surge in demand when a new product is introduced. The previously mentioned Big Three automaker could have handled its surprisingly large demand for minivans by building volume flexibility into its new plant. Defense contractors are required by the government to be able to increase production in a national crisis quickly and dramatically.

Such problems are exacerbated by the dynamics of production-distribution systems with long lead times. Production must be adjusted by more than the given change in sales due to unavoidable impacts on inventory levels. Time-based competition by reducing lead times helps eliminate some of the uncertainty which creates a need for volume flexibility.

Maintaining or increasing market share, the associated strategic objective, is particularly significant in mass production industries. Witness General Motors' costly efforts during the mid 1980s to stimulate the market and maintain capacity in the face of declining sales. Subsequently, in the face of a declining dollar, Japanese auto manufacturers stubbornly refused to raise prices correspondingly, preferring decreased profit margins to a lower market share.

Rerouting Flexibility. Downtime uncertainty has a short-run aspect, which arises from the necessity to cope with equipment or quality problems, and a long-run aspect, which occurs when machines are refurbished off-line to accommodate major product design changes. Rerouting flexibility, defined as the ability to adjust the sequence of machines through which a part flows, compensates for these uncertainties. In a flexible manufacturing system the parts routing program may automatically reroute parts if a machine breakdown occurs.

The range aspect reflects the extent to which a part can be rerouted and the variety of parts for which rerouting occurs. The time aspect considers how long it takes to make the adjustment plus how long it takes to perform the rerouted work minus how long it takes to perform the work with the original sequence. Sometimes the second time is greater than the third reflecting a less efficient alternative. Rerouting flexibility is associated with the strategic need to meet due dates specified by customers.

To the extent that uncontrollable factors temporarily take machines out of operation some uncertainty is given and a degree of rerouting flexibility is indispensable in the manufacturing process. Rerouting, however, is a built-in redundancy, a nonvalue

---

* A graph of sales versus time for a three month period where sales are level for the first two-and-a-half months and sharply increase in the last two weeks resembles a hockey stick.

added alternative, when it is used as an excuse for not eliminating controllable causes of downtime. Unfortunately, the borderline between what is controllable and what is not is imprecise. Maintaining equipment and solving quality problems should eliminate all but the uncontrollable uncertainties from a given manufacturing process. Yet there remains the possibility of having vendors design more reliable machines or even performing equipment design in-house.

Material Flexibility. Uncertainties in the composition and dimensions of the materials being processed may occur at the supplier or in a company's upstream activities. They create a need for material flexibility—the ability of human or automated adjustment mechanisms to correct or adapt to unexpected variations in inputs. The range aspect refers to the extent of an unanticipated variation such as a deviation in length of 0.001 inch vs. .01 inch. It also refers to the number of different kinds of variations susceptible to adjustment. A second aspect is the time required to make an adjustment. Since material flexibility contributes to reducing defects, it facilitates the strategic objective of product quality.

Material flexibility's role in manufacturing is in dispute. Traditionally, American companies overrelied on the flexibility of human operators in the manufacturing process to handle unanticipated deviations in materials. Assembly in particular was used as a screen to find defects since it is the point at which components come together for the first time (Whitney et al. 1986). In auto body framing, for example, workers could readily locate bent metal and straighten it so that welding could be done properly. However, assembly typically occurs at or near the end of the manufacturing cycle. All the upstream activities, including product design, purchasing, fabrication, and aspects of quality control, were under less pressure to do their jobs right the first time. It is tempting to avoid searching out the root causes of technical problems, to eschew rigorously adhering to standards, and to be overly tolerant of mistakes when adjustments can be made to the resulting crises.

Once firms begin to experiment with computerized automation the situation quickly changes. As human operators in fabrication and assembly are replaced, material flexibility drops. Gerwin and Tarondeau (1989) surveyed eighty-one French companies using some type of programmable automation in fabrication. Forty-four percent registered either a decrease or a large decrease in this type of flexibility as a consequence of introducing the technology.

To deal with the resulting breakdowns and quality problems manufacturers reduce the compositional and dimensional variations in materials and redesign the manufacturing process. For example, since welding robots used in auto body framing cannot recognize bent metal or do something about it, upstream activities must reduce the deviations they produce and assembly must introduce palletized parts and part feeders. Business units with FMSs significantly differ from those without FMSs in placing more emphasis in future plants on vendor quality, purchasing management, and a zero defects policy (Tombak and DeMeyer 1988).

Computerized automation therefore helps redress the balance between control of and adaptation to uncertainty. Introducing it has the same effect as reducing inventory; uncertainties in the manufacturing process are revealed and eliminated. This hidden advantage, however, is an equivocal blessing. One may find that after correcting the problems indicated by the technology humans will be able to do a better job than AMT. One American company, for example, redesigned a product to make it easier for robots to put together only to find that they had enhanced the abilities of assembly workers to do the job even more. Why then wait until the technology is purchased to clear up problems? Eliminating the problems beforehand may obviate the need for the technology in the first place.

There is also a danger that a firm will move too far too fast in substituting AMT for human resources. First, it is very expensive. Second, it creates new kinds of complex technical and managerial dilemmas. If it is accomplished faster than the ability to improve upstream activities the problems are exacerbated. General Motors, by installing its first computerized body-framing unit before redesigning for automation the components to be attached by it, compounded the difficulties in assembling them. The old designs were not readily put together by the new equipment. Third, going too far too fast swings the balance between accommodating and controlling uncertainty too much in the direction of the latter. Since all process uncertainties cannot be eliminated, some degree of material flexibility must be preserved. Besides being ruthless about eliminating process uncertainties, the Japanese utilize multiskilled workers who stop the line when trouble occurs and cooperate to solve the problem. Especially where very tight tolerances are required, such as in aerospace or defense work, human monitoring of metal cutting operations is indispensable. When tolerances are just a little wider than a machine's capabilities, only a thin margin exists to accommodate variations in materials and cutting tools. Machine operators have the crucial role of detecting tool wear and faulty materials.

What will be the impact of more sophisticated generations of AMT such as flexible humanlike robots? For example, consider the revolutionary FMS developed by a large U.S. defense contractor and the Department of Defense. If successful, it will be the first FMS with totally automated material-handling capabilities. Robots equipped with special end effectors and three-dimensional vision systems are meant to load and unload over eighty different parts whether they are in the form of raw billets, parts machined on one side, or completed parts. The robots must be able to adjust to different sizes and shapes, locations, and obstructions due to chips, coolant, or burrs.

Experiments such as this one will help resolve whether it is technically and economically feasible to substitute automation for humans in tasks requiring an ability to sense deviations and make adjustments. Designers of the project acknowledge that this is the most risky aspect of the entire system. Continuing technical difficulties elsewhere are leading some engineers to believe that creating robots with human flexibility is not possible (Whitney 1986). From an economic viewpoint, making a robot more competent and versatile can increase its cost to the point where it is not as efficient as a factory worker. If success is achieved eventually designers must avoid a rush to install liberal amounts of automated material flexibility into AMT. It may reduce pressures on upstream activities in the manufacturing process to eliminate controllable deviations.

Implications of the Flexibility Classification

Embedding flexibility within a conceptual framework goes a long way toward unraveling some of the mystery surrounding the concept. Through its connection with strategic objectives, flexibility is a potential adaptive response to uncertainty or an aid in controlling uncertainty. It is also a multidimensional concept with each different type possessing a range and time aspect. Most of the types identified here have been recognized elsewhere as being significant (Browne, Dubois, Rathmill, Sethi, and Stecke 1984, Donner 1988, Slack 1988).

Classifying flexibility in this way helps to relieve managers' confusion about the capabilities of a manufacturing system. Slack (1987), for example, studied a company that had recently installed a new flexible machining center. The company dramatically lowered throughput times but also decreased the range of products that it could produce. The marketing manager whose primary concern was with product variety could not understand why the new flexible technology was less flexibile than the old conventional equipment it had replaced. His confusion lay in not distinguishing between the range and temporal aspects of the mix dimension.

The flexibility constructs discussed here are meant to apply with suitable adjustments at most levels within an organization including the individual machine, manufacturing system or cell, plant, or strategic business unit level. Another approach, hierarchical in nature, defines flexibilities which operate at different levels and assumes that those at one level are required to have those at the next highest (Sethi and Sethi 1990). Eleven flexibilities with emphasis on range aspects are divided among three levels. At the level of machines or other components of a manufacturing system there is machine flexibility, for example, which refers to the types of operations a machine can perform without a prohibitive effort in switching from one to another. Mix, changeover, volume, routing, and other kinds of flexibilities occur at the manufacturing system level. The aggregate, presumably factory, level includes production flexibility, the universe of part types capable of being produced without adding major capital equipment, and market flexibility—the ease in adapting to a changing environment. Note, however, that machine, mix, and production flexibility are based on essentially the same concept operating at different levels.

The hierarchical approach to flexibility has at least two limitations vis-à-vis the approach advocated here. When the hierarchical approach assumes that flexibility is a way to adapt to changing environments, it neglects the concept's proactive function. By defining specific flexibility dimensions in terms of their range characteristics, it does not consider their time aspects. These limitations, however, are not inherent in the approach. Ways may be found to overcome them.

Are the six uncertainty, strategy, and flexibility combinations independent of each other? Although empirical research is needed to answer this question, it is tempting to speculate about a complex web of secondary effects. First, uncertainties may impact on flexibilities. Machine downtime and material variations increase manufacturing lead times, changeover times, and modification times. As product life cycles shrink, firms are committed to larger varieties of spare parts for previous offerings thus increasing the need for mix flexibility. Second, flexibilities may affect strategic objectives. Modification flexibility permits minor design changes that improve quality. A product quality objective is adversely affected by rerouting flexibility if the emergency sequence does not ensure as precise machining or assembly as the primary sequence.

Is it possible to design AMT so that all six types of flexibility are present or does having one type preclude another? Designers of the IBM Lexington plant, which manufactures typewriters and printers, consciously sacrificed indiscriminant modification flexibility in order not to hamper mix and changeover flexibility. It requires careful study of the technical and cost implications to implement a minor product design change instead of quick and informal agreement among the affected parties (Swamidass 1988). Other firms are trading off the time aspects of changeover and modification flexibility by introducing completely new products less frequently and updating existing products more frequently. Some companies are decreasing the range aspect of mix flexibility in order to respond faster on the time aspect of this dimension. That is, they are trading less product variety for shorter delivery times.

The conceptual framework implies that a given plant will have flexibility priorities determined by the nature and intensity of the uncertainties it faces. Since industry and strategic variables affect uncertainties, plants in different situations should exhibit different flexibility priorities. Swamidass (1988), for example, suggested that needs for flexibility vary with volume-variety situations. High volume, low variety requires a little changeover flexibility in order to introduce new models and a little mix flexibility to offer some variety. Low volume, high variety requires high changeover, mix, and modification flexibility. Schonberger (1987), however, argued that in high volume, low variety situations changeover flexibility is more important than mix flexibility because shrinking product life cycles are more of a problem than variety.

The conceptual framework takes a static viewpoint. A given set of manufacturing uncertainties determines particular strategic objectives and flexibility priorities that result in a fixed AMT design. From a dynamic perspective the nature and intensity of the uncertainties facing a given plant may shift over time. A company with *strategic adaptability* can change the nature of its objectives to meet the new reality. It must also have *flexibility responsiveness*, the power to alter existing capabilities for flexibility through reconfiguration of the manufacturing process. As priorities change one wants to adjust the limits within which flexibility exists, that is, significantly increase product variety, reduce the time to alter the volume of production, or decrease the extent of alternative routings. Reconfiguration is facilitated by modular system designs, machinery that does not require special foundations, and material transport that is easily rerouted, such as automated guided vehicles (Carter, 1988).

3
DESIGN AND JUSTIFICATION OF AMT

A company determines its requirements for manufacturing flexibility by understanding environmental uncertainties and establishing strategic objectives. The next step in the conceptual model deals with providing these requirements. Although there are a number of ways to build flexibility into a plant, investment in AMT is undoubtedly one of the most popular, at least in North America. Requirements for flexibility enter the design process for AMT as technical specifications. To ensure that these specifications are met managers must have enough knowledge to oversee the process. This chapter begins by indicating in a general way which design characteristics AMT, specifically computerized manufacturing systems, must have in order to score high on each of the six flexibility dimensions. The financial evaluation of AMT is a related issue. How can one justify the need for the required flexibility when it is so difficult to assess the economic impacts? The latter part of the chapter accordingly explores different approaches for handling the intangible benefits and costs of flexibility.

DESIGNING CAM FOR FLEXIBILITY

According to sociotechnical systems theory, discussed in this book's section on organization, a manufacturing system contains both technical and social aspects. Technical considerations include the nature of the hardware and software as well as the equipment's layout. Social factors include the kind of supervision, the degree of task specialization for workers, and the amount of planning responsibilities possessed by workers. In keeping with the sociotechnical approach, information on how each type of flexibility affects hardware, software, and workforce decisions was gathered from the literature and conversations with engineers. Emphasis is placed on production and assembly lines as opposed to other types of manufacturing processes.

As a general rule, flexibility is both expensive and time consuming to design into a manufacturing system. Managers should worry about providing it only where uncertainties are significant. It is not necessary, for example, to think in terms of flexible manufacturing for an entire product. Stable components of the design are appropriate for dedicated automation, whereas parts with a lot of variety or subject to frequent change are candidates for CAM. In a dishwasher the internal parts are standardized while there are different, changing types of doors.

Managers who want CAM technology with mix flexibility, the capability of producing a variety of products, should heed the following guidelines.

1. Workers must demonstrate versatility in handling different products and must have ready access to materials and parts.
2. General purpose machines are emphasized over dedicated ones. However, if a special design or handling requirement makes dedicated equipment necessary, there is a special work station for the part (Gould 1986).

    A variety of fixtures and tooling is transferred into and out of the system. It is this variety that is so hard to handle in computer-controlled mass production. Once more, tooling migrates among the machines or the machines need large tool-storage capacities.
3. Due to the production scheduling difficulties associated with a variety of parts, software aids for schedules are necessary (Graham and Rosenthal 1986b).

How is CAM technology designed to quickly and efficiently change over from one product to another?

1. Workers are willing and able to continually learn new operating procedures as the result of a heavy investment by the company in training. In order to move as fast as possible down each new learning curve, management encourages workers' suggestions (R. W. Hall 1987).
2. Most important, the amount of hard tooling is kept to a minimum. One trick is to use general purpose machines modified for a special purpose since it doesn't take much work to remodify them for a new use (R. W. Hall, 1987).
3. Software is designed to be readily changed when a new product appears. It is, first of all, generic with respect to a particular product family. Providing the computer with the specifications of new members of the family is sufficient to create their parts programs. Second, it is modular so that any changes for new products are accomplished by removing and inserting subroutines (Gould 1986). The software also facilitates rerouting existing products while work stations are converted off-line.

Most companies want to avoid shutting down a line entirely while it is being changed over. Continuing to build the existing product allows demand to be met until the new offering is ready to replace it. In one American company an existing type of compressor was phased out completely and a new type was put on the market. Since its functioning proved to be inadequate under certain conditions, the compressor had to be redesigned. During this time it was not possible to continue to manufacture the original compressor.

Where feasible, part of the process is converted off-line while the rest of it continues to produce. The premiere computerized assembly unit of one French automaker has three parallel body framing lines of modest size. Two are used to manufacture existing models and one is left idle. When the time arrives to introduce new models, engineers set up and prove out the new tooling on the idle line. The other two lines continue to produce the existing models until demand tapers off sufficiently for one of them to be converted. When the remaining line is no longer needed to produce the existing models it is idled until it is used to set up and prove out the equipment for future models.

Changeover flexibility is undoubtedly enhanced when backward integration of the manufacturing process is curtailed. The arguments for doing so are most compelling where rapid technological change exists, as in the computer and electronics industry. A firm can take advantage of new product and process technologies as they come out by switching suppliers if necessary. It does not have to worry about previous capital investment in its own manufacturing process. According to Hayes and Abernathy (1981) the U.S. auto industry's large investment in automating the manufacture of cast iron brake drums delayed by five years the transition to disc brakes.

Limiting backward integration raises the important issue of whether a firm will lose its manufacturing prowess in the long run. As more and more components and subassemblies are provided by outside suppliers valuable skills are lost, which increases the difficulty of manufacturing the remaining items. Interaction of the manufacturing and research and development units is also impaired if the former loses the knack of building prototypes or is unable to provide sound advice on manufacturability. A vicious cycle can ultimately occur, which drives a firm to become essentially a final assembly operation whether it wants to or not. The amount of value added to a product in assembly, however, is decreasing as the number of parts per product and labor content go down (Bolwijn et al. 1986). There were 450 components in a Philips fourteen-inch color television set in 1975 as compared to 175 in 1984. Simultaneously, the value added in component manufacturing is increasing under the impetus of technological breakthroughs meant to improve a product's performance. Enterprises concentrating on final assembly may therefore find themselves in a weak competitive position.

Modification flexibility enhances redesigning a product, and is associated with the following characteristics:

1. The work force quickly adapts operating procedures to meet somewhat new conditions.
2. Either there is a minimum amount of hard automation or some work stations are flexible enough to accommodate the design changes. Fixturing mechanisms do not have to be drastically altered to handle the redesigned product.

3. Software is easily reprogrammed to aid in changing machine routings.

What are the design characteristics for volume flexibility, the ability to increase or decrease the aggregate production level of a CAM process? Before mentioning them it is useful to point out that when sales fluctuations are extremely high, computerized automation is probably too expensive a means of achieving this type of flexibility (Bolwijn et al. 1986).

1. Workers possess skills that are utilized elsewhere when CAM production is low, and they are returned as needed to preserve continuity of staffing (Graham and Rosenthal 1986a). Alternatively, workers are guaranteed a certain number of working hours over the course of a year and in return they accept fluctuations on a daily or weekly basis (Bolwijn et al. 1986). Skilled technicians are willing to work on second and third shifts if needed.
2. The equipment has high capacity limits, or adjustable limits as when capacity is added or removed in modules.

    In mass production, computerized transfer line technology, such as an auto body framing system, creates high capacity limits since the smallest unit of capacity that typically can be added is another line. Volume flexibility is achieved at the cost of unused capacity. When breakdowns occur however high limits facilitate a quick return to normal production and in-process inventory levels.

    An alternative is to construct a manual line next to the transfer line. It requires less investment and is easy to set up and take down, but has higher operating costs and perhaps lower quality.

    Use of flexible manufacturing systems in batch production permits modularization, the ability to incrementally raise or lower capacity over time as needed. One implication is that when demand is increasing capital investment is delayed until it is absolutely required. This is however not just a matter of adding more of the same equipment. More specialized technology is in order as volume increases. When demand turns down many modules can be converted to other uses. Over the entire life cycle capacity is more or less evenly matched to demand (Hutchinson and Holland 1982).

    Reserving empty floor space is critical in batch operations which achieve volume flexibility by adding more machines or modules. Space needs are reduced by purchasing an FMS instead of conventional equipment which performs the same task. A plant manufacturing reagents for use in medical instruments saves space by having all the steps in the process share a single room at different times. As the equipment for the next step is brought in the equipment for the previous step is cleared out. Of course, this approach requires storage room and setup time.
3. Production scheduling software creates slack time to allow for unexpected occurrences.

As part of the JIT philosophy, many companies are transforming their functionally organized shop floors into continuous flow production lines. A flow-line, especially when it contains CAM, is more sensitive to the disturbances of a batch environment.

For example, the integrated complexity of CAM technology reduces its reliability as compared to conventional equipment. Breakdowns are more frequent and last longer due to the difficulty in diagnosing and solving problems. Rerouting flexibility, a capacity for adjusting the sequence of machines through which a part flows when a breakdown occurs, is enhanced by the following design characteristics.

1. People are trained in advance to handle anticipated contingencies. A good deal remains to be done on the spot, however, when a machine breaks down. Workers therefore have an intimate knowledge of the system to prevent damage and reroute production. A group structure facilitates cooperation. The group may include maintenance people.
2. Equipment redundancy exists in one or more of a variety of forms. An overlapping work envelope is illustrated by the situation in which robots on either side of a downed one can do its work in addition to their own. The cost of duplicated equipment is avoided but the work rate is reduced while repairs go on (Groover et al. 1986).

    Some factories have parallel lines, which make the same items or parallel work stations within a single line. An example of the latter is the assembly line of a midwestern manufacturer of lawn and garden tractors. It contains a sequence of work islands each consisting of four stations performing the same functions. Under computer control an automated guided vehicle transports a model finished at one island to a kitting area where components for the next assembly operation are loaded and then to the station at the next island which will be available first.

    To avoid the added expense and complexity of redundant computerized equipment, conventional equipment requiring significant human inputs is sometimes used as a backup. Workers must have sufficient chances to operate the conventional machines to retain their skills. However, mere availability of a conventional backup may reduce commitment to the computerized system if the latter is perceived as ineffective (Boddy and Buchanan 1986). Once more, artificial cost differences between CAM and conventional methods sometimes hinder the use of the latter as a backup. One American defense contractor is experimenting with an FMS that has automated loading and unloading. Since no direct labor and hence no overhead is charged to the system, it appears to be very inefficient to use conventional methods.

    Accompanying redundant equipment, tool migration or large tool-storage capacities ensure that rerouted parts can be machined or assembled.

    The material-handling system features automated guided vehicles, which are the most flexible type of transportation devices. It can be operated in a variety of ways depending on the extent of the breakdown: in fully automatic mode, using local programmable controllers, or manually (Gould 1986).
3. Software aids exist for rearranging the production schedule.

The system may contain other important features especially in mass production. High capacity limits allow workers to quickly restore production after a breakdown is repaired. Large in-process inventories keep production going as long as possible after a breakdown occurs. Both features appeared in the two American auto body framing lines studied by Gerwin and Tarondeau (1986). These features were considered to be required in the long run as new kinds of breakdowns were always occurring. JIT production may therefore not be very successful where large-scale integrated CAM technology exists.

Material flexibility, which allows a manufacturing process to handle unanticipated variations in inputs, relies on the following characteristics.

1. Workers are trained in quality control principles so problems are quickly detected. They have the authority to stop the line when a problem occurs. A team structure facilitates their working together to solve problems. Workers' suggestions for improvements are encouraged, which reinforces the team concept.
2. There are automated monitoring devices and coordinate measuring machines under computer control. In many situations they back up and support workers' efforts rather than substitute for them. The state of the art of on-line monitoring does not permit sophisticated geometry checks, and measuring machines, which are used at the end of an operating sequence, may detect errors too late to avoid damage to expensive parts and equipment.
3. Software provides diagnostic information including traces of a part's route through the system.

CIM Facilities

Although most of what has been said about CAM design also applies to a CIM facility, there are some special considerations worth mentioning. Unfortunately, they are not based on extensive empirical studies as the number of CIM projects is still low. Some instances exist of vulnerability to unexpected events in the marketplace such as a downturn in sales, failure to obtain or keep a contract, or problems with breaking into a new market. The result is usually a glut of excess capacity that is difficult to handle in an *ad hoc* manner, a sign of lack of flexibility.

Why have certain CIM facilities not been able to adjust to unexpected downward market shifts?

- They possess high changeover flexibility within a given envelope, but that envelope is rather narrow. It is not always technically feasible to identify substitute products. Although there is some improvement over dedicated facilities in this regard, it is not a great deal.
- The cost implications of excess capacity eliminate some products that can be produced. As production decreases the relatively large overhead costs of a CIM plant are allocated over fewer units driving costs per unit up.
- In hi-tech industries some existing products are unassignable because by the time workers would have moved down the learning curve, the products would be ready for phasing out.

- Lack of volume flexibility in terms of being able to scale down productive capacity augments the problems. CIM's *raison d'etre* is to integrate design, manufacturing operations, and control. A plant may be so closely integrated, however, that it is virtually impossible to shut down one segment without affecting several others.

Due to the large investment in, and high risk of, CIM projects it is critical at the outset to consider the implications of a possible steep drop in demand. Answers are needed to questions such as "What else could the facility do?" "At what cost?" and "What are the chances of a market disaster occurring?" To provide answers, planners can judgmentally weigh the costs of expanding the envelope against the costs and chances of having to shut the plant.

Other suggestions to consider include the following.

- Make the facility a reserve producer of an additional product family, the bulk of whose manufacture is carried out elsewhere. Then if demand slumps for the main family perhaps production can be increased for the reserve products.
- Do not allocate full investment costs to the facility's initial products for managerial decision making purposes. Some of this overhead expense can be reserved for products to be made in the future.
- The plant should, insofar as possible, be designed on a modular basis with decoupled subsystems and parallel activities in order to facilitate closing down only part of it.
- Consideration should be given to having two small plants rather than one large one since closing the latter shuts the firm out of the particular market.

CAPITAL JUSTIFICATION AND THE VALUE OF FLEXIBILITY

When something as new and radical as AMT is initially considered by a firm the justification decision has two aspects: determination of whether the basic concept is worthwhile and evaluation of whether a particular set of equipment should be selected (Boaden 1989). Concept justification, directed toward senior management, is largely strategic in nature but must also involve financial issues. Consciousness raising is a necessary ingredient, getting top managers to understand the new approach without resorting to detailed technical arguments. Often this type of justification, which has elements of an informal advocacy process, takes a considerable amount of time as managers struggle to grasp the wider implications of the technology for their company. Its significance lies in providing senior management with a valuable educational experience which paves the way for a more reasoned evaluation of specific projects. Equipment justification, directed at financial officers, is mainly economic in nature but includes strategic considerations as well. It is necessarily formalized, detailed, and technical in nature.

Very little research exists on concept justification, whereas there is an overwhelming literature on equipment evaluation, including the particular example of AMT. We do not consequently have a basic understanding of the former including what kinds of information, methods, organizational roles, and problems are involved. It is not known if justifying a concept occurs before equipment evaluation or whether they go on simultaneously. Most important, does the former represent the decisive step with the latter being a mere technical afterthought? If so, the research emphasis on equipment justification for AMT is badly misplaced; we need to learn how to improve concept justification.

Equipment evaluation has structural, process, and technical aspects. The justification literature, however, treats only the last component. Little consideration is given to meshing techniques with supporting organizational arrangements. If, for example, it is decided to justify AMT on a strategic basis, final authority should be in the hands of a manufacturing vice-president instead of a financial executive. Accounting people should review the proposal at the plant level and provide just one set of inputs in the ultimate determination. Designing techniques in isolation from accompanying organizational supports reduces the chances of successful applications of those techniques.

Virtually all of the AMT justification techniques in the literature presuppose a given equipment configuration as input. The implied separation of design and evaluation appears to be reasonable; one cannot analyze benefits and costs without the existence of a proposed system. Some good arguments exist, however, for developing models that treat design and evaluation as occurring more or less simultaneously (Kulatilaka 1988, Primrose and Leonard 1986). First, it is necessary to learn during design whether overall strategic and technical objectives are economically feasible. If not, aspirations are lowered. Second, alternative designs for the same capital alternative may emerge and one wants to discover which meets the objectives most efficiently before a great deal of time is invested. Ideally, design and justification continuously evolve, the results of one feeding the other as plans become more concrete.

Primrose and Leonard (1986) suggested a joint process that contains some of these characteristics. The first step is to establish overall strategic objectives for some problem requiring a solution. Perhaps sales are being lost due to poor delivery performance. A capital project is identified to solve the problem and a preliminary financial evaluation is conducted to see if it is economically feasible. Engineers establish technical specifications to achieve the objectives and prepare a detailed design. A detailed justification is then conducted.

Recent efforts to formalize the design process for manufacturing systems could in principle be extended to consider financial issues. Donner (1988) proposed a computer model that would translate knowledge of required flexibilities and other factors into a hardware configuration. The model's basic units are manufacturing activities such as milling, turning, inspection, and assembly. Parts to be manufactured are defined in terms of an ordered sequence of activities. The design process is hierarchical in nature. Activities are aggregated into machines that are combined into cells that are formed into systems. Information on flexibility is included among the given attributes associated with an activity. It is also inferred at steps in the design process using certain evaluation mechanisms. Rerouting flexibility, for example, is determined from the number of alternative paths a part may take through a cell or system, information that depends on which machines the part can flow between.

Intangible Benefits and Costs

Why don't conventional discounted cash flow techniques suffice when evaluating a proposal for AMT? As Gold (1982) and Meredith and Hill (1987) pointed out, they work best for proposals with limited objectives, such as increasing capacity or making performance more efficient, and localized impacts such as direct labor savings. AMT is usually a solution to a strategic problem and has implications for an entire business unit. It consequently produces significant intangible benefits and costs that are considered out of bounds in a traditional financial analysis.

Two alternative scenarios ensue. AMT is successfully justified on the basis of a strict financial analysis, but its potential is never achieved. Managers confuse it with traditional automation; they perceive its primary benefit as direct labor savings rather than strategic capability. Organized labor sees it as a threat to job security and resists its implementation (Primrose and Leonard 1986). More likely the technology cannot be justified and the company is deprived of an important asset, assuming net intangible benefits are large. These benefits would have eventually showed up as additional profits even though they could not have been traced to the technology.

AMTs significant intangible benefits arise from a number of different factors. Many benefits of flexibility cannot be measured. Consider changeover flexibility as an example. How does one assign a value to being able to manufacture some currently unknown part at some unknown time in the future? The ability of a computerized system to act as a centralized source of operating information improves the accuracy and timeliness of data in ways that are hard to quantify. With AMT, production data is collected, recorded, processed, integrated, and displayed as the byproduct of automated control. Traditionally, it is obtained as the result of a laborious investigation in which supervisors cover considerable distances to learn what is occurring at different work stations and then try to piece it all together.

Intangible benefits also arise due to AMTs integrative capacity (Berliner and Brimson 1988, Cohen and Zysman 1987, Gold 1982). Each element of the new technology is just one step along the road to a whole new concept of manufacturing. Although many gains are not achievable until all the steps are in place, failure to incorporate any one step may jeopardize realization of those gains. Decisions on each element should therefore be based on its role in an entire package rather than considered in isolation. An FMS is undoubtedly a necessary ingredient in the eventual installation of CIM. Evaluation of its benefits should include its contribution to making the CIM system a success. This argument makes most sense when a firm has already committed itself to a detailed long-range plan. It makes less sense when a company is proceeding one step at a time on a trial-and-error basis and is not clear where it will wind up.

Even direct labor cost savings, the benefit that is typically quantified when justifying AMT, has important subjective elements. Bennett, Hendricks, Keys, and Rudnicki (1987) indicated some issues arising in the acquisition of NC machines, a member of the AMT family with relatively minor strategic impacts.

1. Ideally, a labor cost is considered direct when it is readily traceable to a product. Some managers, however, feel pressure to identify indirect labor as direct in order to keep overhead rates low.
2. A new NC machine usually replaces a few conventional machines. The displaced operators are probably reassigned rather than laid off. The relevant savings are the often intangible costs avoided by reassignment, not workers' wages.
3. When a new NC machine's production volume is increased up to some limit, it usually can be accommodated without adding workers. However, it is difficult to predict accurately when the production increases will occur, and therefore the labor savings as compared to conventional machines.
4. Predicted direct labor savings may not equal actual direct labor savings if the individual who justifies and acquires an NC machine is different from the person who hires, fires, and reassigns workers.

What of AMTs intangible costs, many of which arise from organizational changes needed to support the new technology? The justification literature has not done as good a job of enumerating them or estimating their values. Most articles pay little attention to them; some avoid them completely. As compared to shorter lead times, less inventories, and high quality, the following factors are not often analyzed:

1. Training and recruiting highly skilled managerial and technical people
2. Development of new systems and procedures in production scheduling, cost accounting, quality control, and maintenance, which have little utility elsewhere in the factory
3. Reorganizations at the shop floor and departmental levels
4. Organizational resistance to all the changes from managers, technical specialists, and workers
5. Large, in-process inventories to keep production going during breakdowns that occur frequently and are difficult to correct
6. The need for increased control over variations in the inputs to the manufacturing process
7. Impacts of upstream and downstream bottlenecks that negate benefits or increase costs
8. The intangible foregone benefits of workers displaced by the new technology, that is, decreases in some dimensions of flexibility when AMT replaces a labor-intensive manufacturing process.

Beatty and Gordon's (1988) study of CAD/CAM installations enumerated some "hidden" costs that did not surface until after the systems were operating, including a loss in productivity until employees mastered the new technology, the need for a catalogue of standardized parts, the hiring of a systems manager, and adding employees to customize the software.

In order to understand better the factors that give rise to intangible benefits and costs, it is helpful to categorize them in the following manner:

1. Zero-order intangibles—essentially tangibles—are used as a reference point here to represent the ideal situation. Their economic value is known with certainty or at least up to a probability distribution. They influence capital costs, annual costs, and/or annual revenues, more specifically unit values such as sales prices or quantities such as the amounts sold of products. They may also influence the timing of costs and revenues, which impacts on discounted values. Cost impacts are typically easier to quantify than revenue changes.
2. First-order intangibles are variables whose economic value is not quantifiable but that can be readily stated in physical terms with certainty or by using probabilities. One may be able to specify that an FMS reduces lead time from nine to four weeks but be unable to quantify the impact on revenues and costs except perhaps in qualitative terms.
3. Second-order intangibles are nonpecuniary factors that can be enumerated but that are not measurable in physical terms except perhaps on a qualitative basis. Consider the experience gained by maintenance people in the intricacies of electronic repairs.
4. Third-order intangibles represent factors producing unanticipated benefits and costs that are typically not measurable. Since these intangibles cannot be enumerated in advance, this category is an empty set during justification. Although an FMS will enhance a firm's strategic capability, it is not possible to know in advance all the strategic changes which will ensue let alone their impacts.

A superb example of the last category is offered by the evolution of the Peerless Saw Company (Meredith and Hill 1987). It was in the high volume steel blade business, a market steadily being captured by foreign competitors offering similar quality at a lower price. The firm's president decided to move into custom blade products, but that required development of a unique CAD/CAM system with untried laser-cutting technology and internally developed software. Although the system could not be financially justified, its implementation was accompanied by an unanticipated net benefit. The new CAD/CAM process was creatively applied to other products leading to the establishment of an entire new division.

Another example provided by Gerwin (1981a) concerns a new FMS that required high quality raw castings within narrow tolerance limits. When management realized that outside suppliers could not meet this need consistently, a decision was made to have castings supplied internally by the foundry. The impact on the foundry, in terms of upgrading operations, was large. Core room facilities were modernized, control over sand quality was improved, and inspection activities were increased. The changes, which were made to accommodate the FMS, had the side effect of upgrading the quality of castings provided to the rest of the division. On the other hand, the foundry became more dependent on the division for the disposition of its castings. A shutdown or slowdown in operations, particularly in the FMS, could affect the foundry's activities within a week.

The particular category of intangibles to which a variable, such as flexibility, quality, or inventory level belongs, is not fixed. First, as a company moves from considering stand-alone equipment through cells to integrated systems and beyond to CIM, new intangible factors arise and others shift into higher-order categories (Meredith and Suresh 1986). Second, if new parts are to be produced along with CAM some of the factors move into higher-order groupings. Third, a particular variable—quality, for example—may be zero intangible with respect to costs but first-order intangible for revenues. Fourth, research efforts may eventually lead to putting an intangible into a lower-order category. Ulrich, Sartorius, Pearson, and Jakiela (1991), for example, have proposed measuring the economic value of reducing the time it takes to introduce a new product as the change in net present value of sales revenues due to their earlier realization, assuming that volume does not change.

Hundy and Hamblin (1988) identified at least three different justification methods that are used to deal with AMTs intangible benefits and costs. Some individuals advocate modification of the traditional discounted cash flow analysis to make it more compatible with the technology's needs (Gold 1982, Kaplan 1986). They argue that the approach's basic principles are sound but that too rigid adherence leads to erroneous decisions. Furthermore, senior management will more readily accept a new radical technology if it is justified on familiar grounds rather than with a completely new approach. Variables traditionally considered as first- or even second-order intangibles have their financial impacts estimated insofar as possible and are admitted into the analysis.

A second approach—multiattribute decision analysis—uses zero-, first-, and second-order intangibles as criteria (Canada and Sullivan 1989). In theory, a project's scores on all criteria are translated into common units of value and the alternative with the highest value is selected. In practice senior managers are often presented with each alternative's scores expressed in different units and a judgment is made. Advocates believe this new approach is needed to properly justify AMT because it is fruitless to try to express most intangibles in economic terms.

Occasionally, a third approach is appropriate. AMT is evaluated on the basis of whether or not it meets a compelling strategic need or opportunity. Formal analysis and economic considerations play a relatively minor role; the potentially significant nonmeasurable impacts of intangible variables matter. Judgment based on experience is used in making decisions.

Our discussion hardly mentions optimization models for capital investment in flexible automation. They are in an early stage of development. Fine (1989) in his review of this work identified three main streams of models: those that determine the economic value of being able to handle market uncertainties, those that analyze the impact on inventories including cycle, safety, and seasonal stocks, and those that analyze strategic benefits when competitors' reactions to purchase of the technology are taken into account. In the latter situation it has been shown that possessing mix flexibility, in terms of an ability to manufacture products for different markets, can make a firm worse off.

Modified Discounted Cash Flow

How is the traditional discounted cash flow analysis modified to allow for AMTs peculiarities? Variables not normally considered, but that are expressible in financial terms with some effort, are included. The cost impacts of a change in inventory levels or the revenue implications of a shift in flexibility are estimated. Variables not translatable into financial quantities are handled in a manner suggested by Kaplan (1986). Assume that management believes an alternative's *net* intangible benefits have a positive economic value, and that net present value is negative when they are not considered. Determine by how much annual cash flows must increase to attain a net present value greater than zero. Then decide whether it is worth paying that yearly amount to have the net intangible benefits.

Traditionally, it is assumed that in the "do nothing" situation to which a new capital alternative is compared, the annual cash flows are unaffected by the decision. If, however, a company decides against AMT while its competition seizes the opportunity, then, due to the technology's strategic ramifications, the firm's competitive position will deteriorate with a consequent reduction in cash flows. In a modified analysis one tries to estimate these cash flow decreases arising from changes in competitive position, a factor normally considered intangible.

Companies that are among the first in an industry to adopt a new manufacturing technology, however, assume great risk and may not successfully implement it. Firms that wait may gain valuable information from their competitors' mistakes, facilitating a quick, smooth implementation later. These intangible costs of a first mover should be traded off against the intangible costs of a late adopter. When this is done the deterioration in competitive position of the late adopter may not commence until several years after other firms adopt AMT, if at all. Starting in the early 1980s General Motors took the lead in the auto industry in purchasing state-of-the-art manufacturing technology. To date, in spite of a massive $40 billion investment, the firm's preeminent position has still not been restored. Once more, it has served as a guinea pig for competitors by ironing out expensive bugs in the equipment before they make purchases (Keller 1989).

It is generally believed that firms use discount rates that are higher than needed to reflect a capital alternative's degree of risk (Kaplan 1986). AMT projects are affected more than conventional ones because their benefits occur further in the future. It may be a few years from the initial major expenditure on an FMS to the commencement of full-scale production. Once more, an FMSs changeover capability allows its use for a new product for years after a transfer line is scrapped because the original product came to the end of its life cycle. If the analysis is modified to permit lower discount rates, AMT will receive a fairer evaluation. On the other hand, admitting intangible variables into the analysis creates more opportunity for advocacy bias, the tendency of requestors to make a project appear more beneficial than it actually is, to creep into the justification. High discount rates serve as a protection against this phenomenon.

Simulation Models. A final modification is the use of simulation models to give a more explicit account of risk and the dynamic nature of justification. AMTs high risk stems from large investments and the uncertainties in estimating benefits and costs. Its flexibility draws attention to the sequence of choices that are possible over time.

The models to be discussed are essentially sophisticated net present value (or internal rate of return) generators. Only a single, financially oriented criterion is utilized because it is assumed that traditional intangibles, such as flexibility, have measurable economic impacts. Implicit economic impact occurs when flexibility is not measured in physical terms but where a cost parameter is lower or a revenue parameter higher for a flexible project relative to a conventional project. Explicit impact occurs when flexibility enters the model as a physically measured parameter that influences costs and revenues directly or through other intermediary variables. The flexibility measures are almost always exogenous; their values are not influenced by other factors operating in the model. Range aspects have received much more attention than time aspects.

Hutchinson and Holland (1982) developed a simulation model that calculates net present value for two specific capital alternatives—transfer lines and FMSs—in a situation of market uncertainty. A given number of products are introduced at random points in time with sales in any period essentially represented by a position on a life cycle curve with random variation. Adaptation to these uncertainties occurs either through the lower operating costs of transfer lines or the flexibility of FMSs. The possibility of using both sets of equipment is not explored. There is no opportunity for management to influence the market uncertainties, which implies that flexibility impacts on costs but not on revenues.

Each alternative's net present value arises from periodic decisions on how much to invest and periodic calculations of revenues and operating costs. Capital investments in transfer lines are made assuming a line's capacity can have any numerical value, but capacity cannot be changed after it is initially determined, and the line can only handle one product. The specific amount is based on an updated forecast of a product's sales. There is a fixed capital cost per unit of capacity; no economies of scale are allowed, which eliminates a possible advantage of transfer lines.

Capital investment in FMSs assumes that more than one product can be handled per system. Each system's capacity is purchased in increments over time up to some maximum amount, an example of volume flexibility. Capacity that is no longer needed for a product in the declining phase of its life cycle, can be converted to handle another product reflecting changeover flexibility. Decisions on new and changeover capacity are made using the updated sales forecasts and a simple product assignment algorithm. Unit capital costs are the same as for transfer lines and there is a unit conversion charge.

The primary advantage of transfer lines are in unit operating costs being half of those for FMSs. Since actual sales of a product may turn out to be greater than the amount forecasted, either capital alternative is allowed to produce the excess at an added unit operating cost, which is less for transfer lines. Sales revenues for both capital alternatives are always the same, therefore financial evaluations are made by comparing costs.

Experiments with the model were conducted using four criteria: expected discounted total costs and costs per product, expected undiscounted total costs and costs per product. Risk, in terms of the variances of the four cost distributions, was not assessed. Parameters that were allowed to vary included the number of products that are introduced and the standard deviation of a product's actual sales distribution, factors that should place a premium on flexibility. Increasing the number of products resulted in a statistically significant improvement in the advantage of FMSs over transfer lines on three of the four measures. Surprisingly, increasing the standard deviation of sales produced no significant effect on any measure.

The grand mean of discounted total costs was about 6 percent lower for FMSs as compared to transfer lines. An analysis of its cost components throws light on flexibility's economic impacts, which are, of course, dependent on the model's assumptions.

- Changeover flexibility, the ability of an FMS to convert excess production capacity instead of having to add new capacity, exerts an economic impact through the unit conversion charge, which is less than the unit capacity cost. It creates a cost advantage for FMSs.
- Volume flexibility, the ability to incrementally add capacity to an FMS, saves interest costs by deferring capital investment into the future rather than having to make it all initially.
- By allowing a better match of capacity and demand in each period, volume flexibility results in the production of fewer units of overcapacity by an FMS, but this effect is canceled out by a transfer line's smaller unit overcapacity charge.
- Mix flexibility exists since more than one product is manufactured on an FMS, but it has no economic impact in the model. For one reason, unit capital costs are the same for both types of equipment and capital costs have no fixed component. The investment in an FMS with n products is the same as for n dedicated lines.
- The lower operating costs of transfer lines almost counterbalance the economic advantages of flexibility.

Azzone and Bertele's simulation model (1989) determines net present values for a general class of capital projects that might include conventional equipment, FMCs, and hybrids. Market uncertainty is represented through a product's sales, which in any time period are determined from a position on a life-cycle curve and random variation. The components of certain revenues and costs are influenced by strategy and flexibility, but the latter two factors have no direct links with each other. Strategic parameters are independent of the projects considered; nevertheless, they create differential financial implications. Flexibility parameters, which vary among the projects, mainly impact on capital costs.

Three types of capital costs may be incurred in any time period. First, the cost of fixtures and software for a new product varies among projects indicating an economic impact of changeover flexibility. The number of new products introduced at any time is determined from a probability distribution representing the firm's strategic position on this issue. Second, inventory costs are influenced by two other strategic factors: required lead time and the level of service (that is the probability of not exceeding the lead time). Third, the cost of new machines depends on how much additional capacity is needed.

Machine capacity depends on the three aforementioned strategic parameters. It also depends on mix flexibility as measured by setup times, rerouting flexibility in terms of the ratio of expected production to the production of a fully operational system, and changeover flexibility represented by the probability that existing manufacturing technology will be able to process a new product. Finally, machine capacity is influenced by each product's demand.

Sales revenues are a function of the strategic parameters; a decrease in required lead time, for example, increases average demand and raises price. To this extent strategic factors influence market uncertainties. Operating costs include a unit charge multiplied by average demand and a fixed charge for each machine type multiplied by the number of machines of that type.

Azzone and Bertele discussed an application of the model to an Italian auto firm that was considering an FMS in lieu of existing stand-alone equipment. The estimated strategic parameters were common to both alternatives. The estimated values of mix, routing, and changeover flexibility in terms of new fixtures and software favored the FMS, whereas the value of changeover flexibility in terms of the probability of technological compatibility favored the existing equipment. Based on a consideration of expected returns and risk the model recommended purchasing the FMS. Its expected net present value was greater than zero and the probability of occurrence of a net present value greater than zero was .98.

In contrast to using simulation, Kulatilaka (1988) developed a stochastic dynamic programming model for the financial evaluation of an FMS when market uncertainty exists. His formulation is of interest because it addresses a consequence of flexibility not explicitly handled by the previously discussed simulation models. The range aspect of a flexibility dimension, say having a choice of which among a number of products to manufacture in each time period, creates an intricate web of alternative sequences of decisions over time. Assuming it is costly to switch among products, a choice made in the current period influences which choices can be made later. Thus, current decisions should be made in the light of their effects on future ones. An FMS may provide a temporal sequence of decisions with higher economic value than the sequence that is more or less forced upon a transfer line. Using numerical methods, Kulatilaka arrived at solutions that demonstrated under some restrictive assumptions the conditions favoring a greater profit for an FMS versus rigid technology.

Multiattribute Approaches

Multiattribute approaches assume that it is not feasible to reduce all intangibles to economic terms. First- and second-order intangibles each expressed in their own units are instead considered on an equal basis with financial criteria. One may formulate the attributes in terms of a hierarchy in which nested under each attribute at one level are clusters of more specific ones at the level below. Decision makers find such an arrangement compatible with their own thinking processes. Alternatively, the selected criteria all exist at the same level.

To make choices, various scoring models have been developed that convert a project's outcomes on the disparate criteria into a single overall value on a scale common to all projects. Three examples of this approach are discussed: the basic weighted additive scoring model, a qualitative hierarchical model specifically designed for evaluating advanced manufacturing technology (Wildemann 1986), and the analytic hierarchy process (Saaty 1980).

In the weighted additive scoring model, as discussed by Canada and Sullivan (1989), the attributes are judgmentally selected. Normalized weights are determined for the attributes in a process which establishes a preference order among them. Although certain techniques exist for extracting the necessary information from a decision maker, the weights are often hard to estimate and vary from person to person.

The kinds of outcomes achievable on any attribute are translated into scores on a value scale that are arbitrarily chosen to run from say 0 to 10. The translating functions, which may differ from one attribute to another, implicitly establish trade-offs. For example, it may turn out that one is indifferent to trading a reduction in customer lead time of a week for a 10 percent reduction in the defect variance; both reductions represent the same change in value. Determining the functions is a judgmental task that is almost as formidable as estimating economic worth. The overall value of a project is assumed to be a weighted average of its values on the individual attributes and the project with the highest overall value is selected.

Canada and Sullivan (1989) discussed the assumptions necessary for a project's overall value to be a linear function of its values on the attributes. Roughly speaking, the preference order among any two attributes must not change with variations in the outcomes of the others. If quality is judged as more important than lead time, this holds true whether the outcomes for all the other attributes are the highest possible, the lowest possible, or somewhere in between. The trade-offs established among any two attributes must also be independent of the outcomes for the others. One is indifferent between a one-week reduction in lead time and a 10 percent reduction in the defect variance regardless of the outcomes for all the other attributes. As Canada and Sullivan pointed out, it is frequently impractical to attain these types of independence or to even know whether or not they have been attained.

One example of the weighted additive scoring model is the method developed by the Cost Management Systems program of Computer Aided Manufacturing-International (CAM-I) to evaluate investments in advanced manufacturing technology (Berliner and Brimson 1988). It has three sets of attributes: financial, such as net present value and amount of investment; quantitative, such as process yield and throughput time; and qualitative, such as potential for technological obsolescence and degree of radicalism of the products to be produced. These categories, which correspond respectively to zero-, first-, and second-order intangibles respectively, are useful for introducing flexibility into justification decisions. Initially, it enters the qualitative set in a move aimed at consciousness raising. When senior management becomes accustomed to its presence, flexibility enters the quantitative category and eventually some aspects may be accepted into the financial set.

Wildemann's (1986) strategic approach for justifying AMT is based on the hierarchy of attributes depicted in Figure 3.1. An overall objective of deciding whether or not to purchase the technology depends on its attractiveness and the firm's preparation or readiness for it. Attractiveness in turn depends on the technology's strategic opportunities and risks. Preparation, opportunities, and risks each subsume attributes idiosyncratic to the company and the technology. For CAD/CAM, preparation might include financial capability, relations with vendors, and experience with integrating systems. Opportunities could involve cost reduction, flexibility, and the effects of an integrated information flow. Risks might include the chances for long-term support by the vendor, technical obsolescence, and the reliability of cost-benefit calculations.

The choice procedure works from the bottom up. Each attribute related to strategic opportunity (risk) is assigned a value of high, medium, or low. The respondent subjectively determines overall opportunity (risk) using one of the three values in

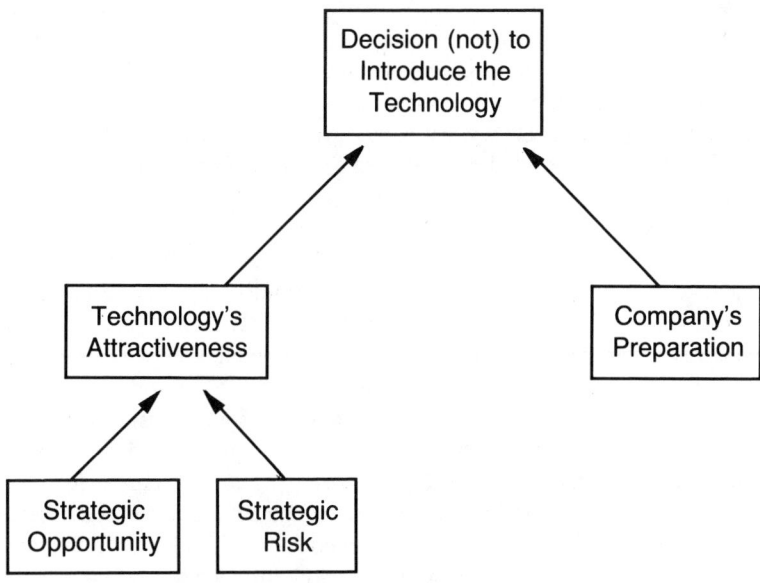

FIGURE 3.1. Wildemann's hierarchy of attributes.

the light of his or her assignments for the relevant attributes. Attractiveness is rated high when overall opportunity is high and overall risk is medium or low, and medium when opportunity and risk are high or when the former is medium and the latter is medium or low. Attractiveness is low in all other situations. The firm's overall preparation is determined in light of the values assigned to its relevant attributes.

A decision is made to introduce the technology if attractiveness is high and preparation is high or medium, or if the former is medium and the latter is medium. The technology is not introduced if attractiveness is medium and preparation is low, or if the former is low and the latter is low or medium. In all other situations additional study is required. Note that two or more AMT alternatives can be eligible for introduction so that other considerations presumably are used to break ties.

Wildemann's method puts the three basic elements of any capital justification, benefits (opportunities), costs (preparation), and risk, into a structured framework to which decision makers can easily relate. Once more, the attributes used in these three categories are empirically based; Wildemann investigated decisions concerning FMS, robotics, and CAD/CAM in forty-five West German firms. Preparation explicitly includes the intangible costs of needed organizational improvements, items that do not receive enough attention elsewhere. A new technology's attractiveness is consequently tempered by the realization that the firm is not prepared to handle it. Determining that preparation is too low is really a blessing in disguise if it leads to developing organizational supports necessary for AMTs successful implementation in the future.

The analytic hierarchy process (AHP) (Saaty 1980) is a particular type of weighted additive scoring model that is appropriate for hierarchically organized attributes. One could use the AHP to convert Wildemann's method from a qualitative to a quantitative basis, assuming the necessary additional information was available. It employs pairwise comparisons among attributes to establish weights and the same kind of comparisons among alternative projects to determine their values on the attributes. There is a consistency test for the weights that checks into the order and degree of stated preferences. It has not been finally resolved, however, whether the resulting rankings are free from arbitrariness (Dyer 1990). Applications of the AHP to FMS justification appear in Canada and Sullivan (1989) and Wabalickis (1988).

All three multiattribute models are essentially deterministic in nature. The risk associated with a project's overall value, which arises from uncertainties in the estimates of key parameters, cannot readily be calculated. One can, however, introduce risk as an attribute and conduct sensitivity analyses.

In practice, the weights assigned to an attribute, or a project's value on the attribute may vary over time. Changeover flexibility should perhaps receive more weight in the future when the current product comes to the end of its life cycle. An FMSs score on quality may steadily increase over time as more is learned about the system's capabilities and limitations. The weighted additive scoring model and Wildemann's approach, designed as static models, could not readily handle these extensions. The AHP can incorporate multiperiod differences in weights and scores without difficulty.

The Strategic Approach

Circumstances exist in which strategic criteria predominate over financial objectives. Usually there is some pressing competitive need or opportunity, little hope of computing meaningful financial impacts, and/or a single decision rather than a continuing series. Choices are based on managerial judgments instead of formal analyses; experience replaces computation. The technology is adopted if it seems to meet the need or exploit the opportunity. Rough calculations suggesting economic feasibility are sometimes also a factor.

During a crisis the overriding concern is to find a capital alternative that will help the firm survive. Recall that the Peerless Saw Co. invested in a unique CAD/CAM system in order to save itself even though the system did not pass financial tests. A second example is the farm tractor division of a large diversified manufacturer (Gerwin 1981a). It was designing a new product line and replacing its plant's machinery to stem declining sales and profits, which threatened to shut it down. One critical choice was between a hybrid transfer line and an FMS under conditions of substantial uncertainty in the sales forecasts. If the new product line failed or needed to be redesigned, a transfer line would require considerable additional investment for retooling. The FMS could be put to other uses or accommodate redesigned models with little extra investment. The division and headquarters decided on the FMS, in part because no further large amounts of funds were available. An FMS could give the unit a new lease on life if disaster struck.

Hundy and Hamblin's (1988) sensitivity analysis of their simulation model also indicates the preeminence of an FMS over conventional equipment under market adversity. They compared internal rates of return for these two types of alternatives under pessimistic conditions, that is, the product is unsuccessful, no new versions are introduced, the selling price is 5 percent lower than expected, and materials' prices are 5 percent higher than expected. The FMSs internal rate of return was 17.6 percent versus 0.9 percent for the conventional equipment in part because the former could be assigned to completely different work. Hundy and Hamblin's analysis also showed that the FMSs advantage diminishes as environmental conditions improve.

The following other rationales exist for a strategic approach, some of which have been discussed by Meredith and Suresh (1986).

- The AMT project will help achieve a significant competitive advantage in terms of say lead time or quality, and it is not feasible to estimate the financial consequences.
- The technology goes into full-scale operation but is viewed as an experiment since management desires to learn about its capabilities and limitations, provide experience for technical people, and stimulate diffusion throughout the company if it is successful.
- A firm designs and puts into full-scale operation an AMT project with an eye toward marketing it or some aspect of it in the future. The technology not only provides useful learning, it acts as a showcase for potential customers.

Evaluation of Current Approaches

To evaluate modified discounted cash flow, multiattribute models, and the strategic approach, first consider the demands made on an enterprise's information processing capabilities. The two formal approaches make unreasonable impositions. As their advocates have stipulated they require knowledge of *all* significant benefits and costs. However, unanticipated consequences are the rule rather than the exception in organizations as signified by the category of third-order intangibles. Ingersoll Milling Machine, a pioneer in the use of CAM, found that the actual benefits often came from factors not considered in the justification (Kirton 1986). The formal methods also force individuals to be precise about uncertain quantities: financial implications, probabilities, values, and weights. While the strategic approach requires less heroic information processing, it offers hardly any precision at all. Its frequent application might hinder achieving profitability.

All three approaches are subject to advocacy bias, the tendency of individuals in the justification process to bias information according to their predispositions. It is most obvious for the strategic approach but holds as well for the formal methods. Although the modified discounted cash flow and multiattribute approaches are designed to remove subjectivity from justification, they merely shift its locus. Occasions for choosing which intangibles will be considered, estimating financial impacts, selecting attributes, or providing information on weights and values are opportunities for advancing preconceived notions.

These problems shift attention to ways of improving organizational structures and processes for gathering, analyzing, and communicating data needed for justification decisions. Just how *is* the marketing department to produce a valid and reliable estimate of the impact on sales of a reduction in lead time? Although the AMT literature has hardly addressed questions such as this one, something can be said with respect to the modified discounted cash flow approach.

Recommendations are often made for companies evaluating AMT to rely on the experiences of others. Specific estimates of financial savings for discounted cash flow analyses are available from articles reported in the trade literature. One must regard this information, however, with a healthy scepticism for the following reasons.

1. It is well known that successes are reported much more frequently than failures so that a biased sample exists.
2. There are no well-accepted standard definitions of terms like FMS, CAD/CAM, and CIM. Reporting firms probably attach a more integrated label than justified to their technologies. Once more, the technologies are highly customized and constantly changing.
3. Many technical and organizational changes occur simultaneously in a plant. It is impractical to isolate the specific effects of AMT.
4. The alternative to which AMT is compared is sometimes inappropriate, outmoded existing equipment rather than what competitors currently have.
5. A dramatic change in performance may occur but only in a limited part of the factory so that the overall effect is small (Boddy and Buchanan 1986).

6. Reporting companies are aware that their accounts will be read by potential customers and competitors.

Firms also search internally for the required data but how are they to do it? Boaden (1989) recommended a procedure initiated by a small group that conducts a literature search. They compile an initial list of sources of benefits and costs for the particular type of AMT being considered. A questionnaire is developed asking key managers whose departments are affected to have the financial impacts estimated. The questionnaire is discussed with all concerned parties and revised prior to dissemination. Conferences with respondents after administration provide feedback on the estimates' rationales indicating which ones are based on solid foundations. A justification is prepared based on the acquired information. Although this procedure is by no means a panacea, it involves key managers in justification and commits them to achieve the estimates they and their staffs have projected.

There are no clear-cut choices when deciding on a justification method except perhaps in extreme situations. If a firm is considering an individual robot or stand-alone machine, discounted cash flow will undoubtedly suffice. If a very large CIM project is contemplated with mainly intangible benefits and costs, then a strategic approach is needed (Meredith and Suresh 1986). For the rest, a possible solution adapted in part from some ideas of Keen (1981) is sketched out here under the label of experimental AMT justification. Ingersoll Milling Machine apparently follows some of these principles (Kirton 1986).

Experimental AMT Justification. The basic idea is to adopt a gradual incremental approach to implementing AMT through a series of modest steps taken one at a time. There is no detailed preconceived plan for a final system to which evolution is directed. Although an outline may exist, the evolutionary path can change at any time in the light of previous results and new technological and market considerations. Each step represents a full-scale operational project requiring a formal capital justification. It also functions as an experiment with respect to future steps providing valuable information on how to proceed. A future step may make an advance or correct a problem.

Using this approach reduces the information processing requirements of the two formal methods of justification. Each step is justified mainly on the basis of financial benefits and costs; intangible considerations are minor. Intangible contributions to some preconceived final system are not included since it is not clear where evolution will lead. The experience gained in justifying previous steps combined with observations of results through postaudits feeds into the justification process for the next step creating more accurate financial estimates.

There are other advantages to an experimental approach.

1. Advocacy bias is moderated. Since requesters are not proposing one large system for a go/no-go decision, they do not stand to lose everything all at once. Once more feedback on the results of previous steps act as a reality check on biased estimates.

64   DESIGN AND JUSTIFICATION OF AMT

2. Risk is decreased by limiting the investment at each step and giving the organizational support system a chance to develop. Once more the entire process may be cut off at any point.
3. The approach encourages specifying new and modified design criteria as experience is gained and conditions change. Design and justification therefore proceed together with one influencing the other.
4. There is a continuous evaluation of the nature and value of the evolving concept based on feedback information rather than gut feelings.

What are some of the problems? For one, decision makers are susceptible to erroneous arguments based on sunk financial and psychological costs. After a great deal of effort and money has already been expended, it is convincing to hear that only one more modest investment will bring the substantial benefits for which everyone has been waiting.

In sum, limited information processing and advocacy bias characterize capital justification within any firm. These problems extend to environmental sources of data. Vendors eschew dispassionate feedback; they offer performance information collected under idealized conditions. Reports from the trade literature must be treated with caution. Even the research literature on AMT justification displays more concern about benefits than costs, perhaps out of the conviction that the technology is good in itself. Interjecting formal techniques into this situation allows them to get caught up in the problems. Instead, ways of altering the structure and process of justification are needed that will reduce information processing and the motivation to bias. The experiential approach outlined here is one possibility worth exploring.

4
PERFORMANCE MEASUREMENT

Once AMT has been put into operation it is necessary to measure its performance. Traditionally, this next step in the conceptual model has stressed financial evaluation, an approach that is being called into question (Berliner and Brimson 1988, Johnson and Kaplan 1987, Kaplan 1990). Consider the following consequences of an emphasis on financial performance measures.

- In advanced manufacturing environments managers concentrate on reducing direct labor costs even though they account for a small fraction of the total. A small decrease translates into a large savings in allocated overhead expense due to the existence of unrealistically high burden rates for automated activities.
- Allocating overhead on the basis of direct labor shifts these expenses from automated to labor intensive operations in a factory. One extreme example is an experimental unmanned FMS recently developed by a large military contractor in conjunction with the U.S. Air Force. The absence of direct labor implies that no overhead costs will be assigned to the system according to the accounting procedures required by the government. Overhead costs actually generated by the FMS will be allocated to the other operations in the shop.
- Measuring performance against static budgets and historically determined standards discourages efforts to instill a philosophy of continuous improvement. Reference points against which to judge results need to continually increase as success is achieved.
- An emphasis on periodic financial evaluation contributes to a short-term orientation. No balance is achieved with long-term strategic considerations.
- Often financial evaluations are produced too late and in too aggregate a form to be useful in controlling operations.

- Manufacturing costs arise from underlying activities such as setting up, using machines, and receiving orders. To control costs one must measure directly the occurrence of these activities.

In order to overcome these deficiencies, it is necessary to supplement financial indicators with physical measures of quality, delivery performance, and flexibility. This chapter deals with incorporating the flexibility dimensions into performance measurement. One critical element is a mechanism for evaluating discrepancies between what is desired and what exists. The initial section explains how to perform a discrepancy analysis for flexibility and indicates ways of closing any gaps. The next section deals with operationalizing flexibility, one of the major roadblocks to the extensive use of the concept. It discusses some major problems and some commonly used scales. Without this knowledge performance measurement is impossible. The chapter concludes with the results of research on the flexibility in practice of an important kind of AMT, computer-aided manufacturing equipment.

DISCREPANCY ANALYSES

A performance measurement system should distinguish between required, potential, and actual flexibility for each of the previously identified dimensions. The first, a constituent of the conceptual model, represents strategic management's judgment of how much is needed of a particular type of flexibility. It is based on a decision to be proactive or adaptive in the face of a given level of uncertainty. The second represents what can occur given the existing system design if conditions external to the system are appropriate. This is a theoretical notion and is evaluated based on personal knowledge of the design. The third arises from utilization of the equipment and is determined on the basis of experience.

Limitations on flexibility arise from misalignments among the required, potential, and actual conditions. There are six ways of ordering the three concepts, each representing a different set of misalignments. The situation in which a required condition is greater than a potential one, which is greater than an actual condition is undoubtedly the most common. Here, the difference between what is needed and what is obtained is factored into two kinds of discrepancies, required greater than potential and potential greater than actual. Managers should consider these discrepancies as signals to uncover the factors causing them. Some factors emphasized in the CAM literature will be discussed.

Required Greater than Potential

What factors cause required flexibility, the amount needed, to be greater than potential, the amount designed into the system? Understanding these factors provides an appreciation of the limits within which flexibility can be achieved so that attempts are not made to go beyond what is feasible.

Either required flexibility is too high or potential is too low. In the former situation, a company does not realize that some dimension is relevant or there is at best a hazy idea of how much is needed. A specification is made only after AMT is in operation and then it is too late. Even if required flexibility is accurately specified in advance, an increase in manufacturing uncertainty after the design is finalized, causes it to shift up. Various methods for reducing uncertainties can be employed here to close the gap.

Relatively low levels of potential flexibility are due to:

- Limitations on financial resources
- The state of the art in AMT
- Strategic management's decisions on selection and design
- Constraints in the value-added chain identified prior to selection and design

The last three causes will be examined in detail.

AMT has not evolved to the point where it can compete with the flexibility of humans. Some engineers question whether that will ever be possible (Whitney 1986). Changes in flexibility accompanying AMTs introduction are therefore influenced by the nature of the conventional system being replaced (Gerwin and Tarondeau 1989). Factories starting from a labor-intensive technology already have a high degree of flexibility. Installing AMT is likely to produce mixed results. Plants starting from a rigidly automated manufacturing process do not initially have much flexibility. Installing AMT is likely to improve the situation.

There is little that manufacturing managers can do to overcome AMTs technical constraints. Instead, ways must be found to get around them. One suggestion is to lower aspirations concerning what AMT can deliver to more realistic levels. To avoid aspirations being inflated by the excessive zeal of vendors, managers should check out the accuracy of their claims with previous customers. Second, managers should not assume that AMT is the sole means for achieving flexibility. Organizing the work force into teams of multiskilled individuals, who rotate tasks without being closely supervised, is a proven method of meeting unexpectedly changing conditions. Other alternatives include modularized product components, subcontracting, and adjustable plant capacity limits.

Strategic managers' decisions on the selection and design of AMT are often unavoidably biased against flexibility. No way exists to explicitly consider trade-offs between the concept and other manufacturing criteria. Managers must, for example, make an overall judgment concerning a system's flexibility and productivity since these criteria often pull the design in different directions. Since the efficiency benefits of productivity are more measurable than the strategic benefits of flexibility, the former may be emphasized over the latter.

This inherent bias has unfortunate implications for design. Focusing on productivity creates a need for more hard automation; deemphasizing flexibility means less programmable and human elements. The decrease in flexibility, however, is not necessarily in proportion to the increase in the fraction of hard automation in the system. AMT is only as flexible as its most rigid components will allow; a rigid subsystem is a flexibility bottleneck.

Consider, for example, computerized auto body welding systems. They combine rigid fixturing technology with programmable robots. It is much more time consuming and expensive to adapt the fixturing mechanisms to model changes than the robots. If it takes three months to adjust the fixtures and one month to reprogram the robots, the entire system is out of operation for three months. The former's rigidity impedes the latter's changeover and modification flexibility from being fully utilized.

In another example, Willenborg and Krabbendam (1987) found that changeover flexibility in FMSs was overestimated because *every* element of the system, including machines, material-handling devices, tooling, and fixtures, had to be able to handle a new product design. Limitations stem more from the last two components than the first two (Gould 1986).

Manufacturing managers can have great impact on potential flexibility through their decisions on the selection and design of AMT. To make these decisions wisely they must know their company's flexibility requirements and understand in a general way how AMT is designed to meet them. It is not necessary to have a technical education. They must set overall constraints, such as the trade-off between flexibility and productivity, rather than try to second guess the detailed technical choices of engineers. Due to the lack of quantifiable measures of flexibility's net benefits, they must be prepared to make judgment calls; otherwise they will lose control over the balance of hard and soft elements in the system. One rule of thumb is that the proportion of total cost due to elements dedicated to a given product should be directly related to the proportion of the time the system will spend handling the product (Scott 1984).

A good balance between flexibility and productivity is also achieved through proper organization of the selection and design process (Graham and Rosenthal 1986a). The project team should not have a narrow focus on hardware at the expense of software, systems, and human resources. Hardware designers should not have the overwhelming bulk of their experience with dedicated automation. The vendor selected must not be deficient in software experience. Otherwise the following indications of a lack of flexibility will probably appear: a low priority for software development, not enough redundancy, and a low capability for expansion.

An enterprise's value-added chain runs from research and development through manufacturing to sales and distribution. Constraints along the way, identified prior to AMTs selection and design, force a reduction in the technology's flexibility potential if they cannot be altered. There is little sense in building flexibility into computerized automation if upstream and/or downstream activities are not able to handle it. In other words, the concept of a flexibility bottleneck applies to the activities in the value-added chain. The chain's flexibility, and that of AMT embedded in it, is controlled by the most rigid link.

Consider, for example, the gear manufacturing process, which consists of rough forging, machining, heat treating, grinding, and assembly. The latest technology for heat treating is a computer-controlled rotary furnace with different chambers each capable of being set for time and temperature independently. Each chamber can process a different batch of parts that arrive in random order but it is uneconomical to process individual parts randomly. Gear machining, the previous step, is consequently not a prime candidate for the introduction of equipment flexible enough to handle batch sizes of one.

A second example concerns material-handling problems. In factories building products from large numbers of components, such as auto assembly, it is often difficult to get the right parts to the right process at the right time while simultaneously keeping inventories low. Keeping the problem manageable requires putting maximum limits on the number of different components to be handled. This limits mix flexibility in terms of allowable differences in the products being assembled.

Managerial discretion, although limited, has some ability to alleviate this type of problem. Before selection and design of AMT it is advisable to thoroughly scrutinize the value-added chain for bottlenecks, activities whose potential flexibility is a good deal less than the required flexibility of the new technology. It is undoubtedly easier to identify bottlenecks in the manufacturing process since links are harder to define and flexibility is more difficult to measure elsewhere in the chain.

After bottlenecks are identified, a decision must be made on whether to limit the new technology's potential or to remove the bottleneck. The former choice, which implies less flexibility is delivered than is needed, prevents realizing major strategic benefits albeit at a reduced capital investment. Sometimes due to technological constraints, as in the heat treating of gears, this is the only option. The latter choice usually means a much larger change than originally intended. Capital investment, time to complete the project, and disruption to the organization, will all increase. For these reasons it may be difficult to sell the expanded project to strategic management.

Potential Greater than Actual

The next significant discrepancy is between potential flexibility—the amount designed into the system—and actual—the amount obtained in practice. Understanding the reasons for this type of misalignment provides information on which controllable factors have the greatest leverage in realizing a system's potential. Constraints in the value-added chain uncovered after the new technology is working and management's decisions on how to operate the system are two factors that reduce the level of actual flexibility.

Due to problems in determining flexibility bottlenecks beforehand, many instances are recognized only after the system is operating. Activities in the value-added chain outside of manufacturing are likely candidates, but their precise impacts are not easy to determine. The marketing department may not be able to learn fast enough about the changing needs of customers or product engineers may not know how to design for computerized automation.

Recognizing bottlenecks leads to the same kind of choice associated with the first type of discrepancy. A decision to maintain the level of actual flexibility prevents realizing important strategic objectives. It also contributes to the technology missing its targets, which may impede future efforts to extend AMT throughout manufacturing.

A decision to remove the bottlenecks augments the amount of change to which the organization must adjust.

Attaining AMTs potential for flexibility is hampered by the way it is operated (Graham and Rosenthal 1986a, Jaikumar 1986, Reich 1983). American managers steeped in the values and assumptions of high volume standardized production are not always concerned with the technology's strategic potential. They use it to produce high annual outputs of a few parts rather than to take advantage of its flexibility and emphasize performance measures such as machine uptime and productivity as opposed to meeting changing market needs. Managers rely on traditional organizational structures designed for stable production rather than new approaches meant for changing conditions.

These arguments explain the results of surveys which have inquired into managers' perceptions and use of AMT. Farley, Kahn, Lehmann, and Moore (1987) interviewed twenty-nine senior manufacturing managers from large U.S. companies concerning anticipated expansions of flexible automation. The perceived importance of three major benefits, cost reduction, product quality, and shorter new product lead times was measured using ten-point scales. Cost reduction was viewed as the major benefit with an average score of 7.9, whereas shorter new product lead times, the time aspect of changeover flexibility, was seen as the least important with an average score of 7.0. Jaikumar (1986) surveyed sixty Japanese and thirty-five American FMSs with similar technical characteristics. He found that average mix flexibility in terms of the number of parts produced was ninety-three in Japan and only ten in the United States. Average changeover flexibility defined as the number of new parts introduced per year was twenty-two in Japan and only one in America. Margirier's (1986) mail survey of nineteen users of computerized automation in France found a similar situation. The equipment was selected more to reduce unit costs than to increase product diversity or decrease lead times.

Data subsequently collected by Ettlie (1988a), however, has more optimistic findings. In the twelve plants from his sample that utilized FMS, there were 187 average number of parts scheduled on the new technology. Either American managers are learning over time to use computerized automation more effectively or these results are due to differences in the companies, parts, and manufacturing technologies surveyed.

Graham and Rosenthal (1986a) attributed gaps between potential and actual flexibility to use of organizational forms meant for mass production. First, traditional functional structures produce organizations for FMS that do not include all the roles needed for effective operation. The time taken to change over to a new product goes up when software is written by programmers who are based elsewhere. Only a few companies in their study assigned maintenance people to the new technology. Second, most of the companies trained lower level managers inadequately and underestimated the extent of responsibilities operators should have.

Jaikumar (1986) provided clues on how to operate FMS technology better in order to get the most out of its flexibility. Managers should shift from directing day-to-day operations to developing systems and structures necessary to provide support for the hardware. One example is creating and managing a small team of engineers, which designs and operates the technology until it is running smoothly. Second, emphasis must be put on long-term planning for manufacturing process improvements and training employees. The result will be a manufacturing system capable of responding quickly to shifting markets and unanticipated process problems.

Other Discrepancies

The next most important sets of misalignments are potential greater than actual, which is greater than required, and potential greater than required, which is greater than actual. These situations in which potential is greater than required may arise in a number of different ways.

1. Manufacturing considerations are more important than marketing issues. Purchasing an FMS with random processing is done primarily to reduce inventories rather than to handle market uncertainties. It may cut production lead times beyond the levels required to meet customer needs.
2. There are organized efforts to reduce uncertainties that have not been meshed with the plans to select or design AMT.
3. It is difficult to estimate what is required before selection or design is finalized.
4. A phenomenon known as "creeping flexibility" troubles the firm. Under pressures from certain departments potential has been gradually acquired in some dimensions over and above what is required. Product lines have multiplied as the result of the marketing department's short-sighted fixation on meeting customers' needs.

Having potential greater than required creates an opportunity to explore proactive or defensive strategies. A firm can immediately employ excess flexibility to its advantage by redefining market uncertainties, or it can decide to bank flexibility for later use. Alternatively, it may reduce potential flexibility and distribute any liberated resources elsewhere.

Due to the current emphasis on increasing flexibility, it is easy to overlook the substantial savings that can occur from reducing potential (and/or actual) flexibility to a level commensurate with requirements. Some companies have found they can create more focused factories, that is, decrease the range aspect of mix flexibility, especially if they respond more quickly on the time aspect of this dimension simultaneously. By rationalizing their product structures companies achieve smaller product variety. Existing sales are analyzed to identify the most frequently purchased items in a product family. The remaining items are discontinued, offered to customers at a price or delivery penalty, or replaced by frequently purchased items having greater specifications than desired. In the process, resources are freed up that are then used to shorten delivery times. Another possible advantage is a significant reduction in the number of parts handled. Through its product structuring efforts, one business unit of a large American corporation discovered that 92 percent of its offerings were made from only 31 percent of its parts.

One specific approach to reducing product variety is to assemble to stock only the most frequently demanded basic units such as chassis or central processing units in combination with a number of standardized bundles of options. Customers or dealers are able to have these items rapidly. If units with additional options are desired they are produced to order and take much longer. Some firms are trying to convince dealers to put on additional attachments. These practices avoid the traditional costly and time-consuming policy of building completed products to stock on the basis of unreliable forecasts and then reconfiguring them to meet actual demand.

Another approach is to decrease the number of offerings within a range of product characteristics while simultaneously satisfying demand for the entire range. For standardized products the trick is to sell customers, at no extra cost, an item that more than meets desired specifications. For example, a customer wanting an electric motor with an unobtainable horsepower will get the next largest horsepower motor that is offered. For less standardized products a computerized system might compare an order's specifications against items that have been manufactured in the past. A customer might receive, at no extra cost, something more than is needed, a practice that is cheaper than having applications engineers prepare a sample. Their time is freed up to concentrate on major customer opportunities. Software that aids customers in developing product specifications also keeps them from regressing out of a given product structure. It can guide them to select the first offered item that more than meets their needs.

The three last sets of misalignments all possess the self-contradictory condition that actual is greater than potential and therefore are of little practical interest. Managers who find a balance between required, potential, and actual flexibility should not be complacent. One of the most cogent arguments against the use of flexibility in manufacturing operations is that it encourages laxity in eliminating controllable uncertainties. Analysis of uncertainties often reveals ways in which engineering change orders, machine breakdowns, or quality problems can be reduced, thus decreasing the requirements for a certain type of flexibility. Managers then have the options of redefining market uncertainties, banking the excess potential, or reallocating the resources devoted to maintaining it.

To conclude this section, one significant implication deserves some emphasis. Discrepancy analysis gives a new twist to the issue of focus versus flexibility. Since companies have no specific guidelines on whether to contract product variety or expand it, they are always susceptible to the current fad in their industry. From the viewpoint of discrepancy analysis a choice between the two strategies depends upon certain contingencies facing a firm. The first is the state of the market as mirrored in required flexibility and the second is the state of a company's manufacturing process as reflected in potential (actual) flexibility. Provided measurements exist one can determine whether required is less than potential and actual in which case more focus is an option, or whether required is greater than potential and actual in which case more flexibility can be considered.

MEASURING FLEXIBILITY

What attempts have companies and researchers made to measure the flexibility dimensions? Measuring has to do with finding scales to express the dimensions in quantitative physical units. Mix flexibility, for example, may be indicated by the number of different parts handled by an FMS. A second aspect involves using the scales to determine how much flexibility is needed or possessed. A third aspect, determining the financial implications of a given amount of flexibility, is discussed elsewhere in this book since we do not want to measure the concept in terms of its economic impacts. For one reason, it is useful to be able to investigate how changes in flexibility relate to changes in economic value.

Taking advantage of the manufacturing flexibility concept in decision making is hampered by problems in measurement, especially where AMT is concerned. Unless managers can make explicit tradeoffs between flexibility and other criteria the former does not receive the weight it deserves in manufacturing choices. Short-run financial considerations are emphasized over long-run strategic implications. Producing low variety and high volume is stressed over meeting changing market needs.

Two key decisions that are usually affected are the purchase of new manufacturing equipment and the evaluation of implemented technology. First, an inability to quantify the benefits of AMT including flexibility is a significant limitation to its purchase (Rosenthal 1984). Internal diffusion is probably hindered more than innovation (Bennett et al. 1987). Top management is more likely to accept a qualitative justification for the initial technology than for expansion. Second, a lack of performance measures inhibits effective use of AMT that has already been purchased by impeding knowledge of its extent of flexibility. It is not possible to determine whether gaps exist among required, potential, and actual types or to judge whether improvements in manufacturing processes are closing the gaps.

AMT vendors are limited in creating designs that meet the flexibility needs of users when those needs are not operationalized. With flexibility scales available it may be possible to develop general manufacturing system designs to meet frequently expressed combinations of needs. The designs can be used as guidelines in working out detailed plans for specific cases. Flexibility scales could also be used to compare the capabilities of American and foreign manufacturing technology and to indicate where we need to make improvements.

Conceptual and Methodological Issues

Measuring manufacturing flexibility depends on resolving some basic conceptual and methodological problems. One issue is that the concept may be investigated at a number of different organizational levels including the business unit, factory, manufacturing system or cell, or individual machine. Management needs to decide which levels are appropriate. At the business unit level the needs of the R&D and marketing departments may conflict with the development of manufacturing flexibility. Once more, a lack of flexibility in these departments may hinder obtaining the benefits of manufacturing flexibility. At the factory level and below manufacturing flexibility is the primary focus of attention, but production technology exemplifies only one means of achieving it.

Methods of achieving flexibility at one level may not occur at another level or may constrain efforts at lower levels. A source of mix flexibility for a business unit is having each plant specialize in its own product line. Mix flexibility achieved in this manner at the business unit level limits the degree of the same kind of flexibility at the factory level. Mix flexibility at the manufacturing system or cell level cannot be achieved by having individual machines or modules specialize in different parts. Since there is no integration a system does not exist.

A second issue is whether to investigate required, potential, and/or actual flexibility. This choice has implications for data collection. Ideally, one wants to operationalize flexibility using objective information compiled from existing records or *ad hoc* sources. Only where this is not feasible should recourse be made to the subjective evaluations of experts since they are more likely to be contaminated by perceptual biases.

Required flexibilities may be determined from a variety of sources running from mainly objective to mainly subjective. Perhaps the most objective information comes from customer surveys inquiring into, for example, the product range and lead time wanted. Respondents, however, tend to provide information on existing incremental needs rather than future radical needs. They may not be sensitive to ways in which the firm can redefine needs for them. A useful supplement to market surveys involves top functional managers visiting a small number of key customers in each major product line to speak with people who are looking toward the future.

Von Hippel (1988) suggested that a firm identify "lead users" of their product, especially in industries where market needs change rapidly. A lead user faces new needs months or years before they are generally recognized. Often it has conducted useful analyses of problems and experimented with alternative solutions. Von Hippel tested his ideas by studying CAD systems for designing printed circuit boards. He focused on the problem of increasing board density, that is, the amount of chips and circuits on a board. Lead users were identified through a questionnaire mailed to a random sample of user firms and a subsequent cluster analysis. They were designing very high density boards, were dissatisfied with existing CAD systems, and built their own systems. Five companies were selected from this group to participate in a group discussion that led to a new concept for improving the CAD systems. Through a questionnaire sent to the entire lead user group, preferences were determined between this concept, the system being used by the respondent, the best commercial system available at the time, and a special application system included to see if respondents preferred exotic over generally applicable software. Seventy-nine percent of the sample chose the concept developed by the five lead users as their first choice. Their preference was maintained even when it was assumed to cost more than the competing concepts. Although this research does suggest that a firm can identify market requirements for flexibility from lead users of their products, lead users may not in all situations want to reveal their analyses and solutions out of fear that competitors will learn about them.

If direct contacts with customers are not helpful, the conceptual framework indicates more subjective bases from which inferences on required flexibility can be drawn. Information on strategic objectives is useful provided it is precise enough. Knowledge of manufacturing uncertainties can help if it can be determined. Calculations of average product life cycles help pin down the required range and time for changeover flexibility. If all else fails, one can fall back on subjective evaluations of requirements from experts in the company's major functional areas.

Potential flexibilities are likely to come from the subjective evaluations of the company's and vendor's engineers below the factory level and the company's strategic managers at the factory level or above. Often capabilities even at low levels are not clearly known due to technical complexities, and individuals will bias their evaluations in ways that produce favorable outcomes for themselves. In a few instances objective data are available as when the range of mix flexibility is determined by the envelope specifying the dimensions of the parts that a FMS handles.

Actual flexibilities are determined objectively when performance data exist such as the average range of products handled or average lead times. This information, however, is possibly affected by environmental conditions during the time period investigated. When demand is weak the ranges of the four market-oriented flexibility dimensions will be constricted and the time aspects will be brief in duration. It is unlikely a full range of products will be demanded and it is easy to guarantee short lead times to customers. If a correction is not made the effects of uncontrollable market factors will contaminate differences between potential and actual flexibilities. When performance information is unavailable or difficult to compile it is necessary to resort to the subjective evaluations of experts.

Since there are alternative measures for a flexibility dimension, a third issue is to establish criteria such as cost effectiveness, validity, and reliability to help choose among them. Cost effectiveness is facilitated when there is a modest expenditure of time and financial resources for data collection. The number of indicators is kept to a minimum, which implies the same measure holds at a number of different organizational levels. The frequency of collecting data and computing scale values is controlled but is consistent with the need for timeliness. Data are readily available from existing records instead of being compiled from new *ad hoc* sources. Setup times, for example, are obtainable from costing reports and production lead times from the material requirements planning system.

Validity has to do with whether an indicator actually reflects the underlying concept it is desired to measure. Content validity, a specific aspect, refers to whether the scale takes into consideration a concept's entire domain. For most managerial concepts, including flexibility, there are no rules for specifying the complete domain and therefore no rigorous methods for assessing this type of validity. Instead, the subjective evaluations of knowledgeable individuals tend to ensure that relevant items are included and irrelevant items are excluded from a scale. Interviews with people of varying perspectives, such as customers, marketing managers, manufacturing engineers and managers, design engineers, and representatives of vendors, are helpful.

Construct validity, another specific aspect, is the extent to which a flexibility measure relates to measures of other concepts in a manner consistent with theory. If theory indicates a positive or negative relationship, this can be investigated using correlation analysis. It is believed, for example, that product diversity and product turnover should correlate negatively with machine utilization (Ettlie 1988a; Jaikumar 1986). A similar test is to see if a flexibility scale does not correlate significantly with measures of concepts with which it is supposed to be unrelated.

Unfortunately, not much theory has been developed to explain flexibility's interactions with other variables especially where AMT is concerned. Although product diversity and turnover are held to be inversely related to productivity for conventional equipment there is not much evidence to indicate the nature of these two relationships for AMT. Given the dearth of applicable theory recourse can be made to expert judgment. Suppose several pieces of equipment are ranked according to their score on a flexibility scale. Based on written descriptions of the equipment and an understanding of the flexibility concept being measured, each member of an expert panel can also produce rankings. Assuming interrater reliability is established the experts' rankings are compared to the scale's. If a positive relationship exists, it can be assumed that the scale has construct validity.

Reliability, which also has a number of aspects, involves whether a scale consistently indicates whatever is being measured. It is not much of a problem when indicators are based on objective data but is an issue when they are developed using perceptions. Interrater consistency is checked by taking repeated measurements from a number of individuals. Where large discrepancies exist, discussions should be held to resolve them. Then the purged data can be averaged. Interitem reliability means consistency among the different items used to measure a specific flexibility dimension or all the dimensions. A test is conducted using coefficient α in which a sufficiently high value of the statistic indicates the items tend to measure the same concept (Nunnally 1978).

Lack of correspondence between strategic manufacturing objectives and performance indicators, caused by measurement problems, is a fourth issue. It is a widespread problem in American industry especially with respect to flexibility. According to the 1987 Manufacturing Futures Survey (Miller and Roth 1987), the four market-oriented flexibility dimensions were included in managers' rankings of required competitive capabilities. Yet rankings of the top ten manufacturing performance measures did not include any direct measures of flexibility. Foster and Horngren (1988), who interviewed twenty-five English and American vendors and users of FMSs, found that a major concern in two user firms was that performance measures did not reinforce the strategic reasons for adopting the technology. Although flexibility was a motivation for purchase, there were no explicit measures used to assess it.

A practical approach to dealing with this problem is to construct measures on a hierarchical basis in which ends at one organizational level are broken down into means of accomplishing them at the next lower level. Consider the performance measurement system at Wang Labs in which strategy is connected to operations through a four-level hierarchy of ends and means (Cross and Lynch 1988/89). Corporate management develops a broad overall strategy that is supported by individual business units through marketing and financial objectives. Within each business unit, business operating systems, that is, functionally oriented groups of activities such as production, new product introduction, or sales administration, have goals which support the objectives. Marketing objectives are advanced by customer satisfaction and flexibility goals, whereas financial objectives are sustained by flexibility and productivity goals. Each department's operational criteria support the goals within a business operating system. Customer satisfaction is maintained through quality and delivery criteria, flexibility by delivery and processing time criteria and productivity by processing time and cost criteria.

Fifth, to help determine how much improvement is necessary in an indicator, its value must be expressed relative to some reference point or base line. A customer-based standard is required flexibility, the use of which implies that improvement efforts are based on a discrepancy analysis. Actual lead times may, for example, be expressed as fractions of the times by which customers want delivery. This type of measure is appropriate whether a firm's posture is defensive, that is to just meet customers' needs by closing a gap, or whether it is proactive, in other words, to exceed customers' current needs by creating a gap. For many indicators, however, required flexibility is only available in the form of subjective judgments.

Competitor-based reference points are a second alternative. When a firm is unsure about customer wants it is forced to use competitors' flexibility as an indirect measure of those wants. In another scenario it is not economically feasible for any company in an industry to meet the existing levels of customers' needs. Then success depends on remaining ahead of or catching up to competitors' levels of flexibility. Unfortunately, the information necessary to compute reference points may not exist; experts' perceptions have to be used. As a last resort a firm falls back on matching its current levels of flexibility against those occurring in the past.

Finally, will the performance measures reflect control or learning? In the former situation they facilitate evaluation by providing information to managers at upper levels for judging the units reporting to them. In the latter instance they enhance self-improvement by producing feedback on the consequences of decisions for individuals at lower levels. Once more, to prevent comparisons from being made, data are not communicated to higher levels or are aggregated before being sent. No concern is given to constructing generalized measures the purpose of which is to compare diverse businesses throughout the firm. Decision makers, however, may pay little heed to measures on which they are not evaluated.

Flexibility Scales

No truly appropriate flexibility scales have been developed in the research literature. Hickson, Pugh, and Pheysey's (1969) Workflow Integration Scale includes a subscale called workflow rigidity defined as the extent to which a manufacturing process is in an invariable sequence, has a limited use, and operates as a line. Its coverage of the domain of flexibility is confined to the range aspects of the rerouting and mix dimensions. Gerwin and Tarondeau's (1986) measures of the six flexibility dimensions indicate changes, not levels. Based on technical managers' perceptions these measures could suffer from perceptual biases. Swamidass and Newell (1987) also tapped managers' perceptions in asking them to rate their companies' goals for flexibility over the preceding two years using a ten-point scale. A mix flexibility scale ran from narrowest product variety in the industry to widest. A changeover flexibility scale ran from least frequent new product introductions in the industry to most frequent.

Kumar (1987) used the entropy concept to suggest measures for the range aspect of the flexibility dimensions. He assumed a manufacturing process' flexibility to depend on the number of options available and on the freedom or opportunity to select a given option. The mix dimension, for example, might depend on the number of different parts that are made and the fraction of the process' total output due to each part. The routing dimension might be contingent on the number of alternative paths and weights representing the availability of each path.

Suppose n options exist and x_i for $i = 1, 2, \ldots, n$ is the freedom to select option i where the x_i are normalized weights. A flexibility measure in situations of this type is given by the standard expression for entropy in thermodynamics,

$$S = - \sum_{i=1}^{n} x_i \ln x_i$$

Among the measure's desirable properties is that it has its minimum at $S = 0$, representing no flexibility, when all but one of the $x_i = 0$, that is a forced choice exists. For n given it attains a maximum when all the x_i are equal.

Most efforts to develop flexibility measures, particularly by companies, have concentrated on the mix and changeover dimensions. In line with the growing significance of time based competition the time aspects of these two dimensions have been emphasized.

Mix flexibility's range is often measured by the number of different kinds of outputs at a hierarchical level, products from a factory or parts from an advanced manufacturing system. Ettlie (1988b), for example, measured the number of part families and the number of parts scheduled on FMSs. Buzacott (1982) suggested using the number of parts handled by an FMS divided by the number handled by the entire factory.

A more sophisticated measure is the range of some major defining characteristic of the outputs. Examples include the wheelbases of autos from an assembly plant or the dimensional envelope within which parts can be processed by an FMS. The measure takes into consideration that, although a small number of outputs are handled, they may be very different in kind from each other. It is also a practical alternative for the situation in which there are many outputs. Then it is difficult to count the number or to readily distinguish between them.

Margirier (1986) developed a measure of mix flexibility that considers the degree of variation among products,

$$N \left(\frac{1}{3} a_1 + \frac{1}{3} a_2 + \frac{1}{3} a_3 \right)$$

Here, N is the number of products manufactured and the expression in parentheses is a coefficient of diversity. The quantities a_1, a_2, and a_3 are, respectively, the ratios of the maximum to minimum: batch sizes, manufacturing times, and parts' dimensions. When these ratios are around 1.0 there is little variation among products and the value of mix flexibility is around N. When these ratios are large a great deal of variation exists and mix flexibility is correspondingly higher in value.

Firms adopting the JIT philosophy use an indirect measure,

$$\text{Average setup time} = \frac{\text{total setup hours}}{\text{number of complete setups}}$$

over some time period. As this quantity decreases it is possible to run more different kinds of parts frequently and in small quantities. The significant reduction in setup times associated with FMSs permits their random processing feature.

Measures of mix flexibility's time aspect are often based on the reaction time concept, that is, the interval between the receipt of a signal and a response to it (Bolwijn et al. 1986). At the business unit level this amounts to total lead time as measured by the average interval between receipt of a customer order and its delivery. An indirect measure is the fraction of units produced during a given time period that are reconfigured. As lead times diminish, sales forecasts become more accurate. Production is better matched to demand. Less completed units for which there is no need are torn down and rebuilt into units for which a need exists.

At the plant level production lead time is the interval from receiving an internal order to produce, and shipment. A typical measure is the average lot close out time, the average time from start to completion of all jobs closed in a given interval. Average setup time is one of the most frequently measured components of production lead time.

Where high product variety exists, it is not easy to measure the average production lead time. Extraneous uncontrollable factors such as changes in product mix influence the calculations. It is also time consuming and expensive to collect information on all jobs. A computerized material requirements planning system may help or a sampling procedure can be devised. A firm can also turn to simpler measures such as the percentage of orders or units made to schedule. These are not adequate alternatives, however, if scheduled completion times represent what the company has promised rather than what customers want. Although the measures may indicate excellent performance, customers will still be dissatisfied.

Gunn (1987) suggested a measure which focuses attention on how to make improvements, the ratio of value-added lead time to total production lead time. It recognizes that production lead time is the sum of times to perform value-added activities such as metal cutting or assembly, and nonvalue-added activities like waiting, transportation, or defect detection and correction. Gunn believed that the ratio's value should be greater than .5 for a world class manufacturer, but conceded that for many U.S. manufacturers it is probably less than .1.

The range of changeover flexibility is measured in different ways. Perhaps the simplest is a count of the number of major design changes during a time period. The number of new parts introduced per year on an FMS (Jaikumar 1986) is an approximation. A correction also should be made here for the degree to which the new and old outputs differ from each other. Even if the number of changes is low, outputs entering the mix may be very different from those leaving the mix. A second type of measure is based on turnover such as the fraction of products manufactured at the start of a time interval that are still produced at the end of the interval. A third alternative is to calculate the ratio of that investment in a manufacturing system which is applicable for future parts to the total investment in the manufacturing system (Gustavsson 1984).

The time aspect of changeover flexibility is also a reaction time. At the business unit level it is the interval between when a signal is received to start development of a new product and when the response of full-scale production is made. At the plant level it is the time devoted to manufacturing's activities in new product development. Due to the growing significance of simultaneous engineering, more and more manufacturing activities are carried out in parallel with R&D activities. The manufacturing time remaining on the critical path is the most crucial to consider. Once more, efforts to rationalize the new product development process may cause some traditional R&D activities, pilot development, for example, to be transferred to manufacturing. The resulting increase in manufacturing's share of development time should not be mistaken for a sudden reversal.

Typical measures include the average new product development time at the business unit and plant levels. In the former situation it is often difficult to pinpoint an initiation date so the first well-specified time is used.

A complementary factor to consider is the product life cycle as measured by the average time between the introduction of new products. When used in conjunction with new product development time it can signal important information to management. A high ratio of the average product life cycle to average product development time forewarns of customer dissatisfaction. It suggests that new products are not being introduced frequently enough. Customers are more aware of lengthy periods between new products than of short development times especially in industries producing standardized items. As the ratio decreases to a value of one, management must devote more effort to reducing the denominator.

It is often useful to break down a measure into separate determinations for new models and updates. Separate calculations of development time are needed because there are different activities or at least different times for the same activities. Separate life-cycle calculations are helpful because different frequencies of introduction are preferred. Decreasing the frequency of new models and increasing the frequency of updates is in line with meeting market changes more rapidly.

Measurement of the other flexibility dimensions has not received the same amount of attention. Modification flexibility's range is measured in terms of the number of engineering change orders (ECOs) handled during a given time period, but this is very simplistic. It does not consider the type or extent of each change. One also wants to distinguish between ECOs determined by market forces beyond a firm's control and those which, resulting from sloppiness in activities upstream from manufacturing, can be avoided. It is in the company's interest to eliminate the latter. The temporal aspect reflects the average amount of time it takes to make the changes.

Volume flexibility's range can be operationalized in terms of the average amount of fluctuation in aggregate production over a given time period divided by the capacity limit. Lim (1987) suggested using the average range of batch size per part. This dimension's other aspect reflects the amount of time taken to switch from one production level to another.

Rerouting flexibility has a long-term component that is salient when machines are taken out of production to accommodate major design changes and a short-term aspect arising from the need to cope with machine breakdowns. A direct measure for each component's range should reflect the parts that are affected and the machines that are stopped. Once more, each particular part-machine combination may be associated with a different number of alternative routings. Buzacott (1982) suggested an indirect measure that avoids these complications, the average drop in production rate when a machine stoppage occurs. The time aspect, which involves the average duration to return to normal production, includes the period during which a machine is down and any subsequent period of surge production.

Material flexibility's range can be measured by the fraction of key dimensional and metallurgical properties as well as the extent of variations in those properties handled by the equipment. The dimension's second aspect reflects the time it takes on average to adjust to variations.

FLEXIBILITY IN PRACTICE

Many enterprises have placed huge bets on AMT to help them increase flexibility without sacrificing cost efficiency. AMT partisans have, in fact, prophesized a future in which the fully automated, computer-controlled factory will increase productivity, improve quality, reduce lead times, and decrease inventory levels. Simultaneously, the factory will be manufacturing just about everything from gum drops to tractors by pushing a button.

Information concerning the flexibility of AMT in practice has not been as optimistic. For example, after assembly of minivans began in Ford's revamped St. Louis plant, a substantial change had to be made to a heavier but less strong vehicle floor pan because the computer-controlled equipment was not flexible enough to handle the original floor pan (Nag 1986). A second example is a comprehensive survey which indicates that U.S. flexible manufacturing systems exhibit an "astonishing" lack of flexibility. In fact, some perform worse than the conventional equipment they have replaced (Jaikumar 1986). In another instance, some thoughtful robot designers have questioned whether robots can ever be made as flexible as the humans they replace (Whitney 1986).

Just how flexible then is AMT in practice? Based on a pilot study in auto fabrication and assembly and a large sample mail survey of French manufacturers, it is possible to give some answers for one major aspect of AMT, computer-aided manufacturing technology.

The Pilot Study

The American automotive industry is the largest component of our manufacturing sector in terms of employment and value of shipments. This archetypal mass production industry has been forced to adjust to a dramatic shift in competitive conditions. Finding ways to create and maintain flexibility has become critical. The industry must learn how to bring out a new model in two or three years rather than six or seven and to do so it must replace its inflexible capital equipment which has kept its product strategy captive to its operations technology (Skinner 1984).

Automobile production falls into the two general categories of fabrication and assembly. These are two different kinds of operations employing different kinds of process technology. Fabrication is characterized by a rigid technology incorporating transfer lines and automated machine tools. Assembly has a flexible technology in which manual labor plays an important role. In order to answer the questions raised here it is necessary to study fabrication and assembly separately.

The research was conducted in two American and one French auto company. Each company was represented by a single plant. Within the plants key steps in manufacturing a car were covered. These included three engine fabrication activities (EF1, EF2, EF3), two engine assembly activities (CA1, CA2), and two body assembly activities (BA1, BA2). Three of the activities were in the American plants (EF1, BA1, BA2) and four were in the French plant (EF2, EF3, CA1, CA2). Table 4.1 provides detailed information on the nature of the activities, the kinds of programmable automation employed, and the titles of the respondents.

Comparisons were made between programmable and conventional approaches to the same activity in the same plant by asking about the extent of change in the six flexibility dimensions of the conceptual framework. The programmable process could be run in parallel with the conventional process, or it could have replaced the conventional operation. Ideally, these comparisons should be made for manufacturing processes producing the same kind of parts, but, practically speaking, it was necessary to accept departures from the ideal. A detailed account of the computerized and conventional processes in the American plants is provided in Appendix 2 at the end of the book.

Data were collected using a structured questionnaire, which appears in Appendix 3. The initial section asked for descriptive information on the nature of the manufacturing equipment and the components produced by the equipment. The body of the questionnaire searched for differences between programmable and conventional processes on flexibility and other manufacturing characteristics. Each

TABLE 4.1 Sample Information (Pilot Study)

Code Name	Country	Function	Respondent
BA1	United States	Automated body framing using PCs,[a] guided vehicles, robots, transfer lines	Manager of new model programs
BA2	United States	Automated body framing using PCs, guided vehicles, robots, transfer lines	General supervisor of tooling
CA1	France	Engine assembly using PCs, conveyors, programmable machines	Unit manager
CA2	France	Gearbox assembly using PCs, conveyors, programmable machines	Plant manager
EF1	United States	Engine parts fabrication using PCs	Supervisor of master mechanics
EF2	France	Cylinder cover machining using PCs	Unit manager
EF3	France	Cylinder block machining using PCs	Unit manager

Source: Gerwin and Tarondeau (1986).
[a] PCs = programmable controllers.

question employed a five-point scale asking whether a particular characteristic had decreased a lot, decreased, not changed, increased, or increased a lot. These answers were assigned values of -2, -1, 0, $+1$, and $+2$, respectively. The respondent would indicate the change on the scale and provide a verbal explanation.

Information was also obtained on the mechanization or automation levels of the original and new manufacturing processes. As Kaplinsky (1984) pointed out, mechanization can be applied to different components of the production process such as the transformation of materials, the transfer of materials, or the control of operations. Programmable automation based on digital control systems seems to have its greatest impact on the third component. Therefore, its mechanization potential should be appropriately captured by Bright's scale (1958), which is based on increasing sophistication of the control system. It runs from 1 (lowest) to 17 (highest).

The methodology employed is characteristic of exploratory research in an area in which virtually no previous work has been conducted. The data represent the respondents' subjective perceptions rather than objective information from company records. Only one respondent was interviewed per manufacturing activity, making it impossible to study the data's reliability. The validity of the flexibility scale has not been established. Consequently, the study's results should be considered as preliminary.

Table 4.2 provides changes in the six manufacturing flexibility dimensions for each of the fabrication and assembly activities. It also has three combined measures, the ranges of which are from -2 to $+2$ as for the individual measures. An activity's

TABLE 4.2 Changes in Flexibility (Pilot Study)

Flexibility	Assembly					Fabrication				Overall
	CA1	CA2	BA1	BA2	AVG.	EF1	EF2	EF3	AVG.	AVG.
Mix	0	2.0	1.0	2.0	1.3	0	1.0	2.0	1.0	1.1
Changeover[a]	−1.5	−1.0	−1.0	1.0	−0.6	0.5	0.5	0.5	0.5	−0.1
Modification[a]	−1.0	−0.5	0.5	1.5	0.1	1.0	1.8	0.5	1.1	0.5
Volume	−2.0	−1.0	0.7	1.0	−0.3	0	1.5	1.0	0.8	0.2
Rerouting	0	0	−2.0	−1.0	−0.8	0	0	1.0	0.3	−0.3
Material	−1.0	−1.0	−1.5	−2.0	−1.4	0	0	0	0	−0.8
Market	−1.1	−0.1	0.3	1.4	0.1	0.4	1.2	1.0	0.9	0.4
Process	−0.5	−0.5	−1.8	−1.5	−1.1	0	0	0.5	0.2	−0.6
Global	−0.8	−0.3	−0.7	−0.1	−0.5	0.2	0.6	0.8	0.5	−0.1

Source: Gerwin and Tarondeau (1986).
[a]The measure of this type of flexibility was obtained by averaging the scores of two questionnaire items.

market flexibility is the average of its scores on the mix, changeover, modification, and volume dimensions since they are associated with market uncertainties. Process flexibility is the average of an activity's scores on the rerouting and material dimensions since they are associated with uncertainties in the manufacturing process. Global flexibility, the average of an activity's market and process scores, represents an overall measure.

Averages have also been computed across all seven activities, the assembly activities, and the fabrication activities for each of the individual and combined dimensions. Qualitative information from the American interviews which helps explain the reasons for the observed changes appears in Appendix 4.

Detailed task effects between assembly and fabrication turned out to be stronger than country effects between the United States and France. The methodology employed for this determination reflected the paucity of data. The observations for each type of flexibility were arranged in a 2 × 2 table based on the task variable's two levels and the country variable's two levels. A calculation was made of the difference between the means for the two task levels and then for the two country levels.

Comparison of the magnitudes of the two differences for each flexibility aspect indicated that four task differences were larger than, one was equal to, and one slightly smaller than their corresponding country differences. The global task difference was a good deal larger. Further computations that also took variances into consideration yielded the same results. Once more, part of the country effect is due to an unavoidable confounding with a detailed task effect. The two French assembly activities are for engines while the two American assembly activities are for bodies.

Consider initially the overall average changes in the individual and combined flexibility measures. There was an increase in mix, a small increase in modification, and a negligible increase in volume flexibility. Changeover flexibility virtually remained the same, rerouting flexibility had a very small decrease, and material flexibility decreased. There was a small increase in market, a small decrease in process, and virtually no change in global flexibility. Due to the need to examine assembly and fabrication separately, these results have little import for the pilot study. They will prove useful later in making comparisons with the mail survey.

For assembly the average changes in the individual dimensions do not always agree with what occurs at the individual activity level. Mix flexibility increased due to a greater number of items being handled. However, in the American plants at least, the range of component characteristics produced decreased (-1.0 in BA1 and BA2). In one plant, for example, the new system can only accommodate from a small two-door to a small five-door body, whereas the old system could manufacture a small three-door to a medium-size station wagon body. Although the computerized technology handles a larger number of items, they are more similar to each other.

For changeover flexibility, which is based on combining two questionnaire items, the ability to accommodate major design changes has decreased and the time to accommodate them has increased, due primarily to the greater amounts of hard automation in the new assembly systems. In one instance, however (BA2), the ability increased a lot and the time remained the same because the old process had a great deal more specialized fixtures than the new. Modification flexibility, which is also based on combining two items, has hardly changed reflecting an increase in ability and a decrease in time for body assembly, but a decrease in ability and an increase in time for engine assembly. Volume flexibility exhibited a slight decrease due to a gain in body assembly being roughly canceled out by a loss in engine assembly.

Rerouting flexibility has decreased at least in body assembly due to the incorporation of single lines, more process disruptions, and more difficulty in finding the causes of disruptions. Material flexibility has decreased due to a reduction in human intervention, which had previously been responsible for handling unexpected process variations.

With respect to assembly's combined measures, there was virtually no change in market flexibility as the mix dimension went up, the modification dimension virtually stayed the same, and the changeover and volume dimensions decreased. Once more, an increase in body assembly was canceled by a decrease in engine assembly. The process measure went down reflecting decreases in rerouting and material flexibility. The average change in the global measure represents a small decrease.

For engine fabrication average changes in the individual dimensions reflect shifts at the individual activity level. Mix flexibility in terms of the number of items handled has increased as programmable controllers facilitate the storage and execution of multiple operating sequences for different sizes of parts. With respect to changeover flexibility the ability to handle major design changes has remained the same due primarily to the lack of change in the dedicated nature of transfer and transformation equipment. However, the time needed to adjust to a major design change has been reduced due to the ease with which operating sequences can be altered, and due to improved diagnostic capabilities. For modification flexibility the ability to accommodate minor design changes has increased and the time to accommodate them has decreased. Minor design changes tend to affect programmable process elements more than dedicated process elements. It is easier to reprogram than to change the hard-wired switches of the conventional equipment. Volume flexibility, the ability to accommodate shifts in aggregate production, has gone up in part because capacity limits have become less rigid. Programmable controllers facilitate the incorporation into the process of additional machines to relieve bottlenecks.

There has been some increase in rerouting flexibility due to the new technology's improved diagnostic capabilities. No change was observed in material flexibility, since the ability to deal with dimensional and metallurgical variations lies more with mechanical and hydraulic features than with control systems.

The average changes in engine fabrication's combined measures are fairly characteristic of their values at the activity level. Market flexibility has increased due to shifts in the mix, changeover, modification, and volume dimensions. Process flexibility exhibited virtually no change as the rerouting dimension hardly changed and the material dimension stayed the same. Overall, the global measure showed a surprisingly small increase.

In sum, the overall changes in flexibility indicated mixed results with some individual dimensions going up, others going down and still others hardly changing at all. Among the combined measures, market increased a little, process decreased somewhat, and global stayed about the same.

In assembly, the replaced conventional processes were relatively labor intensive. Using Bright's scale the average original automation level (OAL) of the four activities was 4.0 (four in BA1, four in BA2, five in CA1, and three in CA2). Here the changes in flexibility were mostly negative. Four of the six individual dimensions decreased. Mix flexibility might have decreased if it was measured differently and the modification score was virtually zero. As indicated in Table 4.3, which will be compared to the mail survey, market flexibility remained virtually unchanged. Differences existed, however, among its individual dimensions and activities. Process flexibility decreased and the global measure went down somewhat.

In engine fabrication, the replaced conventional processes consisted of rigid automation. Bright's scale scores for the OALs of the three activities averaged out to a value of 12.3 (fifteen in EF1, eleven in EF2, and eleven in EF3). Here the changes in flexibility displayed a clear tendency to increase. Five of the individual dimensions went up, whereas one remained unchanged. As portrayed in Table 4.3, this resulted in an increase for market flexibility and virtually no change in process flexibility. The global measure increased somewhat.

The Mail Survey

The mail survey was for a large sample of French manufacturers who have adopted some form of programmable automation. Difficulties in identifying a population of users made it infeasible to select a random sample. Consequently, the results of the subsequent statistical tests must be interpreted with caution. Over 800 alumni of the Ecole Nationale Supérieure des Arts et Métiers identified as holding manufacturing, production, methods, works management, industrial manufacturing, or director positions were chosen for the sample. They received an abbreviated version of the pilot study's questionnaire, which asked them to concentrate on one example of computerized automation in their firms based on its importance or representativeness. Responses were received from 163 companies of which 107 had experience with the new technology.

Data were analyzed from the eighty-one firms that use programmable automation in fabrication. It is not possible to determine how representative this sample's

TABLE 4.3 Average Change in Flexibility versus Original Automation Level (OAL)—Pilot Study

OAL	4.0	12.3
Market flexibility	0.1	0.9
Process flexibility	−1.1	0.2
Global flexibility	−0.5	0.5
Sample Size	4	3

TABLE 4.4 Changes in Flexibility (Mail Survey)

Flexibility Dimension	Proportions of Firms with Indicated Scale Scores					Average Change	Proportion ≤0
	−2	−1	0	+1	+2		
Mix	.03	.05	.29	.25	.39	.93[a]	.37
Changeover	.06	.29	.39	.19	.06	−.10	.75
Modification	.04	.11	.31	.23	.31	.66[a]	.46
Volume	.00	.04	.27	.41	.29	.95[a]	.30
Rerouting	.20	.37	.23	.14	.06	−.51	.80
Material	.14	.29	.44	.10	.04	−.39	.86
Market						.61[a]	.14
Process						−.44	.79
Global						.08	.47

Source: Gerwin and Tarondeau (1989).
[a] Significant at the .01 level.

characteristics are of the population of French fabrication activities which have programmable automation since no comparable population figures exist. Most companies were in the small- to medium-size range with 23 percent having less than 200 employees, 40 percent between 200 and 1000 employees, 25 percent between 1000 and 5000 people, 7 percent greater than 5000, and 5 percent unclassified. The companies' products included industrial equipment and components (22 percent), electricity and electronics (12 percent), consumer durables and nondurables (23 percent), miscellaneous mechanical (27 percent), other (7 percent), and unclassified (9 percent). The production systems included 20 percent open job shops, 40 percent closed job shops, 23 percent mass production, 14 percent continuous process, with 3 percent unclassified.

The technology consists of NC and CNC machines (48 percent), flexibile manufacturing systems or cells (7 percent), transfer lines controlled by programmable controllers (17 percent), miscellaneous equipment including robots (20 percent), and unclassified equipment (7 percent). Most companies purchased their technology recently suggesting the results could change over time. Fifty-four percent of the equipment had been in place for one year or less, 35 percent for between one and five years, 7 percent for greater than five years with 4 percent unclassified.

Table 4.4 shows for each flexibility dimension the proportion of companies having the indicated scale scores, the average change, and the proportion of firms that stayed the same or decreased. The last two quantities are also presented for the three combined measures.

Considering individual dimensions first, the average increases in mix (+.93), modification (+.66) and volume (+.95) flexibility were significantly different from zero beyond the .01 level using one-tailed t tests. The first dimension, however, was defined in terms of the number of items rather than the range of item characteristics.

The pilot study suggested that the latter operationalization might decrease in a large sample. Changeover ($-.10$), rerouting ($-.51$), and material ($-.39$) all decreased. The proportion of enterprises for which a given flexibility dimension decreased or stayed the same ranged from a healthy minority of .30 for volume flexibility to an overwhelming majority of .86 for material flexibility.

For the combined measures, market flexibility exhibited the only significant average increase (.61). The overwhelming majority of firms reported some improvement (proportion $\leq 0 = .14$). Process flexibility decreased ($-.44$) with the overwhelming majority of firms registering a fall or no change. Global flexibility, the aggregate measure, hardly changed with about half the firms showing increases and half showing decreases or no change.

The pilot study suggested that average changes in flexibility are negative when the original automation level (OAL) of an activity is low and that they increase to become positive at high OAL values. To investigate, we employed a simplified version of Bright's scale (1958), more suitable for a mail survey. It collapses his seventeen levels into five based on general categories he devised and contains a sixth reflecting types of automation recently developed. The scale runs from 1 (manual) to 6 (capable of recognizing unanticipated environmental changes and adapting to them).

In general, the tendency suggested by the pilot study does exist, but its strength varies depending upon the flexibility measure used. Table 4.5 shows the average value of each combined measure for the firms with the indicated OAL. These data must be considered with caution as few observations exist at OAL = 3,4 and none at OAL = 5,6. The pattern for market flexibility does not conform very well to expectations. It has no negative values and the value for OAL = 1 is out of line. There is, however, a steady increase from OAL = 2 onward. The pattern for process flexibility is closer to expectations with negative values at low OALs that steadily increase to zero at OAL = 4 (and presumably positive values beyond). The global measure displays the anticipated effect but not in a striking manner. Its values steadily increase from a very small negative figure at OAL = 1 to a small positive value at OAL = 4. The magnitudes are surprisingly low.

To provide some statistical support for these results one can utilize regression analysis with each combined flexibility measure as the dependent variable and OAL

TABLE 4.5 Average Change in Flexibility versus Original Automation Level (OAL)—Mail Survey

OAL	1	2	3	4	5	6
Market flexibility	.69	.52	.58	.81	—	—
Process flexibility	−.91	−.22	−.06	0	—	—
Global flexibility	−.11	.15	.26	.41	—	—
Sample size[a]	27	36	9	4	0	0

Source: Gerwin and Tarondeau (1989).
[a] Five observations were unclassified.

as the independent variable. The hypothesis that the slope of the regression line is less than or equal to zero was tested using a one-tailed t test. Rejection supports the contentions since it implies a positive relationship between the variables. Parameters were estimated using the seventy-six observations at the firm level. For market flexibility the slope's value of $-.02$ was not significant at the .05 level. However the values for process flexibility (.39) and for the global measure (.18) were significant beyond the .01 level. Caution is advised in interpreting these results because OAL is an ordinal scale, but we can take some comfort in the high significance levels.

Comparison of the Two Studies

Is it reasonable to compare the results of the pilot study and mail survey? The former involves fabrication and assembly activities while the latter is based on fabrication only. Both sets of data, however, include a range of OALs running from labor intensive through automated, which is a more significant consideration.

Given the data's limitations there is surprisingly good agreement between the two studies. Table 4.6 reproduces the overall average changes in flexibility for the individual and combined measures. The signs are the same in both sets of data except for the global measure. The magnitudes are close except perhaps for volume and material flexibility. Mix, modification, volume, and market increase, changeover, and global stay the same, and rerouting, material, and process decrease. Thus, introducing CAM into a manufacturing process will not improve flexibility per se, although more measures turn positive at high OAL levels.

Average changes in the combined measures as a function of OAL can be compared through Tables 4.3 and 4.5. Scores on Bright's scale of 4.0 and 12.3 in the former table correspond respectively to scores of 2 and 5 on the mail survey's scale. Both studies suggest that market flexibility moves up with OAL and is positive over a wide range, findings that are relevant for the market turbulence of the recent past. Process flexibility increases as a function of OAL, but it remains negative over a

TABLE 4.6 Overall Average Changes in Flexibility

Flexibility Dimension	Pilot Study	Mail Survey
Mix	1.10	.93
Changeover	−.10	−.10
Modification	.50	.66
Volume	.20	.95
Rerouting	−.30	−.51
Material	−.80	−.39
Market	.40	.61
Process	−.60	−.44
Global	−.10	.08

wide range. If, in fact, American companies have traditionally had more process flexibility than needed, the only danger of this result is in going too far, that is in preventing adjustments to process uncertainties that cannot be eliminated. Global flexibility starts off negatively and becomes positive as OAL increases reflecting its underlying constituents. In the more reliable mail survey all three combined measures have higher values at any given OAL than in the pilot study.

Finally, whether one considers the individual or combined measures, the average changes in flexibility are surprisingly small in magnitude. Virtually all are within a range of -1 to $+1$ on the flexibility scale. Perhaps CAM is not as significant a vehicle as previously thought for handling manufacturing uncertainties.

5
EVOLUTIONARY PROCESSES

Our tour through the conceptual model concludes with a look at its feedback processes, the means by which temporal adjustments occur. Changes in environmental uncertainties lead to activation of the adaptive feedback loop that runs through the performance measurement variable. As uncertainties shift, existing strategy becomes ineffective, which is reflected in low performance. Strategic adjustments create changes in required flexibility and production technology. The feedback loops from strategy to uncertainty, along with external influences, are sources of the changes in uncertainties that initiate evolution.

This chapter's initial section concentrates on relationships among the three feedback loops. There is a need to balance investments in uncertainty reduction and adaptation. As firms begin to also redefine market needs through flexibility, it is likely that a three-way balance will have to be struck. The next section on short-run evolution demonstrates how the model's feedback loops provide a basis for changing manufacturing flexibility. Previously discussed concepts are brought together to sketch out a method for creating a flexible factory. The last section on long-run evolution examines theories of how a factory's strategies and technologies change over time. Starting from Abernathy and Utterback's ideas, it traces the likely impacts of AMT on the way in which a plant matures.

RELATIONSHIPS AMONG THE FEEDBACK LOOPS

Although it is dangerous to draw broad generalizations from limited evidence, comparisons of Japanese and American manufacturing practices suggest a need for achieving a balance between adaptation and uncertainty reduction. Then a firm can adjust to uncertainties in the short run while it is working toward their removal in the long run.

Japanese companies seem to excel in neutralizing market and process uncertainties (Hayes 1981, Schonberger 1982). Just-in-Time (JIT) manufacturing depends on uncertainty regulation in order to freeze production schedules, to have parts arrive on time, and to produce high quality items. Japanese flexible manufacturing systems have high utilization rates because they are designed to eliminate anticipated contingencies (Jaikumar 1986).

Simultaneously, Japanese firms are proficient in adapting to market and process uncertainties. With JIT uncertainties may occur unexpectedly or as the result of removing in-process inventories. If something goes wrong, the whole manufacturing operation is quickly notified and works together to solve it (Hayes 1981). This flexibility is achieved through automated monitoring devices, multiskilled workers, and the authority for workers to stop the line.

American firms have historically underutilized adaptation to market uncertainties, putting them at a disadvantage in coping with large-scale, uncontrollable changes in competitive circumstances. Critics of our mass production industries have often made this point in one form or another (Abernathy 1978, Choate and Linger 1986, Reich 1983). Once more, in the 1987 North American Manufacturing Futures Survey (Miller and Roth 1987) respondents were asked to note gaps between existing capabilities and those needed in 1991. The largest gap was in the flexibility to introduce new products and make design changes. Finally, Jaikumar (1986) found that American firms use their flexible manufacturing systems in a much more rigid manner than the Japanese.

Simultaneously, U.S. companies overemphasized adaptation to uncertainties in the manufacturing process, thus reducing pressures to eliminate the causes of machine breakdowns or quality problems. Why, for example, design components for a tight fit when the assembly department is adept at putting loosely fitting parts together? In this situation adaptive mechanisms are nonvalue-added activities that contribute to waste.

In attempting to correct their lack of responsiveness to customers' needs and their wholesale adaptation to process contingencies, American firms must keep the notion of balance in mind. Otherwise, they may well wind up at the other extremes. In too many companies adjusting to market demands is viewed as the only strategy. Potentials for reducing or redefining market uncertainties are not being tapped. At the same time our firms are eliminating machine breakdowns and quality problems with a vengeance. As the need to adapt to process contingencies diminishes, some of them are discarding production workers, a main source of flexibility, too rapidly. The requirement to adjust to process uncertainties will never disappear completely.

One must know at what point adaptation should give way to uncertainty reduction in order to achieve a balance between the two. Attempting to establish some guidelines provides an opportunity to examine the disadvantages of adaptation through flexibility. Consider the following problems:

1. *Mix.* Product variety leads to complexity and confusion, which raises overhead costs.
2. *Changeover.* By the time the current product is out of date, new developments in process technology will make existing flexible equipment obsolete obviating its utility in handling a new product.
3. *Modification.* This type of flexibility reduces pressures on product engineers to get designs right the first time leading to a flood of unnecessary engineering change orders.
4. *Volume.* To have this type of flexibility, heavy investments in excess capacity, empty floor space, and/or slack time in the production schedule are required.
5. *Rerouting.* Alternate routings are redundancies that discourage efforts to eliminate machine breakdowns.
6. *Material.* This type of flexibility reduces pressures on upstream activities to eliminate quality problems.

If uncertainty reduction is to substitute for an overreliance on adaptation, one should find specific methods of control that function in lieu of the flexibility dimensions. As indicated in Table 5.1, some well-known devices exist that embrace a firm's marketing, product engineering, and production areas. Considering each type of uncertainty from top to bottom in the table, marketing approaches include long-term contracts with customers, and product life extension practices such as inducing more frequent and varied use among existing customers, finding new customers, and creating new uses. A cross-functional team is a joint engineering-production-marketing technique. Leveling demand is another marketing method. Production approaches include preventive maintenance and total quality control. Calling on

TABLE 5.1 Uncertainty Reduction Devices

Nature of Uncertainty	Uncertainty Reduction Device	Type of Flexibility
	Product Markets	
Demand for the kind of products offered	Long-term contracts with customers	Mix
Product life cycles	Life extension practices	Changeover
Appropriate product characteristics	Cross-functional design teams	Modification
Amount of aggregate product demand	Leveling demand	Volume
	Process Operations	
Machine downtime	Preventive maintenance	Rerouting
Meeting raw material standards	Total quality control	Material

banked flexibility is still another method with implications for at least the top two uncertainties in the table.

The conceptual framework is, by implication, more than a way of understanding about manufacturing flexibility. By embracing uncertainty reduction as well it subsumes other modern production management approaches. By suggesting how the various methods for control and adjustment relate to each other, it facilitates development of a coherent plan for revitalizing a factory. Integrating the available methods for improving competitive performance is one of the major challenges facing our manufacturing companies. Operations managers often employ a variety of approaches without possessing a clear understanding of their interactions.

More specifically, American manufacturers are busily engaged in uncertainty reduction and adaptation as part of their revitalization efforts, but there is little evidence they realize the compensatory nature of these two strategies. For example, are preventive maintenance (uncertainty reduction) and redundant machines (adaptation through rerouting) typically part of a common plan to deal with machine breakdowns rather than separate piecemeal approaches? Put another way, operations managers must ensure that planned changes in a preventive maintenance program are reflected in decisions on the need for redundant machines in a new FMS. When tackling product market uncertainties, the need for a comprehensive plan is just as acute, but more difficult to achieve, since the adaptive methods may lie in the production domain and the control approaches in the marketing realm.

In the future, achieving a balance between uncertainty reduction and adaptation will become more complex as a third element, redefining uncertainties, enters the equation. Overreliance on market adaptation encourages a short-term outlook (Hayes and Abernathy 1981). Since customers define needs in terms of their immediate situations, serving existing markets takes precedence over creating new ones. Incremental, imitative changes predominate over radical, innovative solutions. An enterprise becomes a follower in the industry as opposed to a pacesetter. Many American firms are merely playing catch-up to the Japanese in the battle over product variety and lead times. Too much adaptation also leads to creeping flexibility, unnecessary product proliferation for example, which wastes valuable resources. Adopting more of a proactive approach in terms of redefining market needs for customers creates a balance. It requires not only preeminent manufacturing flexibility, but an ability to develop imaginative strategies.

What is the role of AMT in controlling uncertainty? It does not make a significant contribution to directly eliminating the market and process uncertainties identified here. In fact, it has been in the past a source of process uncertainties. Most American managers do not yet perceive it as a way of redefining uncertainties, although it has this potential in the future. The new technology also is capable of banking flexibility, a property that generates opportunities for proactive and defensive managerial applications when the need or opportunity arises.

There is another sense in which it is possible to use AMT as a device for moderating uncertainty. Some managers view it as a means of controlling the variabilities introduced into a production process by virtue of human involvement. The threat of strikes, turnover, absenteeism, working to rule, or limiting production is mitigated when AMT replaces workers, contributes to limiting their skills, or closely monitors their activities. To the extent, however, that a growing number of managers view labor as a source of knowledge rather than of uncertainty, the need for this type of control will diminish.

CREATING THE FLEXIBLE FACTORY

Is it possible to sketch out a procedure for changing a factory's degree of flexibility in the short run? This section offers a tentative yes to the question by relying on the conceptual model's feedback loops. The procedure, which has four phases, is currently more a rough guideline needing refinement than a finished product for immediate application. Phase I identifies critical flexibility dimensions requiring investigation. Phase II measures gaps. Phase III selects methods for closing the gaps such as capital investment in AMT. Phase IV discusses continuous assessment.

Two major points should be kept in mind when designing a program of this nature. First, changing flexibility is ordinarily part of a wider improvement program also involving cost, productivity, and quality objectives, which is initiated at the business unit level or above. Decisions on flexibility must take into consideration these other objectives as well. For example, the manufacturing department and marketing unit may conflict over rationalizing the product structure. Any cost improvements for the former must be balanced against reduced product variety for the latter. As another example, domestic sourcing may reduce lead times but lower quality.

Second, in many American firms the current priority is to improve adaptation to market uncertainties, that is, to better meet the changing needs of customers. To accomplish this objective managers concentrate on integrating the activities of the various business functions. The danger is in losing sight of the other strategic implications of flexibility. Attention should also be paid to adapting to process uncertainties, eliminating all types of controllable uncertainties, discarding or banking flexibility, and redefining market needs for customers.

In Phase I senior managers at the business unit level initially decide which product lines to continue; it makes little sense to produce buggy whips in a flexible manner. Using a previously established strategic plan, a decision is then made on which range and time aspects of the flexibility dimensions are relevant for competition in the industry. This choice is different from and prior to determining which aspects need improvement. Business unit managers also select those relevant aspects for which excess flexibility is desired with the intention of redefining market needs or creating a reserve.

In Phase II requirements at the plant level are determined for each relevant flexibility aspect based on an analysis of strategic objectives and knowledge of market and process uncertainties. The business unit's proactive intentions and customer surveys have an important role here. Estimates are also made of the existing values of potential and actual flexibility for each relevant aspect. It is critical to state all three kinds of flexibility in quantitative terms using numerical values or category scores so that any gap can be measured. Based on discussions with business unit managers the gaps are subsequently prioritized to determine the order in which they will be handled.

Slack (1988) provided an illustration of how Phase II could be put into action. He employed subjective measures that reduced data collection and analysis but made for less precise results. For each pertinent range and time aspect required flexibility is measured using a five-point scale running from very important, that is, a major element of business strategy, to never important. Potential flexibility is indicated by a seven-point scale running from very much better capability than the nearest rival through lowest capability in the industry. Actual flexibility was not measured.

In an implicit gap analysis the relation between each relevant aspect's required and potential score is used to determine its priority.

1. If requirements are judged to be high and potential is judged to be low on their respective scales urgent action is needed since this aspect is holding the company back from competing effectively.
2. If requirements are low and potential is high an excess of flexibility exists. The alternatives include reducing the amount of flexibility and transferring resources elsewhere, or banking the excess for future use.
3. If requirements and potential are high this aspect represents a competitive advantage for the plant. Nothing need be done except to periodically monitor requirements for any decreases and to ensure that potential doesn't slip relative to the competition.
4. If requirements and potential are low this aspect is currently irrelevant to the plant's success. No action is taken other than to periodically observe requirements for increases and protect potential from further deterioration.

In Phase III the plant identifies methods for handling flexibility gaps such as JIT, CIM, and time-based competition. To reduce gaps the plant may choose methods that eliminate uncertainties thus decreasing requirements, or which raise potential thus adapting to the uncertainties. In order to redefine market uncertainties or store flexibility, it must find ways of improving potential beyond customers' immediate requirements. Major efforts are referred to the business unit for its strategic and financial approval. As with any improvement program financial resource constraints may prevent closing or creating all the gaps, at least initially.

Phase IV involves continuous monitoring of what has been accomplished along with periodic adjustments to changing conditions. The performance measurement system monitors each gap to see if it is opening or closing, and indicates a need for corrective action. Flexibility indicators used for this purpose do not have to try to pinpoint precise values of concepts which are notoriously difficult to measure. Rather, they must be able to identify broad trends. If a trend is favorable the right things are probably being done and motivation to continue exists. If the trend is unfavorable it acts as a signal to make improvements.

The performance measurement system should also be able to respond to changes in customers' needs by shifting the values of required flexibilities. Wang Labs' system incorporates a built-in mechanism for making this adjustment (Cross and Lynch 1988/89). Each department in the workflow has two external measures, quality and delivery, for which requirements are jointly specified with its downstream neighbor. A department also has two internal measures, cost and process time, for which it determines requirements in the light of its external requirements. A shift in customers' needs, therefore, reverberates backward through the workflow as each department communicates new requirements to its upstream neighbor.

Organizational and behavioral issues influence the success of any procedure for changing a factory's flexibility. Decisions must be made on how to organize the improvement process and on how to gain commitment from the people responsible for implementation.

Organization of the improvement process should follow one key principle. Satisfying customers' needs through flexibility is an integrative activity requiring joint efforts. Actions to satisfy customers, previously made by separate functional units acting more or less independently, will be made by teams each of which represents different functional interests. Responsibility for these actions, previously diffused among various functional managers, is now focused in team leaders. As an example, Phase I is best handled at the business unit level by a task force that includes the top managers of all major functions thus setting a pattern for organization at lower levels.

Alternative organizational choices exist for Phases II, III, and IV. In line with the description of these phases each plant may have its own interdepartmental task force to integrate all activities concerning the relevant flexibility aspects. This structure ensures that plans for one aspect do not interfere with plans for another. An alternative organization is a separate task force for each relevant flexibility aspect with responsibility for all the plants in the business unit. Each aspect then receives focused attention and there is a better sharing of information across plants. No matter which alternative is adopted, there remains a need to integrate manufacturing's efforts with those of other business functions.

A significant behavioral issue when changing flexibility is how to secure the commitment of affected individuals. Functional managers may resent the transfer of their decision making power to the task forces. Managers who are supposed to use flexibility indicators for self-improvement may not bother with them. Their scarce time and resources will be devoted to performing well on the short-term financial measures on which they are judged. Managers who are evaluated with the indicators may find ways to circumvent them, especially when they are subjective.

One way to gain commitment is to involve affected parties in formulating a change program. Nutt (1986), for example, conducted ninety-one case studies in service organizations and found that more organizations actually used changes that resulted from participation versus nonparticipation in decisions. When improving flexibility functional managers must be involved in the task forces. People who are evaluated by the flexibility indicators should play a role in formulating them. At Wang Labs neighboring departments in the workflow jointly set performance goals

(Cross and Lynch 1988/89). Factory workers should be encouraged to recommend methods for changing flexibility.

A second behavioral issue involves conflict between the various business functions in determining the size and priority of gaps particularly when the necessary information is subjective. The marketing department is likely to aim for a wider range of products and a shorter production lead time than is the manufacturing unit. Evidence from studies of interfunctional groups indicates that when opposing sides squarely confront an issue the chances for resolution are high (Davis and Lawrence 1977). An ultimatum from above is a second approach when time is running out. Top managers, however, must guard against dictating too frequently to the task forces since their subordinates will believe participation is a sham. A third alternative of burying the issue in hopes that it will disappear is rarely satisfactory.

AVOIDING MATURITY

What is the long-run path by which a factory matures over time? The seminal answer, Abernathy and Utterback's verbal model of evolution from fluid to specific states, is discussed initially (Abernathy 1978, Utterback and Abernathy 1975). It corresponds closely to the conceptual model's adaptive feedback loop. Next, extensions of their basic model that have mitigated its deterministic connotations are explored. Hayes and Wheelwright (1984) took into account that some firms may choose to evolve along paths off of an industry's main line of development. Other researchers have studied the processes by which a firm or industry in a mature stage can attempt to halt its decline (Abernathy, Clark, and Kantrow 1983, Gerwin and Tarondeau 1986, Hicks 1986). AMT has a role to play in both extensions.

The Abernathy-Utterback Model

Abernathy and Utterback's (Abernathy 1978, Utterback and Abernathy 1975) influential developmental theory traces the impacts of gradual decreases in environmental uncertainty on an entity they dubbed the productive unit. Defined as an interdependent product line-production technology combination it emphasizes that in addition to the well-known product life cycle there is a manufacturing process life cycle, and that the evolution of one can't be understood without reference to the other. This evolution is deterministic in the sense that a decline in uncertainty leads to a specific trajectory for the productive unit's development. Management is compelled to make strategic decisions which reflect these forces if an organization is to survive. There is little, if any, opportunity for managerial discretion.

In Abernathy and Utterback's view, as uncertainty decreases various characteristics of a productive unit evolve from an initial fluid state to a final specific state through three intermediary stages. These characteristics include, but are not limited to, product diversity, product innovation, the nature of the equipment, and the nature of tasks and workers. Strategy that commences as product differentiation gives way over time to cost leadership.

In the initial fluid stage of a new industry environmental conditions are highly uncertain. Potential customers, for example, are unsure about their preferences for various product attributes and the extent to which each firm can meet their needs. At the same time there exist a number of diverse product variations, and companies are not clear about their capabilities and limitations. Once more, product attributes can be readily adjusted to create further variations.

Given these conditions, firms and customers experiment with the various product alternatives to gain more information. Uncertainty is reduced as experience accumulates and unattractive product variations are rejected. Eventually, product attributes become standardized enough for a dominant design to emerge, that is, a single product variation captures a large chunk of the market. With most uncertainty now eliminated, the product becomes highly standardized with options for different market segments and ultimately a virtual commodity.

There is frequent radical product innovation in the most fluid stage. The emergence of a dominant design implies product attributes become relatively stable so that innovation turns incremental. Process innovation becomes significant in an effort to make manufacturing technology more efficient. In the most specific state, product and process are inextricably linked, hence little innovation is possible.

The manufacturing process begins as a highly flexible job shop with general purpose equipment and independent operations. As product characteristics are defined process technology becomes more standardized. When a dominant design appears, dedicated production and assembly lines are purchased. At the extreme, the manufacturing process is a highly automated continuous flow line with interdependent segments.

Initially, workers have sophisticated craft skills that permit them to operate and control equipment. During the movement toward a dominant design, the need for skills and training is reduced. A division of labor appears between operation and control. In the final stage, operators have given way to monitors who observe the equipment and intervene during breakdowns.

Evidence for the model comes from studies of auto engine and assembly plants at Ford (Abernathy 1978). The studies cover a period extending from the beginning of the twentieth century to the mid 1970s. Product characteristics hardly reflect the recent upheavals caused by foreign competition and new technologies. Process characteristics are based almost exclusively on traditional nonprogrammable methods. Some consideration was given to programmable controllers and robotics, but their possibly unique impacts were submerged by their early stage of development, concentration on their mechanization versus flexibility characteristics, and the overwhelming amount of traditional processes studied.

In engine plants, the trends followed paths indicated by the model with a few exceptions. Product diversity, the ratio of the number of different engines to the number of engine plants, was initially high. It decreased to its lowest point during the Model T and Model A eras and then hit another peak just after World War II. From that time until the mid 1970s the trend was downward.

Product innovation, measured by the rate of engine model change per plant, has tended to decrease over time, but not uniformly. The maximum rate of change occurred in the initial period. Thereafter the measure decreased only to hit a smaller peak in the 1950s. From then on it continued its downward movement through the mid 1970s.

The trends in equipment and process technology characteristics followed the model's predictions, but a good deal of the changes in their values occurred during relatively brief periods. Presumably to facilitate data collection the largest values of each measure were determined for each point in time as opposed to average values. Transfer span, the number of different machine operations that are automatically linked, started at a value of 1 in 1910–1912 and increased rather slowly until transfer lines were introduced in the early 1950s when its value was over 150. A machine's automation level, measured by Bright's scale (1958) which runs from 1 (lowest) to 17 (highest), had a value of 5 in 1910–1912 representing use of single-function power tools with fixed cycles. It very slowly increased to the early 1950s when it also jumped. By the 1970–1972 period automation level had a value of 16 signifying use of equipment that can correct its own performance while operating.

During the initial period, 1903–1908, skilled workers performed unpredictable tasks of long duration. They were responsible for large components or meaningful subsystems of the engine. As mass production evolved semiskilled workers took over. Tasks were now characterized by repetition, predictability, short duration, and single operations. With the introduction of transfer lines in 1951 knowledge of the process technology became the critical skill. Workers' responsibilities changed to monitoring and intervention under less repetitive and predictable conditions.

Generally, the trends in assembly plants have been as predicted but less marked than in engine fabrication. The relatively smaller changes from initial values indicate that they did not evolve as much. Product diversity, measured by the average number of different vehicles per plant, hovered around a value of 2 from the 1920s until the mid-1960s when it began a steady decrease to a value of 1½ in 1975. The rate of product change, high initially, declined to a fairly constant value maintained until the 1960s and then decreased further.

Assembly plants, initially flexible and labor intensive, became more specialized and capital intensive, but these changes occurred suddenly, not gradually. Transfer span had a value of one from 1914 through the mid 1960s and then rapidly increased to twenty by 1975. From 1914 until World War II automation level had a value of 4 representing hand-controlled power tools. It then increased to 12, which indicated equipment which changes speed, position, and direction according to signal. By 1910 semiskilled operators largely replaced craft workers and, for the most part, remained predominant over the rest of the time period. During the 1908–1909 period task duration as measured by cycle time was 514 minutes. When the assembly line was introduced in 1914, cycle time went to 1 minute and remained at that value through the mid 1970s.

The general applicability of the Abernathy-Utterback model has not been subject to a great deal of research. To a reasonable degree it seems to reflect the long-run development of the domestic auto industry, but will it hold for other American industries as well? In the chemical industry evolution occurs very rapidly with new products being manufactured with highly automated equipment in the initial stage.

For commercial aircraft evolution seems to have stopped at a rather fluid stage of development (Abernathy 1978).

Porter (1983, 1985) identified a number of conditions required for the model's evolutionary path to occur. In the initial state there is product differentiation, a characteristic hard to find in the basic metal and mineral industries for example. Product performance is the main competitive advantage as opposed to cost reduction, which could also be instrumental in gaining a new product's acceptance. For a dominant design to emerge product innovation must occur and the innovation must spread throughout firms in the industry. Moreover, there is low intrinsic market segmentation as buyers' needs merge over time. In specific states process improvement finally becomes possible because a dominant design has taken hold. In addition, sales volume becomes large enough to support mass production and economies of scale exist.

The role of national variations in determining the model's applicability has not been explored. Will it hold in foreign countries for the same types of industries to which it applies in the United States? Casual evidence suggests the answer may be no although admittedly thorough investigations are needed. It is generally believed that the Japanese auto industry has exhibited less product diversity than the American but that its task and labor characteristics are more fluid. In assembly there are less distinct models, but workers have more varied and responsible jobs.

Once more, Abernathy et al. (1982) noted that the model does not provide a good explanation of the development of the European auto industry through the mid 1970s. The European market remained in a relatively fluid state with technical performance more critical than cost reduction except in the low cost segment. Consumer preferences stayed sufficiently diverse to support a number of design options even among firms in the same country.

Extensions of the Model. Subsequent efforts to build on the model have had the effect of reducing its deterministic orientation. The product-process matrix of Hayes and Wheelwright (1984) has four columns representing stages of product evolution from custom items to commodity status and four rows signifying steps in process development from job shop to continuous flow. Movement along the diagonal from upper left to lower right stands for matched development of product and process as strategy shifts from overall product differentiation to overall cost leadership. It is therefore synonymous with the path from fluid to static conditions followed by a productive unit in the Abernathy-Utterback model.

The product-process matrix also takes into account that a firm may choose a developmental path off the diagonal by pursuing a niche strategy. Evolution above the diagonal represents product differentiation in a niche, whereas movement below the diagonal aims for cost leadership in a niche. Consider the example of Rolls Royce, which by positioning itself above the diagonal in the auto industry, makes a narrow line of cars using a manufacturing process with some of the characteristics of a job shop (Hayes and Wheelwright 1984).

In the Hayes and Wheelwright formulation, strict determinism gives way to managerial discretion. Strategy makers have an opportunity to choose from among various overall and niche strategies each of which guides development along a different path. Depending on management's acuity any number of them may be viable. A firm wanting to compete through product differentiation, on an overall or niche basis, must decide when to drop a product line that is becoming mature. A company emphasizing cost leadership must avoid stepping into a market too early or too late. Environmental uncertainties rather than dictating a path put limits on available alternatives. Paths too far away from the diagonal are considered to be nonviable, although it is probably more realistic to assume that the chances of a path being successful decrease as it moves away from the diagonal.

Gunn (1987) proposed that firms with CIM technology will have more viable paths through the product-process matrix than companies without the new technology. The former will have alternatives in the lower left and upper right corners not available to the latter. Since hard tooling is replaced by software, production of standard or custom items can be handled by the same manufacturing process.

A second extension is due to researchers who questioned the model's implications for industries already in specific states of development (Abernathy, Clark, and Kantrow 1983, Ayres 1984, Reich 1983). If the life cycle of a productive unit is inexorable the result is extinction for a mature industry. Ultimately, built-in rigidities make it vulnerable to externally developed product variations. The model's logic implies, for example, that the American auto industry, subject to tight integration between product and process, which prevents substantial improvements in either, will continue to lose market share to foreign imports, or will shift most of its production to low cost foreign locations. New industries, however, will grow to take its place in the nation's economy.

Abernathy, Clark, and Kantrow (1983) argued that a productive unit does develop along the lines of the model but that reversals can occur, a process they referred to as dematurity. Individual firms in a mature industry can move backwards to more fluid stages if environmental conditions are right and if management is ready to seize the opportunity. Environmental uncertainty may suddenly reassert itself through a shift in consumer preferences due to foreign competition or the impact of an external technological breakthrough. It is then up to a firm's management to meet the new conditions by stimulating product and process change to an earlier developmental stage. Managerial discretion, however, is limited by an organizational context that has been fine tuned for mass production.

Although it was limited, a classic example of dematurity occurred within Ford as the result of General Motors' challenge in the 1920s (Abernathy, Clark, and Kantrow 1983). From 1908 until 1925, Ford's engine plant steadily advanced to a relatively specific state due to the Model T's success as a dominant design. Extensive price reductions and productivity gains signaled this development. By the early 1920s, however, market share began to slip as G.M. took advantage of changing consumer tastes and new technologies. The Ford plant finally shifted back to more fluid conditions. Product standardization was abandoned as the Model A and then the V-8 engine were introduced. Due to the tardiness of these efforts General Motors became the industry leader.

Abernathy, Clark, and Kantrow (1983) believed that another, more extensive reversal is under way in the auto industry. Consumer preferences have shifted from comfort and style to fuel efficiency and reliability under the impetus of foreign competition and new technologies. Their evidence established that customers are now willing to pay a premium for innovation. It also indicated that U.S. auto makers have responded by increasing product diversity and its visibility through advertising. No evidence, however, was presented of corresponding changes in the manufacturing process, a necessary condition for dematurity to occur.

Evolutionary Paths

Two features of theories based on conventional manufacturing processes suggest a need to study how AMT affects evolution. First, the Abernathy-Utterback model reinforced the generally held view that product innovation and cost efficiency are inimical. As a productive unit develops, product diversity and change give way to cost reduction as a competitive advantage. Dematurity, backward movement along the development curve, therefore, implies more product innovation at the expense of productivity. Existing international competitive conditions place a premium on both. To what extent does AMT facilitate an alternative renewal process in which neither are sacrificed?

Second, Abernathy (1978) using data from Ford indicated the special role of a conventional manufacturing process in hindering dematurity. He found that the manufacturing process' stage of development at any moment of time serves as a lower limit below which the product's does not penetrate. From 1905 until the mid 1930s, the manufacturing process at Ford's engine plant advanced relatively slowly with little change in direction from very fluid conditions to an intermediary state. The engine line's evolution was initially much more rapid and then subject to two significant reversals in direction. It remained, however, in more specific states than the process except during two brief periods.

Abernathy's evidence implies that AMT may have substantial benefits over dedicated automation in facilitating a productive unit's renewal. It will allow product characteristics to revert back to a greater extent when the need arises. Providing a bank of flexibility that is not completely used in a mature state will allow product characteristics to revert back even more.

There have been just a few studies of industrial development which concentrate on AMTs impacts. Gerwin and Tarondeau's (1986) pilot study of auto engine and assembly plants in France and the United States focused on various types of CAM suited for mass production. Hicks (1986) investigated small- and medium-size metal working plants with NC or CNC machines. Both studies contended that CAM did have impacts on evolution in general and renewal in particular. Each however identified a different renewal process which was also distinct from that of Abernathy, Clark, and Kantrow.

The implications for evolution of Gerwin and Tarondeau's pilot study are explored in Figure 5.1. It indicates average changes in automation level and global flexibility for the four assembly activities and three engine fabrication activities when computerized automation is substituted for conventional equipment. Recall that the change in automation level is an activity's score on Bright's scale recorded before and after the substitution. An activity's change in global flexibility is tantamount

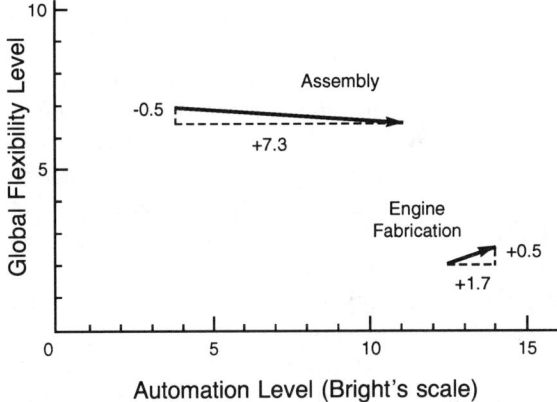

FIGURE 5.1. Trends in flexibility and automation.

to a weighted average of the recorded changes in the six individual flexibility dimensions.

The mean change in automation level is the difference between the average original automation level (OAL) and the average new automation level (NAL). For the assembly activities the average OAL is 4.0 and the average NAL is 11.3 representing an increase of 7.3. The corresponding figures for engine fabrication are 12.3 and 14.0, which amounts to an increase of 1.7.

Since Bright's scale is ordinal, it is possible that the automation increase in assembly is not as great as the one in fabrication. On the other hand, the technical change in the former activity involves substitution of a sophisticated automated control system for largely manual controls. The latter's change is a refinement in an already automated control system. Programmable controllers improve the possibilities for real time quality control and automated setup. On this basis it is assumed that the larger numerical increase is in fact greater than the smaller one.

The mean change in global flexibility is computed from the corresponding changes occurring in individual activities. Average global levels before and after the substitution of equipment cannot be determined since the relevant scales measure only changes. Without these values the mean change cannot be anchored in the plot. Some reasonable assumptions were therefore made, which have little impact on the conclusions to be drawn. For assembly activities the mean original value is assumed to be relatively high, and the mean new value is assumed to be greater than that for fabrication. The observed average change from the study is $-.5$. Correspondingly, for fabrication activities, the assumed average original value is relatively low, and the assumed average new value is below that of assembly. The observed mean change is $+.5$.

The plot indicates that engine fabrication and assembly activities have become more similar. There has been a dramatic convergence in automation level with the difference diminishing from 8.3 originally (i.e., 12.3 − 4.0) to 2.7 currently (i.e., 14.0 − 11.3). The difference in flexibility has gone down by 1.0 on its scale. The net result is a movement to the upper right of the graph by engine fabrication and to the lower right by assembly. Both types of activities may soon possess the same levels of flexibility and automation.

Previously, assembly had been developing in line with the Abernathy-Utterback theory toward more product standardization and process rigidity, although it was not in an advanced stage of evolution. Recent production technology decisions have contributed to its continued maturation. Programmable automation is decreasing flexibility and increasing automation. The result has been higher labor productivity but reduced ability to handle new product technologies. In general, the habitual sacrifice of innovation for efficiency is being continued.

If the advocates of the dematurity theory are correct about a resurgence of product diversity, then assembly activities may be becoming too rigid. There is a need for programmable automation that will maintain flexibility while augmenting mechanization. The new technology needs to be designed to interact with human operators rather than to substitute capital for labor.

The very small amount of research on final assembly, the steps after body framing, does not seem to contradict these implications. Carlsson and Trygg (1987) compared final assembly for the Opel Kadett and Volkswagen Golf based on plant tours and interviews. The former process is manual, whereas the latter is, for final assembly, very automated. The Golf is put together in the famous Hall 54 in Wolfsburg, Germany, where a very high 25 percent of the operations are done by robots or specialized automation. Although the authors presented no hard evidence, they indicated that the Opel's process is more flexible, particularly with respect to changeover flexibility.

Of particular interest was the finding that the plant with manual final assembly had more efficient subassembly than the plant with automated final assembly. For the study six parts of one Opel and one Golf including the front door, back seat, and fuel pumps were dismantled and reassembled by experienced assemblers. It took 55.7 minutes to put together the Opel's parts and 85.0 minutes for the Golf's parts. The assembly times of five of the six items were smaller for the Opel. Multiple regression analysis indicated that the reasons were less components and less complex fasteners.

Previously, engine fabrication had developed following the Abernathy-Utterback model into an advanced stage of product standardization and process rigidity. Recent manufacturing technology choices are leading to the renewal of this activity. The renewal is not in line with the dematurity theory, which predicts backward movement along the development curve. That theory is wedded to the traditional assumption that automation and flexibility are inversely related. This new type of competitive response traces a trajectory through continuing upward shifts in the life cycle. Programmable automation is increasing automation and flexibility simultaneously. The result has been an increased ability to handle diverse product technology without reductions in labor productivity. In general, it has been possible to foster innovation while maintaining efficiency.

Hicks (1986) viewed industrial renewal as taking place within an industry but more at the inter- than intrafirm level. Older, sluggish companies give way to newer dynamic ones even if the industry is mature. Newer firms are more competitive because they utilize upgraded production technology; specifically NC or CNC versus conventional equipment or CNC as opposed to NC. Although older firms have the opportunity to renew themselves with advanced equipment they will not take advantage of it. Their obsolescent plants, lack of physical space, and outmoded managerial practices are poorly matched to the new technology's needs.

To test his ideas, Hicks collected data from about 1200 small- and medium-size U.S. metal working plants comprising over 200 industries at the four-digit SIC level. Each plant had less than 250 employees and possessed at least one NC or CNC machine. Over 70 percent of the plants represented single-site firms. No data were collected on companies that had gone out of business or that did not employ NC/CNC.

The data suggest that major interfirm adjustments have occurred in the metal working sector over the last 200 years. There appears to have been a continual process of cohort renewal in which successive generations of new plants have replaced older ones. To summarize detailed birth cohort figures, only 2 percent of the plants existing in 1980–1982 began production in the nineteenth century. Only 21 percent were initiated in the first half of the twentieth century. The rest, 77 percent, began production after 1950.

These results, however, reflect the metal working sector as a whole rather than particular industries. Plants involved with motor vehicles, farm equipment, and construction equipment are lumped together with those involved with computers, aerospace, and aircraft. The observed effects could be due to interindustry adjustments in which factories from mature industries wither away, whereas those in new ones increase. For a more definitive test it is necessary to investigate the cohort renewal process at an industry level. A less crucial point is that the data do not allow a determination of the extent to which the renewal process is characterized by additions rather than replacements.

It is not possible to directly test whether new plants have more NC/CNC than old ones since information exists only for survivors with the technology. Some indirect tests were made using additional data. First, companies have not elected to search for lower labor costs, a logical alternative to upgrading manufacturing technology. There has been for example no large scale migration from older to newer geographic regions for single site or multiplant firms. Second, union efforts to protect workers from new technology are less likely to occur in new plants. The proportion of nonunionized plants in a cohort steadily increased from forty percent for before 1920 to 92 percent for the decade of the 1970s. A third test used data on the year in which factories in each birth cohort first adopted NC/CNC. Hicks found that plants in a wide range of cohorts adopted during the same relatively brief period, the early 1960s to the late 1970s. Period influences, specifically the need to remain competitive and the availability of smaller less expensive turnkey equipment, may be more critical than the age of a plant. Fourth, data on the extent to which NC/CNC was involved in each plant's range of metal working operations and total workload were not analyzed by age of plant. The overall lack of in-depth use, however, suggests that the new technology has not yet become a significant competitive factor for factories having it.

In sum, Abernathy, Clark, and Kantrow (1983), Gerwin and Tarondeau (1986), and Hicks (1986) all question whether new industries replacing old ones is the only way for revitalization to occur. This deterministic process assuredly has long-run economic benefits but may cause social disruption in the short run. Although all three studies agree that mature industries can be revitalized, they offer different processes by which it takes place. Abernathy, Clark, and Kantrow stress the intrafirm level where managerial decisions on new products and new conventional manufacturing technology leads to backward movement along the life cycle. The result, however, is a sacrifice of cost efficiency for product innovation when both are currently necessary. Gerwin and Tarondeau believe that introducing CAM into productive units with dedicated automation shifts the life cycle upward. At a given level of productive efficiency more product innovation is supported. Hicks discussed interfirm adjustments in which new companies with NC/CNC technology replace old ones in the same industry. His is a thought-provoking alternative that requires additional data and more detailed analyses to verify.

The debate over the form of intraindustry revitalization hinges on whether or not companies in specific states can renew themselves. As acknowledged by advocates of intrafirm processes this is exceedingly difficult. It is more than a matter of increasing the R&D department's budget and installing computerized automation. Organizational systems and procedures, structures, and cultural values, fine tuned for high performance in a static environment, must also change. Some researchers believe that once these organizational characteristics are fairly rigid little can be done to help a firm adapt to more fluid conditions (Argyris 1972). Others have suggested that ways exist to loosen organizational rigidities (Kanter 1983, Lawrence and Dyer 1983). For many large U.S. corporations still struggling to meet foreign competition the final answer on which view is correct is still not in.

PART II
ORGANIZATION

6
ORGANIZATIONAL STRUCTURE AND THE MANAGEMENT OF UNCERTAINTY

Until recently, the context and the structure of manufacturing organizations were considered stable. Raw material costs were predictable as were the hourly costs of production personnel. Processing equipment, configured for a particular application, was sufficiently dedicated that adaption to alternative uses was not cost effective. If changing production technology, varying marketplace demands, or fractious labor-management relations threatened the core technology with undesirable uncertainty, it was usually buffered by insisting on firm sales forecasts; by maintaining large inventories of raw materials, parts, and in-process inventories; by signing long-term labor agreements to stabilize wages; and by manufacturing in long runs with no penalties attached to the resulting high finished goods inventories.

THE EFFECTS OF UNCERTAINTY AND INTERDEPENDENCE

Uncertainty in the environments of manufacturing organizations has permanently changed this relatively placid scenario. Established markets are fragmenting and product life cycles are shortening as products become increasingly customized. Customers are putting pressure on manufacturing firms to lower cost and reduce lead time without sacrificing quality and performance. Previously acceptable trade-offs between cost, quality, lead time, and performance now have to be reassessed as higher levels are demanded in all these areas (SME Sociotechnical Study Committee 1989). Table 2.1 lists four types of product-market uncertainty faced by manufacturing departments today, uncertainties that are increasingly difficult to buffer at the boundaries of the firm (Thompson 1967). This is the dominant form of uncertainty experienced at the plant or organizational level of a manufacturing organization.

When the level of analysis is moved down to the department or group or team, uncertainty from the external environment is still a strong factor because of the difficulty of buffering it from the department or group level. Speed of response to changes and other types of demands from the external environment is critical to competitiveness and any attempt to contain the uncertainties at the plant or organizational level will slow down the reaction at the department or group level. However, the dominant problem to be solved at this level is internal, not external. It arises from interdependencies (See Figure 6.1).

Advanced manufacturing technology environments can be highly integrated through information technology that links different manufacturing technologies and their planning and control systems and materials handling systems. These environments can also be relatively decoupled if the manufacturing strategy has settled on islands of automation as its manufacturing philosophy. In such cases, required coupling between the islands must be accomplished through organizational coordination rather than via information technology. A range of in-between variations can also exist, with tight or loose coupling, via information technology and/or through organizational coordination.

Some of these organizational and technological interdependencies are within departments while others are between departments and/or groups. "Because the overall flow of production is accelerated and more tightly linked, technical interdependence is greatly increased" (Cummings and Blumberg 1987). Still other interdependencies are with units external to the production domain (e.g., design engineering that is linked through CAD/CAM technology and design-to-manufacture activity or suppliers that are connected into JIT systems). These interdependencies become amplified as organizational boundaries increasingly collapse (Blundell 1990). This wide range of interdependent relationships is experienced as complexity that must be managed at the level of departments or groups. If it is not managed, either through information technology or organizational coordination, then the interaction effects become unpredictable and are manifested as different forms of uncertainty.

For instance, if the complexity of the design-to-manufacture boundary is poorly managed by the firm, the ability to innovate new product extensions will be constrained. Changeover flexibility would be characterized as low. If supplier interfaces function poorly and the complexity of the JIT system is undermanaged, only a high degree

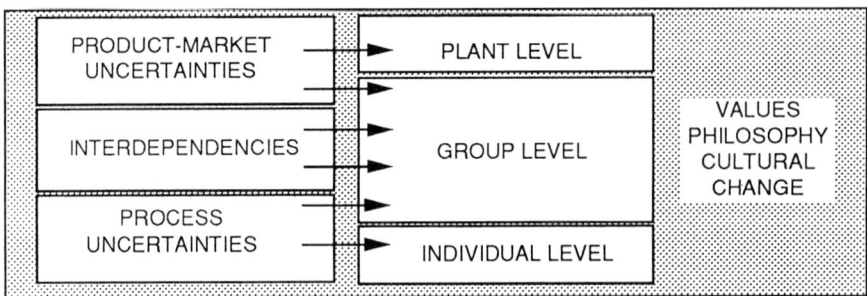

FIGURE 6.1. Primary impacts on different organizational levels.

of rerouting flexibility will provide manufacturing with the capability of meeting customer delivery dates.

When the level of analysis moves still lower, to that of the individual and the support systems that are closely coupled to individual behavior (e.g., training, selection, rewards), process operations uncertainties (Table 2.1) become the dominant problems to be solved. Process uncertainties also have impact as the departmental or group level (Figure 6.1), but their impact is subordinated to that created by internal interdependencies because the department or group level can usually exercise some control over process uncertainties (e.g., by rerouting work when machine breakdown occurs). The individual, for the most part, cannot.

Machine breakdowns that create process uncertainties are largely a result of the computer-based characteristics of AMT. Davis (1983–84) described traditional technology as deterministic in that its ways of breaking down were understood and corrective actions could be appropriately specified and planned. This condition has been supplanted by new technologies that break down stochastically (i.e., the breakdowns are random and the corrective actions are no longer predictable).[1] They can only be determined after diagnosis and analysis by the operator(s). One result is that more automated systems increase rather than decrease the dependence of the organization on its members for effective operation.

Hirschhorn (1984) illustrated one way in which this dependency, in turn, contributes to internally generated uncertainty. With increasing use of sociotechnical designs, more responsibility is delegated to teams and individuals, which reduces the number of rules and specified actions. This increases the experienced uncertainty and tension when actions need to be taken for which clear specifications have no longer been prescribed.

Changing values, new work philosophies and related changes to the corporate culture (Figure 6.1) have a significant impact on individuals and their support systems as well as on the departmental and organizational levels. Such changes are slow to take place. They are experienced as a form of uncertainty that is qualitatively different to those cited in Table 2.1. They create uncertainty in that they demand new behaviors from individuals and groups for which very little previous experience exists (e.g., participative management styles). Changed values, new philosophies of work, and different corporate cultures have important consequences for organizational arrangements and individual behaviors in AMT environments, but the time horizons are generally long term and the undesirable consequences can usually be mitigated by effective education and training and by well-thought-through implementation.

This chapter and Chapters 7 and 8 address organizational issues in AMT environments, issues that are stimulated by the uncertainties and interdependencies illustrated in Figure 6.1. This chapter discusses the impacts at the level of the structure of the organization. Chapter 7 addresses the group level, and Chapter 8, the level of the individual. Product and process uncertainties and increasing interdependencies require different types of flexibility (Table 2.1) if the organization is

---

[1] Stochastic technology is described in detail in Chapter 8.

to be adaptable. This, in turn, means finding structures and processes that can support the types of flexibility needed.

Conventional approaches to controlling and motivating the members of the organization to achieve organizational goals are losing their effectiveness. As a result, the organization is becoming highly vulnerable and dependent on the commitment of its members, a commitment that requires new work values, explicit operating philosophies,and significant changes to the organizational culture. This new paradigm of work, referred to here as sociotechnical systems theory, informs the discussion in these three chapters. It is described in detail in Chapter 7.

STRUCTURAL ADAPTATION TO UNCERTAINTY

Several factories with CAM environments have developed hybrid organizational structures that integrate both functional and product oriented forms. They divide their basic manufacturing operations by functional categories, but within each functional area they structure the production processes and equipment, the work force and the manufacturing methods in team-oriented or product-focused forms of organization. Examples from two different CAM-based engine assembly operations illustrate this structural arrangement.

GE Canada's engine parts fabrication plant (which manufactures sets of high tolerance compressor airfoils—blades and vanes—for jet engines) in Bromont, Quebec, divided the manufacturing area (about 500 people) into five businesses they referred to as PMEs (*petites et moyennes entreprises* or small and medium enterprises). The five PMEs are forge, pinch-and-roll, small rotor machining, vane machining, and large rotor machining. Each PME is a manufacturing subfunctional area of 100 or fewer employees. The senior management team (SMT) is also organized functionally, although across a broader spectrum of manufacturing functions (production and manufacturing engineering, process and quality engineering, materials planning and customer satisfaction, purchasing and maintenance and plant engineering, tooling and tool design, information systems and finance, and human resources). Each of the PMEs is also a responsibility for one of the SMT members.

As a self-contained unit, each PME is expected to cut across the functional organizational lines within its own area to move the different products quickly through the steps needed to complete the required series of operations for that area. For instance, within a business such as forge, a series of different heating, shaping, tamping, and cooling operations are necessary to process a piece of metal into a blade before it can leave the forge PME and pass on to the pinch-and-roll PME. These operations are carried out by product-oriented teams that integrate these functional operations to move the product(s) quickly through the PME. In the same manner, a series of operations take place within the pinch-and-roll PME before the blades move on to the range of operations they will encounter in one of the machining PMEs.

The PME concept, the idea of conceiving of one's subfunctional area as a self-managed, self-contained business unit, was explicitly intended to sensitize the PME members to the idea that they had to take responsibility for events that occurred in their PME's environments, both the internal ones (within the PME itself as well as with the other PMEs, the staff resources and the SMT) and the external ones (the customers and the suppliers).

The Jamestown Engine Plant (JEP) of Cummins Engine Company machined parts and assembled them into 240 to 300 horsepower diesel engines for direct shipment to distributors and original equipment manufacturer (OEM) customers for use in trucks, buses, and other equipment. It also machined high volume engine parts for use in other Cummins plants as well as for their parts distribution network. The JEP, a plant of about 900 personnel (700 on the factory floor), was organized into five businesses that ran autonomously as "miniplants": three machining businesses, one assembly business, and one test and final parts mounting business.

In a structural arrangement similar to the GE Bromont plant, each business reported one level up to one of the plant directors, each of whom also had an overall functional responsibility (assembly and test, purchasing and materials, manufacturing services, reliability, finance, personnel, organization development, and training). Within the businesses, the workers were organized into teams to cut horizontally across the functional lines of the particular business. In addition to his or her assembly or machining skill, each team member also had a "vertical task" (e.g., materials management, accounting control, training). In effect, the business units were organized as matrix structures in which team members reported to both the team and team manager for assembly or machining operations which were oriented toward the output or product task, and to a functional specialization manager for the vertical task function they also exercised (Kolodny and Dresner 1986). Vertical tasks were rotated every six to eighteen months, depending on the complexity of the task.

Both plants are in environments that subject their manufacturing organizations to the uncertainty of unpredictable demands. For example, the JEP can produce 1700 possible variations of the basic engine that it assembles in the plant. Other Cummins engine plants, which are recipients of most of the machined components produced by the JEP, may place demands on the JEP for components for the different engines they themselves fabricate and assemble. Each of their engines, in turn, have a large range of possible variations so that their demands on JEP for variety in components also is large. Cummins' extensive sales and service network, with its enormous variety of older and current engine designs in the field, also places unpredictable demands on the plant for an extremely wide variety of parts in relatively small quantities. These requests may arise to service a customer's existing diesel engine that has broken down and requires a quick response, even if that particular part is only to be produced in a quantity of one.

The GE Bromont plant faces a different kind of environmental uncertainty. It is, to some extent, "a victim of its own efficiency." The sets of airfoils it produces are assembled into different engines the General Electric Corporation supplies to its wide range of customers and their still wider range of aircraft. Since the Bromont plant is flexible and adaptable, it is the recipient of a high rate of change in volume demands, in delivery schedules, even in part parameters from the assembly plants it feeds. In effect, the plant serves as a flexible buffer for design, production, and schedule changes in jet engine manufacturing at its customer, the Aircraft Engines Business Group, changes that manifest themselves as a high degree of unpredictability for the Bromont plant.

Both plants demonstrate the need for structural arrangements that are more responsive than traditional functional designs. Their solutions have been to create product oriented substructures (team structures) with faster response times beneath the functional umbrellas used to organize manufacturing. Functional structures provides a "home" or repository for the manufacturing knowledge and skill needed to operate a factory with considerable CAM technology. The product-focused units (i.e., the teams) overcome the slow response characteristics of functional organizational structures and allow the plant to react quickly to unpredictable events and/or a marketplace that wants fast and/or customized responses.

Product and Process Uncertainty

Longer-term environmental uncertainties must either be adapted to by acquiring manufacturing flexibility through AMT (Gerwin 1989, Sethi and Sethi 1990) or by applying previously banked flexibility against them, or proactively controlled by reducing the uncertainties or by redefining them to steer the firm into more stable products or market niches (Figure 2.1). Such strategies were discussed in Chapters 2 and 5. The uncertainties for which manufacturing flexibility must be acquired have been described as four types of product uncertainty and two types of process uncertainty (Table 2.1). In the shorter run, uncertainties emanating from the marketplace can be absorbed by implementing appropriate organizational structures and human resource practices (Kolodny 1990, Manufacturing Studies Board 1986). Different AMT responses coupled with alternative organizational arrangements provide the organization with flexibility to confront these uncertainties in different ways.

Product Uncertainty. Group technology reduces parts variety proliferation as designers find ways to utilize previously designed parts in new products. This improves producibility which, in turn, reduces the uncertainty resulting from engineering changes (Susman and Chase 1986). Cell arrangements also reduce cycle time (Lindholm and Norstedt 1975). Hence, product-focused arrangements such as manufacturing cells foster both *mix flexibility* and *changeover flexibility* (see Table 2.1) by reducing the uncertainties of wide product demand and shortened product life-cycles, respectively.

Advanced manufacturing technologies alone do not achieve these flexibilities (Forslin, Thulestedt, and Andersoon 1989). For instance, mix flexibility to adapt to the variety of products needed to accommodate unpredictable product demand requires more than the use of flexible machining cells or lines. It also requires organizational arrangements, such as parallelization[2] in both machining and assembly,

---

[2] Group technology, parallelization, and other product-focused forms of organizing are described in detail in Chapter 7.

so that variations can be introduced into the production process without disturbing other products whose continued production is needed (Aguren and Edgren 1980). Each parallel line, in turn, is likely to be organized into work teams that must become increasingly self-regulating. If lead time reduction is an important attribute of mix flexibility, it will not be achieved in an organizational structure that has a slow moving hierarchical chain of decision making.

Changeover flexibility to accommodate the uncertainty of changing product life cycles by extending the life of the product, for example, may require cooperation across the design-to-manufacturing boundary if dimensional changes to the product affect tooling or performance characteristics changes affect testing or other manufacturing procedures.

Manufacturing cells and parallelization (Kolodny 1985) in the organization of the factory floor are also factors for achieving *modification flexibility* because they allow a variety of minor changes to be implemented on one or several production lines while the remaining lines continue to produce to other requirements. Such change will also require high cooperation between design staff and factory workers associated with each cell, cooperation that results from effective within-team and across-team arrangements.

Volume flexibility is achieved when changes in production levels and the time needed to accomplish those changes are realized by the manufacturing organization so that it can respond to the uncertainties of aggregate product demand. Venkatesan (1990) described how Cummins Engines first converted its flexible factory to product-focused units and then subdivided those units again by high and low volume as a better way to utilize its flexible equipment and systems to simultaneously integrate considerations of design stability, frequency of delivery, and numbers of setups.

Managing high and low volume orientations simultaneously in order to achieve volume flexibility also requires appropriate organizational arrangements. The horizontal lines of communications of project teams, for example, usually imbedded in matrix or project management structures (Davis and Lawrence 1977), can respond more quickly than functional structures. Furthermore, because such teams usually operate with a reasonable amount of decentralized autonomy within the organization, one project team can negotiate with another project around the use of its own and shared production facilities to modify production levels quickly and complement the flexibility that different CAM technologies provide.

Process Uncertainty. Management of the uncertainties contributed by process operations also requires changes in the structure of the organizational arrangements. Machine downtime can be adapted to by changing machine sequences and by rerouting the work to achieve needed *rerouting flexibility*. Self-regulating teams with good interteam networks can facilitate the needed machine sequence changes, either within their own manufacturing cell or through the working relationships they may have established with a nearby cell. Manufacturing and industrial engineers, physically located adjacent to the shop floor rather than in office buildings at the other end of the company facility, can also reduce the time needed to reroute work when machines break down.

Even machine repairs can be expedited by organizational restructuring. TPM (total productive maintenance), for instance, brings a fresh perspective to the perennial conflict between maintenance and operations. Within a TPM program, companies and unions can search for more effective classification systems than those that currently create rigid demarcations between tradespeople and operators (Sheridan 1990). Safety and security must be protected, but innovative organizational approaches have opened up many new opportunities for cooperation that do not prejudice these fundamental requirements.[3]

In the short run, *material flexibility* can be achieved when manufacturing and engineering work together well enough and quickly enough to make effective decisions about variances in the compositional and dimensional parameters of parts received or fabricated or subassembled. Timely cross-functional team work is the basis for such coordination.

In the longer run, material flexibility results from building in feedback processes that can return meaningful problem and performance data from both the manufacturing areas and the customers first to the product designers and then back to the decision makers that influence the values of the firm (e.g., toward quality or performance guarantees). This requires redesign of the total system with feedback processes introduced at every stage: from outcomes back to the operating systems; from operating systems back to the product, process, and organization design activities; from design back to the decision makers who legitimize the design choices; and from these decision makers back to the dominant coalition that sets the values and philosophies of the corporation (e.g., of openness, trust, standards of excellence in quality and performance, the organizational culture) (Kolodny and Stjernberg 1986). The entire "value-added" chain needs to be examined if a continuous improvement perspective towards materials is to be built into the organizational arrangements (Johnston and Lawrence 1988).

Figure 2.1 illustrates that environmental uncertainties that cannot be reduced are transmitted to manufacturing. AMT environments with flexible technology and flexible organizational arrangements can adapt to a large proportion of the remaining uncertainty. Even the most flexible CAM technologies, however, will not completely enable manufacturing to adapt to nor completely cushion it from the steadily increasing rate of uncertainty. Liu, Denis, Kolodny, and Stymne (1990, p. 14) suggest that through the use of open processes and conceptually new integrative approaches, organizations will attempt to manage complexity in the future rather than reduce it:

> Current organizational design applies two principles to avoid complexity. The first one is to divide a real situation into specific aspects: technical, human, economic, etc. The second is to stabilize production processes into routines by buffering organization from environmental turbulence (Thompson 1967). But the "reduction" of complexity leads to some major dangers. The first principle ignores the fact that the sharp edged

---

[3] Presentation by Jan Klein, "Team Involvement and the Maintenance Organization" at the Ecology of Work Conference, Toronto, Canada, May 11, 1990.

frontiers between the different conceptual divisions are the very places where real interactions in work and life settings occur. One example of this first principle is that technical problems are formulated and solved without reference to the social consequences they induce. Thus, organizations are not as efficient as they could be if sociotechnical facts are taken into account (Trist 1981). The second principle leads to the rigid shaping of organizations to fit one given situation; as a consequence, it is prevented from becoming flexible and adaptive.

They go on to say that organizations will manage complexity largely by opening their boundaries directly to environmental inputs. In Chapter 7, Liu et al.'s first principle is addressed by examining sociotechnical approaches that cut across the "sharp-edged frontiers."

This chapter examines manufacturing's responses to their second stated principle—how alternative organizational structures are being used in AMT domains. Specifically, this chapter focuses on organizational structure and the two major factors that affect structure: the context in which the organization operates and the design choices about how the organization is to be structured (Hall 1987). Context includes size, technology, the environment, and cultural conditions. Within a "contingency organizational theory" framework, discussed shortly, technology and environmental uncertainty are examined by referring to practice and to some selected studies that appear to offer the most relevant lessons for managing the organizational aspects of AMT. The issue of level of analysis is raised first because structure must be considered at the work group and the individual level as well as at the organizational level.

Specialization and coordination, or differentiation and integration are the essence of organization. These variables need particular attention in AMT because traditional organizational boundaries are collapsing rapidly. This phenomenon of collapsing organizational boundaries is a key topic in this chapter, particularly with respect to the interface between design engineering and manufacturing. Finally, the chapter introduces structure from the perspective that most managers think about it: organizational forms and arrangements.

Chapter 7 concentrates on the design perspective at the level of the work group and manufacturing units—one level down from the manufacturing division or plant or department level focus of this chapter. Chapter 7 begins with a discussion of sociotechnical systems (STS) and their application to AMT environments, a discussion that is buttressed with some new research that has significant implications for CAM technologies. Since work teams lie at the base of many STS designs, this chapter will also be the location for addressing issues about work teams and support systems (i.e., the middle level of analysis of organizational structure). Chapter 7 then follows with a description of the product-focused forms that have broken manufacturing departments away from their traditional reliance on functional structures. Flexible automated manufacturing technologies are the forces driving the movement towards product-focused forms.

Chapter 8 will discuss organization at the level of the individual, focusing particularly on the nature of new technology and its accompanying new skill demands and their impact on the roles of operators, maintenance workers, supervisors, technicians, engineers, and managers. The coordination requirements and integration characteristics of new technology are discussed, and the chapter also addresses the deskilling-reskilling controversy that accompanies any discussion of new skills. Chapter 8 then turns to the work designs needed to accommodate the new skills and roles associated with computer-based technologies. Finally, the chapter deals with human resource practices relevant to AMT, particularly the issue of training.

Level of Analysis

Although organizational structure is usually conceived of in terms of the larger organizational unit, the company, or division or plant, the concepts of structure and structuring also apply at the level of subunits such as a manufacturing line or an assembly cell or a work group. At each of these levels there are patterned behaviors and interactions, rules, procedures and norms, reporting relationships, fixed configurations of workers and machines and systems, skill sets and qualification levels, and even consistent education and training programs. Structure and structuring also apply at the worker or operator level. Operators tend flexible machining equipment, reprogram computer controls, carry out routine maintenance, load and unload parts and fixtures, and do so in patterned and relatively predictable ways that may be referred to as structured actions or "scripts" (Barley 1986).

The concept of structure can extend beyond the boundaries of a particular organization. A manufacturing line or cell that is part of a JIT system has attributes of structure—patterned behaviors and interactions, rules, procedures—that apply for all parts and members of the larger system, namely, the manufacturing unit in question and the suppliers who feed it and sometimes even lateral departments within the firm such as purchasing that are integral to the functioning of the JIT system. The concept of structure is not limited by traditional organizational boundaries. It is, however, necessary to define the relevant boundaries of the unit under discussion when an analysis of the structure of that unit is undertaken.

Structural configurations change, but usually in predictable ways. A plant that uses self-regulating work groups and cross-plant committees to coordinate its operations has an identifiable configuration or structure, even if team membership changes frequently as employees rotate through positions to gain learning in a skill-based pay system or even as cross-plant committees are reorganized to address new plant problems. However, over time, new ways of coordinating (e.g., diagonal slice committees) may arise to become a permanent aspect of the organization's revised structural configuration.

Structure does not define all the activities that take place. In factories with programmable and flexible equipment we also expect that operators will handle exceptions, deal with unpredictable events, and respond to environmentally as well as internally generated uncertainties. Nor is this nonroutine or nonstructured activity confined to the operator level. Much of the explanation for the increasing use of self-regulating work groups is precisely that they can better adapt to uncertainty than the hierarchical arrangements of functional groups. This will be explored in depth in Chapter 7. Several recent conceptual arguments on future organizational arrangements (Liu et al. 1990, Savage 1989) predict that the management of variety and nonroutine will characterize tomorrow's organizations as more and more of the routine is assumed by expert systems.

ORGANIZATIONAL STRUCTURE

For organization theorists who have studied structure by comparative analysis (i.e., by comparing the arrangements in different organizations at a particular point in time), structure is usually viewed as the result of a particular imperative, with technology and environmental uncertainty being the determining variables most frequently cited. (Size is a third variable that has often been researched but it has had less explanatory power than technology or uncertainty.) The relationship of uncertainty to structure was examined above and a discussion of the technology-structure relationship follows next. Both relationships are complicated by questions about what level of analysis to address in the organization and by an increasing realization that the relationship is more complex than the independent variable-dependent variable models that have dominated so much organizational research (Fligstein 1985, Hall, Richard 1987). In Richard Hall's (1987, p. 122) words: "As the explanations of structure have been considered, none has been totally correct and none totally incorrect. Organizations are structured in a context. . . . There are thus multiple explanations of structure."

Structure has usually been studied by disaggregating it into discrete variables (e.g., complexity, coordination, control, formalization, centralization, configuration) that can be measured across comparable situations, usually from large questionnaire samples that can be analyzed statistically. To date, few significant studies of this nature have been carried out in AMT environments.

Although it has seldom been researched, the relationships between structure and other variables can also be in the opposite direction to those mentioned above. Structure, for example, might severely constrain technology. The beliefs about hierarchy in a traditionally organized manufacturing department may make it very difficult to change from a functionally organized machine shop to one that is organized in product-oriented manufacturing cells. Similarly, existing organizational structures can have a large influence on how managers perceive (Duncan 1972) and/or enact (Weick 1969) their environment(s). An organization with a tradition of autonomy within its divisions might find it especially difficult to get them to cooperate to offer an integrated response even though that may be precisely what its environment requires.

An alternative organizational perspective views structure less in terms of explanation and more in design terms (Gerwin 1981b). This is a systems view of structure. It is the one that prevails in this book. A systems view of the environment includes considerations of how the environment can itself be manipulated, the political processes involved in strategic choice (Child 1972), and the fact that there is organizational choice (Trist, Higgin, Murray, and Pollock 1963) (i.e., the idea that management and workers can choose among structural alternatives and can influence the technological arrangements rather than have them determined for them by technical experts).

This is a holistic perspective that views structure more as a configuration of design parameters (Mintzberg 1983) than as single relationships. Miller refers to this view as ". . . different organizational configurations or adaptive patterns that are richly described by the dynamic interaction among variables of environment, organization and strategy" (1981). As AMT installations tend towards becoming CIM factories, such a holistic perspective on the design of organizational structures will increasingly be required (Haywood and Bessant 1990, Susman and Chase 1986).

Some theorists have suggested an alternative view of structure with more of an action perspective (Silverman 1981). Ranson, Hinings, and Greenwood (1980, p. 3) define structure as ". . . a complex medium of control which is continually produced and recreated in interaction and yet shapes that interaction: structures are constituted and constitutive." Structure, then, should be seen as more dynamic than static, as more action oriented (e.g., as in Weick's references to organizing rather than organization (1969) and to structuring and structuration (1990, p. 18)):

> The important ideas . . . are that systems are built from interactions and rules; that such resources as action are the tools people use to enact their organizations; and, most important, that structures are both the medium and the outcome of interaction. People create structural constraints, which then constrain them. Structuration pays equal attention to both sides of that structuring process (constraining and being constrained), whereas earlier notions emphasized one side and neglected the other.

This dynamic perspective on structure tends to be more individually oriented than organizational (Scott 1990). However, it becomes increasingly relevant when the traditional variables that constitute structure change not only because of a changing set of environmental conditions or changes in technology but when the members of an organization themselves change significantly because they are in continuous learning modes. This is a currently highly desired mode for operators in a range of AMT situations. Pay-for-knowledge systems that support many new organizational arrangements, for example, foster continuous learning. In such "learning environments," the concept of structuring may be more useful than that of structure. The idea of learning environments is not confined to conceptualization at the level of the individual. It will increasingly be a way to think about organizations with high adaptability requirements, such as those based upon flexible CAM technologies (Schuck 1985, Swedish Work Environment Fund 1988).

Technology and Structure

The relationship of technology to structure has been extensively studied by organizational theorists (Barley 1986, Davis and Taylor 1976, Dewar and Hage 1978, Gewin 1981b, Hage 1969, Hickson, Pugh, and Pheysey 1969, Khandwalla 1974, Mohr 1971, Scott 1990, Sproull and Goodman 1990, Weick 1990). However, the relationship is complex. The ability to clearly articulate it is compounded by the lack of clear definitions for both technology and structure, particularly the former (e.g., Gerwin (1981b) cites ten different studies that used eight different measures of technology).

Most research into the technology-structure relationship assumes one overarching causality—that the structure of an organization is directly dependent on the organization's technology (Hickson, Pugh, and Pheysey 1969, Thompson 1967, Woodward 1965). Management's role, under this view, is a passive one (Gerwin 1981) (i.e., the structural arrangements follow once the technology is decided). If technology determines structure, then the choice of the appropriate organizational structure should follow from a clear statement about the characteristics of the particular technology or technologies. In fact, such choice has been quite problematic, not just because of the wide variety of automated manufacturing technology arrangements that can exist, but also because the frequently espoused technology-structure relationship is less and less clear with each additional research study (Gerwin 1981b, Scott 1990).

Emery (1978a), for instance, offers a complex view of technology with respect to organization. On the one hand, it is a boundary condition of the organization's social system ". . . mediating between the ends of an enterprise and the external environment" (p. 41). However, technology is also seen as "belonging" to an enterprise, if only for the fact that it is excluded from similar control by other enterprises. It, therefore, represents the organization's "internal environment." Hirschhorn (1984) takes a position somewhere between Emery and technological determinism. He conceives of technology less as a determinant of structure and more as a constraint on organization design decisions.

Scott (1990, p. 36) recently reexamined the technology-structure relation and suggested that both variables need to be rethought and the relationship reconceptualized:

> While it is always imprudent to be overly sanguine about recent developments in any arena of social science, it does seem to me that the past two decades of work—both theoretical and empirical—examining the relation of technology and organizational structure do provide evidence of some progress. From the early days, when our comparative results were based on the study of a handful of organizations, we can now examine results based on sizable and diverse samples. After much dissension over variables and measurements, we can now point to some convergence on both, so that it is possible to draw some conclusions about the effects of technical uncertainty and complexity on structural features of organizations, on the basis of thirty or more empirical investigations. Contingency models, broadly defined, continue to provide the dominant theoretical paradigm.

A contingency model of organizations states that there is no one best way of organizing (Lawrence and Lorsch 1967). How one should organize depends on several variables, particularly the ones we have already identified such as environmental uncertainty and technology. Contingency models also hold that all ways of organizing are not equally effective. Again, depending upon particular market characteristics or the technology in place or the state of competition, one way of organizing might lead to high performance while another might spell failure for a firm.

Other Influences on Organizational Structure

Research into contingency models has tended to concentrate on the effects of a single variable (e.g., technology or environmental uncertainty) on organizational structure. However, it is in the interaction of several contingencies that the unique structure of each organization is established, for instance, in the way a specific technology is combined with a particular set of job and organization design arrangements, appropriate training and development systems, and all this placed within the unique context of that organization's particular labor market.

National societal factors are another contextual contingency condition that can have strong influence on work organization and skill structure. Schultz-Wild (1990, p. 97) noted that variations in ". . . system concepts and layout (greater significance of the central computer and software programs in France; stronger concentration on processing machines, handling facilities and transport systems in Germany) . . ." not only impact the supply of labor and the professional orientation of engineers differentially in those countries but also affect the division of labor and job design in their respective manufacturing systems. It may only be where highly skilled trades-type production workers exist in quantity (e.g., Germany or Sweden) that it will be possible to create work organizations in which high levels of responsibility and autonomy are assigned to workers.

Structural resistance, or inertia, can be an important factor constraining organizational structure, or in d'Iribarne's (1990, p. 69) words, ". . . the internal structure of each workplace . . . acts as a brake on any evolution not in line with its past development." Liu et al. (1990) provided an elaboration of this perspective. They suggested that when new technologies are introduced, they have immediate effects on the contents of jobs and on skills and qualifications, but for a while they do not modify the organizational arrangements. In time, however, the existing organization design becomes increasingly suboptimal with respect to the possibilities that the new technology creates. Only then does a new form of organization, that was at first marginally applied, become more fully adopted because it provides ways to gain the benefits of the new technology. They cite the example of CAD that is too often used as an electronic drafting board because the design and manufacturing organizations that it should work across are seldom organizationally integrated.

Interdependence and Coordination

Uncertainty that originates in product-markets and/or process operations has been the focus of this chapter. As mentioned earlier, there are other uncertaintylike characteristics that a firm experiences when its technical and internal operations become quite complex, its need to respond quickly increases, its links to customers and suppliers tighten; in effect, its need to manage a much higher level of interdependence increases (Campbell and Warner 1990, Manufacturing Studies Board

1986) (see Figure 6.1). Theoretically, these are not uncertainty dimensions since a firm can match the variety and complexity of high levels of interdependence with appropriate structural arrangements, integrated information systems and appropriately skilled personnel (Ashby 1956). In practice, this is a continuous catch-up exercise, particularly because advanced manufacturing technologies are constantly racing ahead of the organization's structural capacity (Liu et al. 1990). This is experienced within the firm as just that much more uncertainty to deal with.

Although effective organizational coordination is the appropriate structural response to perceived interdependence, the specific structural response varies with the type of coordination needed. Thompson (1967) identified three types of interdependencies—pooled, sequential, and reciprocal—and proposed a different type of coordination appropriate for each—standardization, planning, and mutual adjustment, respectively. Three examples illustrate Thompson's prescriptions. If, in the case of pooled interdependence, a large number of CNC lathes require assistance from maintenance mechanics for relatively similar setup procedures, then maintenance can simplify the coordination of such activity by developing clear setup procedures that operators can be taught to carry out themselves even if they are assigned to different lathes at different time periods. In the case of sequential interdependence, if the assembly of diesel engines requires parts to arrive in a particular sequence in order to minimize inventory stocks, then careful scheduling with the supplier for just-in-time arrival to coordinate with the build schedule is called for. Finally, if a program of continuously reducing manufacturing costs requires frequent revisits to and changes of the design to take account of production experiences in manufacturing the product, then a design and manufacturing team that meets regularly will be an effective way of achieving mutual adjustment between the functions.

Rhenman (1973) suggested that an organization must achieve "consonance" when it experiences dissonance either between itself and its environment or between subsystems within the organization. It has several ways to achieve such consonance: by *mapping*—reflecting the environment in its own structure (e.g., organizing by product teams that will be responsive to each of the product customers); by *matching*—understanding the environment or the neighboring systems causing the dissonance and complementing it (at a remotely located Westinghouse feeder plant, CNC and DNC programmers were difficult to hire so machinists from different group technology cells were retrained as members of a programming unit to serve the entire plant and interact with divisional systems software people); by *joint optimization* and *joint coordination*—using a combination of the previous approaches to map for exploitation and match for cooperation; and by *dominance*—either by (1) acquiring greater variety than that of the dissonance creating situation in order to "force down their variety" (Ashby 1956) (developing an FMS with sufficient flexibility to adapt to any future variability in demand), (2) becoming better able to adapt to changing circumstances than competitors through "survival of the fittest" (which describes the strategies of many firms who attempt to lead in technologies such as CAM, CIM, AI, etc.), or (3) developing a "leading subsystem" that can lead or control others (i.e., finding "lead" variables as opposed to some organizational change variables that are considered "lag" variables). For instance, Beer and Spector

(1985) suggest that pay may be a lag variable that does not have to be attended to early in implementing innovation. Work groups would probably be a lead variable. Too many CIM designers have made the costly mistake of considering the "people" side of the design a lag variable (Bomba 1990, Koelsch 1990).

Kaplinsky (1984) showed that the pattern of progress in manufacturing has been to first *integrate within* what he called a "sphere" and to then *integrate between* spheres. The consequences of this line of reasoning is the argument that technology is continuously increasing in complexity and, as it does, the need for accompanying organizational coordination and integration increases. Haywood and Bessant (1990) illustrated Kaplinsky's reasoning by citing how integration first took place within the different functions that comprised a machine tool operation (i.e., integrating the operator's role then adding a programmable controller and then an automatic tool changer to create a CNC machine tool). A desire for faster cycle times led to a need to achieve integration between different CNC machines assembled in physical proximity to form a machining cell. This is a cycle that continues within CAM technologies. For instance, integration within a cell might add DNC and some robotic handling capability to allow the cell to function partly unmanned. At the next level, integration between cells would call for the addition of AGVs, material-handling systems and supporting technical and managerial staff to create a "factory-within-a-factory" (Lindholm 1975). The cycle can obviously be continued to evolve toward a CIM system.

Interestingly, Kaplinsky's logic can be applied in a parallel way to the organizational side of a CAM environment. A machine operator may choose to integrate the operational knowledge of other machines (and/or of maintenance and/or programming) into his or her skill set with the objective of becoming multiskilled. As such, operators and accompanying equipment are put into cell arrangements and flexible teams evolve that call for coordination between multiskilled operators. Additional levels of integration take place within the team as they absorb some of the scheduling and planning responsibilities of foremen and/or supervisors and as they take on responsibility for other functions such as minor maintenance. Again, the cycle continues.

In Lawrence and Lorsch's (1967) study, organizations in the high change environment had much higher integration requirements because of the high differentiation between functional units than did the low change environment organizations. The theorists concluded that the more effective organizations were those that had a good fit between their structural arrangements and the requirements of their respective environments. In other words, the effective organizations in high change industries had high differentiation and high integration simultaneously. The effective organizations in low change sectors had low differentiation and commensurately low requirements for integration.

Lawrence and Lorsch studied organizations at a time when high rates of change and uncertainty in the external environments of some sectors were still a relatively recent phenomenon. In today's AMT environment, uncertainty is almost the norm for each of the different subenvironments faced by different manufacturing technologies. This condition, where every subenvironment is high in its experienced rate of change, was not one they studied. However, if one were to extrapolate their conclusions to this environment, a firm might experience few differences between the subenvironments faced by different islands of automation (e.g., a manufacturing cell, an FMS) in that all are facing similar high unpredictability (assuming "high" for one technology is similar to "high" for another). With few differences in their subenvironments, required differentiation between the technologies should be small, and as a consequence, not much integration required. Since the different technologies within a AMT environment do not currently have much of a need to have their activities coordinated, this would appear to be a reasonable conclusion. For instance, in factories machining engine parts and also assembling them, despite the desire for low level inventories, component stores is still used as a buffer that effectively decouples the two functions.

However, this low need for integration between automated manufacturing technologies or islands of automation is likely to change. First, it will change because the differences in the subenvironments of different technologies will likely increase (some will be used for high volume production and some for low volume (Venkatesan 1990); some will focus on response time and some will emphasize other flexibility features). Second, as shown in the examples used to illustrate Kaplinsky's concept, computerization and advanced manufacturing technologies combine to create ever-increasing interdependencies, which require increasingly greater levels of integration. CIM is one obvious outcome of such combination. FMS is also, although at a slightly lower level of integration. Factories within factories, as described in Chapter 7, are another outcome of this tendency toward increased integration.

The acquisition of flexibility to accommodate uncertain and unpredictable events depends on multiskilled operators in addition to programmable equipment. Multiskilling, in turn, requires cross-training and job rotation, rotation that usually takes place first through different equipment in one's own cell or line, but after that through other technologies.

GE Canada's Bromont plant, for example, deliberately places technologies that are required by several business units in only one or the other of its PMEs to create the kind of reciprocal interdependence that requires mutual interaction, that is, sharing, job rotation, and cross-training. For instance, the forge PME has tumbling and etching equipment that other PMEs require, whereas the pinch-and-roll PME operates a heat treatment unit that forge and other PMEs utilize. Plant management has allocated particular technologies so every PME will have a dependent relationship with at least one and preferably several of the others.

On one hand, this high level of forced interdependence constrains PMEs and slows down their ability to minimize their cycle time. On the other hand, it achieves a higher level appreciation of a plantwide perspective. Bromont's senior management team wants the PMEs to appreciate that it is not acceptable to the plant as a whole for one PME to perform well if other PMEs are not also performing well. In this seemingly contradictory way, this structural action reinforces the goal of having the PMEs take responsibility for themselves as a relatively self-contained business unit but always with a forced appreciation of the systemic nature of the plant and of the interdependencies associated with such a systemic undertaking.

Reciprocal interdependence requires mutual adjustment if effective coordination is to be achieved (Thompson 1967). In their study, Lawrence and Lorsch (1967) identified several ways in which the effective organizations they studied went about achieving the needed integration. They noted that effective integration comprised several things: appropriate coordinating mechanisms (e.g., teams, integrating departments), balanced perspectives on the parts of the individuals in coordinating roles, and open conflict resolution processes to address the "legitimate" differences. Coordinating structure, then, is defined widely enough to encompass process and individual characteristics as well as the more "macro" variables.

The Collapse of Organizational Boundaries

Rapidly increasing interdependencies between the plant and its external environment are collapsing traditional boundaries at a rapid rate (SME Sociotechnical Study Committee 1989). This breaking down of boundaries will be one of the key organizational manifestations of the 1990s. Many of the environment's uncertainties are entering directly into the organization and are no longer capable of being buffered at the boundary. Nonroutine activities will assume more importance than they did when it was possible to seal off the technical core of the organization from uncertain events (Thompson 1967).

High interdependence is managed by effective coordination. The most inclusive coordination is achieved by removing the actual boundaries between the units requiring integration. This has already manifested itself within the manufacturing domains of a large number of organizations, both within the firm's boundaries (e.g., in the design-to-manufacture area), as well as across them (e.g., via JIT systems). A similar pattern is developing across the marketing department-to-customer boundary areas. To satisfy their customers, producing firms have increasingly involved them in the design of the product at an early stage, via focus groups—for consumer products, or through engineering or the quality systems—for industrial products.

As mergers and acquisitions continue and as the number of alternative suppliers for products dwindles, the differences between products offered will be fewer and fewer. Differentiation to achieve competitive advantage will likely take the form of more rapid responses to customer demands and changes. Shortened product lifecycles and customized responses go hand in hand with more frequent interactions with customers and suppliers. This increased interdepenence requires a level of mutual interaction that is impeded by conventional organizational boundaries. Firms will attempt to get closer to their customers by reducing and even removing the boundaries between them. For instance, one company that recently designed an extension to one of its major products had over 20 members of its principal customer for that product as members of the large project team involved in the two-year design and prototype manufacturing effort. Even at the shop floor, the increased permeability of the boundaries can be seen in innovative firms that send shop floor workers out to customer locations when product or process problems are identified, or to vendor locations to alter designs.

Of more salience is how the boundaries are falling within the organization itself. Here, the pace is frenetic. It is alternatively referred to as delayering, demassing, and flattening out. The entire middle is disappearing. This is the vertical dimension of the change. One could call it delegating responsibility downward, but that might be a too-charitable interpretation.

Cutting costs is a large part of the motivation, but it is not just cost cutting that is driving the activity. It is more that the control exercised in traditional organization by middle ranks of supervision and management is less needed now; partly because workers expect more challenge from their work and need less control, but mostly because the spread of information technology has made control a more easily and less expensively managed process. The data to understand and make relevant decisions is available at the top of the corporation and at the operator level and often simultaneously. Control can be exercised by setting objectives and knowing that those affected will have the data to confirm if they have been accomplished. Control is best achieved by getting the information first to those who are actually doing the job so they can exercise the needed control themselves.

The pressure to break vertical boundaries is not just top down. With the increasing use of teams, sometimes without supervisors, many traditional vertical roles are assumed within the team. It becomes less clear then whether responsibility has been delegated downwards or whether upwards responsibility has been assumed. In Haywood and Bessant's words (1990, p. 81) "integrating technologies require closer functional integration, so they imply shorter hierarchies and greater vertical integration in organisational structure." They suggested that to exploit the full benefits of a rapidly responsive and flexible system it was necessary to create managerial decision-making structures that closely involved the shop floor and that delegated a high degree of autonomy.

In some AMT situations, the boundary between operators and engineers is no longer as clear as it once was. Many companies use engineers as operators for complex CAM technology such as FMS (Jaikumar 1986), particularly in the startup years (Martin 1989). Other companies with CNC machines, either in stand-alone applications or within cells, have trained operators and machinists to perform the programming tasks that still other organizations assign to software engineers.

The Design-to-Manufacture Interface. The interface between design and manufacturing is the internal organizational boundary under the severest pressure at the start of the 1990s. All the current trends in manufacturing—cost reduction, quality enhancement, market fragmentation, shortened product life cycles, and particularly their combination (Dean and Susman 1989)—benefit from better management of the design/manufacturing interface. The pressure is on to collapse this boundary in order to reduce product development time, to decrease the number of engineering changes that traditionally follow a new product into manufacturing, and to improve the quality of the products produced by "getting them right the first time." The focus on this interface has spawned a range of titles for this area of activity (e.g., design for manufacturability, manufacturing by design, manufacturable design, concurrent engineering, simultaneous engineering, producibility, and manufacturability assurance (Dean and Susman 1989, Industrial Management Roundtable 1990, Susman 1989, Whitney 1988)).

Hottenstein (1988) studied thirty-four firms where AMT had been or was being installed and suggested four overall strategies they were using to manage the design/manufacturing boundary: *technology*—particularly the use of CAD/CAM, *parts simplification*—to improve part numbering systems and reduce the number of different parts used in products (Kumpe and Bolwijn 1988), *organizational relationships*—which will be examined in more detail, and *supraorganizational efforts*—which raise the status of manufacturing (Skinner 1978) and change the corporate culture and climate.

In a summary of a special Institute of Electrical and Electronics Engineers issue on the social and organizational dimensions of CAD, Majchrzak and Salzman (1989) concluded that more than technological determinism affects these dimensions. Managers and engineers too often have inappropriate expectations about what CAD technology alone could do to improve the design-to-manufacturer interface. Forslin, Thulestedt, and Andersson (1989) showed that such expectations could only be realized if appropriate organizational changes were made, but the engineers involved only came to appreciate this after attempting to implement CAD through the technology alone. Adler (1989b) studied 13 organizational efforts to integrate CAD and CAM and arrived at a similar conclusion—that to be successful, CAD and CAM integration requires more than a technological solution.

Lee (1989) studied CAD and CAD/CAM implementation in the United Kingdom and identified several factors that determined the outcome of the implementation effort: whether pressures to adopt CAD or CAD/CAM were market driven or technology driven and whether they were carried out in a sociopolitical environment of high or low trust affected management's approach to changing the work organization. That approach could either be one of direct (hierarchical) control or one of granting responsible autonomy, the former leading to a traditional work organization and the latter to a flexible one. Buitendam (1987), examined lateral relations in new technology and reinforced both the significance and the perspective Lee attached to the role of trust. He noted that trust is not an automatic consequence of new technology. Rather, it is the opposite. New technology was likely to be more successful when it was integrated with the changing social situation and "The change of the social structure might, in its turn, depend upon trust relations." (p. 78)

Most efforts to coordinate across the design/manufacturing boundary fall within the model of coordinating mechanisms that Lawrence and Lorsch (1967) and Galbraith (1972) identified as those most commonly used to integrate between functions. However, the high current importance attached to this particular interface has resulted in considerable elaboration of some of these methods, ranging from joint team efforts and adding degreed professionals to manufacturing teams to desktop manufacturing and design software. In fact, an entire industry has evolved to develop "design management software" (King, Treece, and Hammonds 1990).

Ettlie and Stoll (1990) studied thirty-nine factories that were in the process of modernizing their facilities in 1987 and found five actions that summarized their design-manufacturing coordination efforts (p. 86):

1. All members of the cross-functional teams are trained in design-for-manufacture methods.
2. Manufacturing signs off on design reviews.
3. Novel organization structures are used for coordination.
4. Job rotation is practised in engineering functions.
5. Personnel move between engineering and manufacturing—permanently.

With respect to the structural arrangements to achieve design-manufacturing coordination, they reported that in most cases (72 percent), design engineering reported to engineering management and manufacturing engineering reported to manufacturing management. In the other cases, design engineering and manufacturing engineering reported to a common boss, although the authors suggested that performance was best when the common boss was an engineering manager. Dean and Susman (1989) found somewhat similar patterns which they classified into four distinct approaches that each constitute changes to traditional hierarchical structural arrangements.

Manufacturing Sign-off. This gives manufacturing veto power over product designs. These approaches are often augmented by software packages that help design to determine attributes that will be satisfying to manufacturing. These packages, which have even acquired a generic title, "electronic design management," are becoming increasingly complex since some are customized for the particular organization. Included in this approach is the idea of having manufacturing take responsibility for building prototypes, normally a design task.

Integrator. The second approach is to use an integrator to coordinate between the two functions. Integrators need to be perceived to be balanced in their orientations (Lawrence and Lorsch 1967). This perception is usually a result of a succession of decisions they make. Given that integrators usually originate in one of the two functional areas being coordinated, a perception of balance is difficult to achieve. The actual titles of people in these roles varies with the organizational situation. Dean and Susman (1989, p. 30) cite one company's choice of the term "producibility coordinator." Responsibility and authority granted to the integrator also varies considerably.

Cross-functional Teams. This approach advocates early enough involvement of representatives of the two functions for simultaneous engineering to occur. As with the cross-functional teams that have long existed in project management and matrix organizations, interpersonal skills are necessary for effective team functioning (Davis and Lawrence 1977). Unfortunately, such skills are often at a premium. The difficulty of working together is compounded by the variety of backgrounds that make up the membership of a cross-functional team. For example, Ford put together a very successful team to make Taurus a "total car" by bringing all the stakeholders in early. This included not only representatives from design and engineering but also manufacturing, finance, and product planning. Said Mishne (1988, pp. 51–52), "Everyone who was concerned or involved with the project was called in before we started designing the car so that we had their input. This eliminated surprises later down the line when we actually started developing clay models."

Product-Process Design Department. The fourth approach is that of creating an entire integrating department that Dean and Susman (1989) refer to as a product-process design department. There are several variations to this approach, but all can move the organization still further down the path of simultaneous engineering.

The decision to select one of these approaches, and its variations, over another is dependent on the culture in the organization, the history of relations between design and manufacturing, the appreciation of the situation by the senior manager or executive common to both functions, and the support and commitment management is willing to provide for the activity (the organizational and technology costs can be very high), quite apart from the level of coordination needed by the particular product design situation. The payoff, however, can be significant. A manufacturer of telephone equipment created a project team to oversee the enhancement of one of its major switching products. The R&D facility and the primary and secondary manufacturing locations were hundreds of miles from each other. The project manager carefully selected a twenty-five-person core team of people who had reputations for being able to extract cooperation and respect from others in the organization. He himself had been selected because he had alternated his career between manufacturing and design. The core team, with some members alternating, and with many customer representatives joining them and the larger project team (which in time grew to 600 persons), traveled back and forth on airplanes, particularly between the R&D and the primary manufacturing sites, an average of two days a week for over a year. That tight coordination between design and manufacturing reduced the expected time to develop the new product from three and a half years to two years.[4]

ORGANIZATIONAL FORMS

The conceptualization and measurement of the variables that comprise structure dominates the thinking of organizational theorists. Organizational form, however, is the preoccupation of managers who have the responsibility of making structures function in the enterprise. Table 6.1 gives a brief overview of the organizational forms that have dominated in the manufacturing environment, particularly from a North American perspective.

---

[4] From site visits and interviews conducted in 1988.

TABLE 6.1 Organizational Forms: Part I—To the 1980s

Period	Form and Features	Primary Emphasis	Internal Characteristics
To 1950	*Functional* Efficient Developmental	Manufacturing	Division of labor Economies of scale Vertical career paths
1950s–1970s	*Divisional* Decentralized External orientation Diversified	Marketing	Local autonomy Local responsiveness General management
1970s–1980s	*Mixed (Project/Matrix)* Adaptive Complex Variety increasing Two or more orientations	New products Information processing	Lateral communications High interaction Multiple career paths Teams

Senior management or the dominant coalition (Thompson 1967) make decisions about the choice of basic structural forms to organize the company or plant or division. Staff are left to fill in much of the supporting structure in terms of functional systems (e.g., accounting, measurement, cost control, recruitment, compensation, training, benefits, quality, sales administration, production control, etc.), although in practice many of their choices are circumscribed by professional standards, legislation, and norms in their specialist areas. Support systems ostensibly follow the organizational form. Often, however, they become the sources of structural inertia. Even after management has changed an organizational form, the support systems are usually slow to adapt, if they ever do, despite their inappropriateness. Most manufacturing accounting systems, for instance, are still based on direct labor costs even though direct labor costs are decreasing and as a percentage of factory costs are now often less than 10 percent (Kaplan 1983).

Functional Forms

Until after World War II, functional structures prevailed. Manufacturing was the backbone of the economy and organizational arrangements were designed around division of labor, scale economies, specialization, and managerial hierarchies that provided clear career trajectories for aspiring managers. Functional forms are efficient, particularly when the cost of capital is low and inventory costs can be discounted. Relative stability in the marketplace allowed production to have long runs, few setups and learning curve effects that provided scale economies.

The logic of functional organization dominates the layout of equipment in a traditionally organized factory as lathes, milling machines, drill presses, grinders, and polishers are grouped together by common function. As repositories of long-term specialist competence, functionally organized structures have much to offer manufacturing organizations who have a need to stay technologically current in a world where manufacturing competence is increasingly sophisticated. At the same time, the logic for maintaining a functional home for a particular type of machining competence is becoming weaker because advanced manufacturing technologies incorporate more and more machining expertise into programmed controllers. With expert systems, this trend will only continue.

Functional forms are a poor structure for organizations seeking quick responses to environmental uncertainties. Cycle time through the machine shop is slow because parts sit in inventory or on skids on the shop floor for long periods of time waiting their place in the queue of parts to be processed on a particular machine or travel extensively back and forth on fork lift trucks between the different machines required to complete the range of machining operations. Hence we see responses like the earlier cited structures of GE Canada in Bromont and Cummins Engine Company in Jamestown where product oriented structures are subsumed under functional ones in an attempt to both have a place to house manufacturing competence yet be capable of responding quickly to environmental changes.

Divisional Forms

The affluent years from the 1950s to the 1970s were the marketing era. Functional organizations grew in size with the ascendancy of American management (Servan-Schreiber 1968) and economic affluence. The marketplace, however, began to differentiate in response to the diversified preferences of an increasingly affluent consumer population. Within large corporations, divisions were created and granted local autonomy to respond to their local markets, where local could mean geographic or product line or market segment (Chandler 1962, Rumelt 1974). Sloan (1964) created the divisional concept at General Motors and the 1950s saw the idea proliferate (Fligstein 1985). With the creation of divisions, the number of general managers, each as a local division head, grew almost exponentially.

This was the era when the power in the eternal centralization-decentralization struggle of corporate life shifted to the decentralized side as responsiveness to local markets and diversified product lines took precedence over the controls and economies of centralized activities. Within each division, functional organizational forms were replicated and assumed all their characteristics, with the manufacturing focus dedicated to the particular division's products. Divisionalization had the effect of disaggregating what would have been an increasingly complex centralized production process into separated and more manageable production units.

Project/Matrix Forms

From the 1970s to the 1980s, market differentiation continued as new product proliferation engaged more and more members of the organization in managing the boundary between the organization and its environment, particularly on the sales, marketing and new product sides. Project managers and program managers were created to cope with the continuously segmenting markets (Kolodny 1979). At the same time, traditional engineering, production and sales functions faced increasing uncertainty and variety in their own scientific, technoeconomic, and market sub-environments, respectively (Lawrence and Lorsch 1967).

In the GE Bromont and Cummins Jamestown examples, the need to respond to both functional and project needs is handled *sequentially* as product focused units are organized one level down from functional groups. Project management and matrix organization structures develop capacity to *simultaneously* manage functional and project orientations (Davis and Lawrence 1977).

With more environmental sectors engaged with simultaneously, considerably more external variety was thrust towards and gathered into the organization. The increasing variety was reflected in large amounts of information that had to be processed within the organization (Galbraith 1972). The increase in information was accompanied by a commensurate increase in the number of managers to match the variety (Ashby 1956) or fit into it or lead it (Rhenman 1973), but in any case to remove its "equivocality," which Weick (1969) refers to as a characteristic of the informational inputs on which organizations operate, "informational inputs that are ambiguous, uncertain and equivocal" (p. 40).

Galbraith (1972, 1973) created a scale of increasing coordinating mechanisms that were necessary to integrate across the differences between the functional departments (Table 6.2). If differentiation between functional units is high, an organization might have to be further along the coordination scale to achieve the needed integration. Coordination is costly. Effective organizations incorporate no more coordinating mechanisms than are needed for the particular situation. Matrix structures provide the most intensive coordination but at significant cost—in complexity, in increased management, in the number of meetings, and in the stress created by intensive lateral communications.

The earlier discussion on the design-to-manufacture interface illustrates the applicability of this scale of coordinating mechanisms. Integrators, cross-functional teams, and product-process design departments are different coordinating mechanisms

TABLE 6.2 Functional Organization Coordinating Mechanisms

Rules and procedures
Hierarchy (common boss)
Planning
Direct contact
Liaison roles
Task forces
Teams
Integrators
Integrating departments
Matrix structures

Source: Galbraith (1972, 1973).

used to achieve the required level of integration between design and manufacturing departments.

Effective coordination is achieved through a set of coordinating mechanisms and behaviors. Basic manufacturing strategy is the highest level of coordination and subsumes Table 6.2 though it may also be manifested in the rules and operating procedures of the unit in question. The manufacturing or plant or production line manager is another strong source of coordination (see hierarchy, Table 6.2) as is a manufacturing coordinator (liaison role). A production scheduling team or task force or committee or manager constitutes another coordinating mechanism. With organization structures becoming increasingly flat, it is likely that forms of direct contact between the operators and/or their work teams will also be within the set of coordinating mechanisms as will direct contacts with suppliers and clients—both those external to the organization and those upstream and downstream of the unit in question. However, since coordination is costly, then if different production lines in the same plant require different amounts of coordination, the effective plant will be the one that provides the needed amount but not more than what is needed. This could mean quite different coordination structures for production lines within the same plant.

The increased information processing capacity of project management and matrix designs converts these forms more into "logics of change" than the logics of control that characterize functional forms (Kolodny 1979), a condition that makes matrix and project management forms highly adaptable.

In matrix structures, production is usually a centralized functional activity, but more often than not the production area is subdivided for the different projects and programs. In AMT environments, centralized personnel can be deployed temporarily to different manufacturing cells (Manufacturing Studies Board 1986). Manufacturing managers must make daily decisions about what could benefit from being shared and what is unique enough to merit its own production line. For example, in aerospace manufacturing, when several different projects flow through common production facilities, some resources such as acoustic test chambers or test aircraft are costly enough and little enough used to be obvious candidates for sharing across programs, but other functions such as assembly and reliability testing are each conducted in the separate production lines of each project or program (Davis and Lawrence 1977). It is within this context that project manufacturing managers can change production levels quickly enough to realize volume flexibility for the firm.

Table 6.3 lists organizational forms from the 1970s on. The dates are very coarse approximations. For example, some of the Swedish experiments with product-focused forms go back to the late 1960s as do some of the sociotechnical systems designs in the USA. Explicit STS designs in India, Norway, and the United Kingdom took place even earlier. The first two forms listed in Table 6.3—product-focused forms and sociotechnical systems designs—are discussed in Chapter 7.

The 1980s came as a shock to North American manufacturing management. Fierce competition from Japan and the other Asian countries struck at the core of almost every company's operations and exposed a far weaker manufacturing flank than anyone had believed possible (Hayes and Abernathy, 1980). Top management's

TABLE 6.3 Organizational Forms: Part II—From the 1970s

Period	Form and Features	Primary Emphasis	Internal Characteristics
1970s	*Product focused* Flexibility Small size Relative autonomy Flow orientation	Production technology Short cycle time	Computerization Automation Group work Cross-training
1970s	*Sociotechnical* Self-regulation Variety increasing Values driven Systems orientation Joint social-technical	Adaptability People commitment	Boundary management Growth and learning Autonomous groups Participative design Multiskilling
1980s	*Network* Changing alliances Interactive Gain leverage Interdependent	Information technology Knowledge work	Lateral relations Communications Boundary crossing Information networks
1985+	*Integrative* Medium size Sociotechnical design Shared philosophy Plants within plants	Business management People, quality, technology, and information rationalization	Continuous design Participative design Collapsed boundaries Parallel processing

excessive concentration on acquiring capacity to address increased variety at the boundaries of the organization had been purchased at the expense of attention to internal operations. Quality and productivity were the primary victims of the inattention. Slowly, North American organizations began to draw back from many of the variety increasing organizational forms they had adopted such as conglomerate forms, matrix organizations and project management.

In manufacturing, they turned their attention to production technology, to adaptability and responsiveness and to people (see Table 6.3) and organized themselves around product focused forms and sociotechnical systems designs. In the service sector, there was a similar turning, but to information technology, knowledge work and, again, people. Network organizations came to the fore here as they have in some of the R&D/engineering, production and marketing functions of manufacturing companies. Integrative organizational forms have surfaced recently, particularly in AMT.

Network Organization

Network organizations is a loose category for a wide range of horizontally oriented and very flexible organizational forms (Aldrich and Whetten 1981, Miles and Snow

1986). Communications systems coupled to computers that facilitate the sharing of standardized data bases are one of the driving forces creating network forms. Hence, JIT systems with their supplier links are network organizational forms. Communications technology and information technology and the knowledge workers that use the information generated characterize these forms. Although these forms have obvious application in nonmanufacturing sectors, as boundaries between functional organizations continue to collapse they are finding increasing relevance within manufacturing organizations. Many of the arrangements used to organize activities in the design-to-manufacturing area, for instance, are best understood as network forms, particularly because some of these structural entities cross beyond the firm's boundaries to include customers within their arrangements.

In one variation of network organizations, small organizations band together in loose confederations built on trust and confidence in each other to bid as if they are larger than they are. For the next bid, the arrangement may be different and the players different as a new alliance is established. Hence interactions and interdependencies can be high, although sometimes short lived, because the players may soon recombine into another network.

In another variation of the form, one finds an increasing use of consortia of R&D personnel from companies in related and even competitive companies. The consortia are usually restricted to the precompetitive stage. Threats from such consortia in Japan, whether real or imagined, have been the rationale for such activity in the West. Efforts are made to preserve the integrity of each company from the other, but despite them, the boundaries between the companies are less rigid than they were.

Other organizations examine the entire "value chain" of the products they produce and form alliances up and down the chain to make their collectivity more competitive (Johnston and Lawrence 1988). Here we find some conflicting advice with respect to the issue of vertical integration. Hayes and Abernathy (1980) pointed out emphatically how the tendency of U.S. firms to vertically integrate made many of them less innovative and reduced their flexibility by constraining them with sources of supply that could become technologically outdated. Dean and Susman (1989) reiterated this position when they cited vertical deintegration as an organizational structure for firms to adopt to increase their manufacturing flexibility. Kumpe and Bolwijn (1988), however, make a case for manufacturers to reconsider vertical integration. They suggest that the flexibility gained by freeing up the manufacturing organization from committed or owned sources of supply in the vertical network may be purchased at the price of having to buy upstream components from competitors, a situation that might considerably increase the uncertainty for a manufacturing firm.

Some commentators have cast the phenomenon of "hollowing-out" under this category of network organizations (i.e., the phenomenon where American corporations may be subcontracting design work and manufacturing out of the country so that the corporation with its name on the product being produced is nothing more than the shell of a manufacturing company). This has begun to be less of a concern for U.S. manufacturing as more and more manufacturers realize that speed of response is more valuable to the organization than savings in the cost of direct labor (Weiner,

Foust, and Jones 1988). This is particularly the case as the cost of direct labor has become a small proportion of the manufacturing cost. The short-term uncertainties of changing product preferences in the market can be responded to far more quickly by a manufacturing organization that is in proximity to the design organization.

There is no clear classification system to differentiate network forms and the term covers too wide a range of activities. With research and classification, understanding will improve. This, in turn, will lead to better application. The form has particular relevance to CAM environments because the sharing of databases between islands of automation in a factory, between the components of an FMS, between suppliers and customers and equipment vendors, between design and manufacturing and customers are all networklike trends that will increase. We see this pattern in the standardization efforts of Open System Interconnection networks such as MAP (manufacturing automation protocol).

Integrative Organizational Forms

This is a speculative classification of forms currently arising. As with network forms, it will take some time to understand the characteristics of the form, but some of its features are currently clear. Integrative forms fly against much of conventional manufacturing logic to incorporate many of the attractive features of product focused forms, sociotechnical designs, and network organizations in concert with quality management approaches and JIT systems. Savage (1989 p. 60) takes this one step farther to incorporate business issues and suggests that ". . . a clearly articulated strategic business vision becomes the glue of the organization."

Savage (1989) differentiates CIM-II from CIM in a fashion parallel to the difference between MRP-II and MRP. CIM-II is associated with fifth generation management in which functional units are replaced by knowledge center "nodes" in which many responsibilities overlap and redundancy is common. He points out that the focus of CIM-II is not just on the manufacturing function, but on the entire enterprise. Savage suggests that CIM-II is "integrative, not integrated" in that the process by which manufacturers are involved in CIM-II is a "continually evolving integrative process" (p. 60). The difference between integrative and integrated can be related to the situation of an open system that is imbedded in a turbulent environment. An integrated approach would attempt to coordinate across the continuously increasing differentiation resulting from matching the environment's variety in a kind of "brute force" way (e.g., by continuously adding coordinating mechanisms of the type listed in Table 6.2). An integrative approach would assume enterprise level system complexity a priori (e.g., as per the primary emphases in Table 6.3) and design commensurate integrative processes into the organizational arrangements. The technological and manufacturing rationalization necessary to develop and sustain such integrative organization has raised the level of sophistication in manufacturing management significantly.

Several examples illustrate the concept. In the early 1980s, Volvo contended that given the good experiences of its smaller plants, such as those in Ghent,

Belgium, and Kalmar, Sweden,[5] and in contrast to the performance at its large Torslanda plant that it had repeatedly attempted to improve, it was unlikely to build large plants again (Jonsson 1981). Nevertheless, in 1988, Volvo commenced the first stage of its new assembly plant in Uddevalla, which is destined to employ in excess of 1000 people. Volvo's realization, and that of other firms, is that large manufacturing plants can be built that incorporate the learning of the last few years. Small size can be experienced at the same time as scale is achieved. Sufficient parallelization of activities can be structured in that many variations of assembled products can coexist, relatively autonomously, and still enjoy the advantages of being part of a larger organization. In effect, manufacturing objectives, often viewed as trade-offs in the past, can be designed to be simultaneously achievable.

The notion of factories within factories (Lindholm 1975) and plants within plants (Skinner 1974) supports this integrative conceptualization. Hence, Cummins Engine Company has successfully maintained a sociotechnical or high commitment design at its Jamestown plant while maintaining its plant population at around 1000 and at the same time reducing its production costs by 35 percent (Beer, Spector, Lawrence, Mills, and Walton 1985, Klein 1989). GE Canada's sociotechnically designed airfoil blades and vanes plant has watched its size increase from the originally targeted 350 to 700. It refers to itself as a "paperless factory" linked through its information systems. Although General Motors' Saturn plant in Tennessee is reduced from its originally planned size, it still incorporates the philosophies of STS designs in a large plant designed to feel small to those working inside (Gwynne 1990).

The "structuring" of all these plants incorporates participative and continuous design processes throughout. They all are seeking the advantages of scale while at the same time maintaining the flexibility, autonomy, and human atmosphere of smaller units. More significantly, they realize that restructuring is needed to successfully integrate the contribution of people with advanced manufacturing systems. Haywood and Bessant (1990, p. 79) referring to such systems as C/HIM (computer/human integrated manufacturing) rather than CIM, stated:

> . . . that the objective in such systems moves from one in which labour is seen as a necessary evil and a cost item to be reduced or eliminated wherever possible, to one in which it is seen as being an important aid to keeping the utilisation of the system high, and thus to recovering its high capital costs.

Bomba (1989, p. 5) makes an almost identical point in referring to the implementation of FMS as requiring congruence between many things and stating that "rather than being a part of CIM, FMS might be more properly termed HIM: Human Integrated Manufacturing."

---

[5] For several years, Volvo's assembly plants were efficient in inverse proportion to their size (Jonsson 1981).

7
SOCIOTECHNICAL SYSTEMS AND PRODUCT-FOCUSED FORMS

In the late 1980s, the pressure for restructuring for more effective performance was strongest on the middle levels of the organization. This trend will continue throughout the 1990s. Computers and information technology have made the information needed to monitor and control activities available simultaneously to higher management, to supervision, and to individual workers. The information gathering and control functions that have been the preserve of middle management have become less necessary. At the same time, demands from employees at lower levels of the organization to have a say in decisions that affect them have met with a desire from top management to decentralize some of their responsibility and delegate some of their decisions downward. In this environment, hierarchical control behavior and thinking have become dated. The combined result has been the widespread collapse of vertical boundaries in the middle levels of the organization.

The pressure on the middle comes from the sides as well as from the top and bottom. The need to respond rapidly to market, customer, and technological changes has collapsed the boundaries between plants and suppliers, between plants and customers, and between production departments and engineering and design areas. Structural and process arrangements that facilitate lateral relationships and communications are increasingly important aspects of organizing (e.g., the different arrangements to bridge the boundary between design and manufacturing and the different techniques to install just-in-time (JIT) inventory controls).

There is a third source of pressure on the middle of manufacturing organizations and it comes from within. It results from the interdependencies between tightly coupled information technology-based manufacturing approaches that are linked to form integrated manufacturing systems (e.g., communications systems such as local area networks (LANs), materials handling technologies such as automated guided

142 SOCIOTECHNICAL SYSTEMS AND PRODUCT FOCUSED FORMS

vehicles (AGVs), and planning and control systems such as material requirements planning (MRP) and computer-aided process planning (CAPP)). There is a less tightly integrated variation of the foregoing in which manufacturing systems are combined with organizational coordination mechanisms (e.g., teams and coordinators) to link more loosely coupled islands of automation with each other and with the less automated functions on the factory floor.

These interdependencies plus the product-market and process uncertainties illustrated in Figure 6.1 have made it very difficult for the traditional center to hold in manufacturing departments. Dramatic changes are taking place in the way work is organized. The changing social systems and technical systems are creating a context for a new paradigm of work, referred to here as sociotechnical systems. The new paradigm is emerging at all levels of the organization, the plant level, the individual level, and the department or group or "middle" level, but it is at the last of these that it is experienced most strongly. The middle level—the group or department or unit level—is the focus of this chapter.

The chapter begins with an introduction to sociotechnical systems (STS) theory and the STS design principles and design processes that are the basis of the new paradigm of work. The major design principles are described and explained with examples drawn from AMT applications. Another alternative to the traditional scientific management approach that has dominated factory life for the last half-century, is introduced here as "engineered organizational design" (EOD). This approach is manifested in a variety of product focused forms of organization.

The last part of the chapter discusses many of the different EOD-based structural arrangements that comprise product-focused forms. They are structural arrangements widely used in AMT environments. They probably form the bridge between traditional scientific management approaches and the new paradigm of work that is evolving under the rubric of sociotechnical systems.

SOCIOTECHNICAL SYSTEMS THEORY

Technological determinism has been the basis of design of manufacturing organizations, particularly process technologies (Daft 1989). In a continuous process technology petrochemical plant, the process for refining crude oil into its by-products follows a specific chemical formulation that dictates the sequence of operations that will take place. The translation of these sequenced steps into a layout of technical equipment—pipes, valves, pumps and so on—follows a determined solution that is assumed to be the optimal way to carry out the production process. In discrete production, a similar logic applies. Even though the operations are not continuous, the "optimal" sequence for machining a part or assembling a component will follow a path that would look similar across a variety of factories.

A technically determined solution maps out the essential technical design of the production system. The work organization, the job designs, the organizational arrangements, the reward, and other support systems (i.e., the total social system) is then put in place to fit the technical solution and make up the total production system. This is referred to as technological determinism. It dictates that social systems will adapt to technical systems. As the building blocks of the social systems, people will be fitted in as necessary to complete the technical design. At times they may have machinelike roles, as when they are less costly than or more flexible than a machine alternative, whereas at other times they may serve as extensions of the machines.

Sociotechnical theorists advocate an alternative perspective. They suggest that organizations should be designed so that people are in roles complementary to machines rather than being extensions of machines (Jordan 1979). They maintain that there is always more than a single way to design the technical system and some of the alternative ways lead to not just better social system solutions but also result in more effective total organizational designs.

Emery and Trist (1960) developed STS theory from experiences in the British coalfields in the late 1940s. To many colliery managers in the United Kingdom, the solution to inadequate production output was to mechanize the mines by adding new technology as it became available (conveyor belts, better explosives, undercutting equipment) and by restructuring the work away from historically inefficient but cohesive work group arrangements toward scientific management methods that had arisen in the automobile and other industries in the United States. "Longwall" methods were devised that divided up the skills of the previously cross-trained miners into highly specialized job categories and supervisors were installed to coordinate their activities (Trist and Bamforth 1951). Output increased, but not anywhere near the proportions the equipment designers and engineers expected or maintained was possible from the new technology and work structures (Trist, Higgin, Murray, and Pollock 1963).

Managers, in some collieries, with assistance from miners and unions, created what Trist and Bamforth (1951) referred to as "composite" approaches. They utilized the new technology but maintained some of the better aspects of the previous work group structures. A large number of design variations were attempted in different mines. Team size, for instance, ranged from as few as eight people to as many as forty-two across three shifts. Several very effective variations resulted, both in terms of productivity and in the work satisfaction they provided to miners and managers. They became the basis of "the emergence of a new paradigm of work" (Emergy 1978b).

Trist (1981, p. 8) described his observations of these innovations at a new coal seam in a colliery he visited in South Yorkshire:

> The work organization of the new seam was, to us, a novel phenomenon consisting of relatively autonomous groups interchanging roles and shifts and regulating their affairs with a minimum of supervision. Cooperation between task groups was everywhere in evidence; personal commitment was obvious, absenteeism low, accidents infrequent, productivity high. The contrast was large between the atmosphere and arrangements on these faces and those in the conventional area of the pit, where negative features characteristic of the industry were glaringly apparent. The men told us that in order to adapt with best advantage to the technical conditions in the new seam, they had

evolved a form of work organization based on practices common in unmechanized days when small groups, who took responsibility for the entire cycle, had worked autonomously. These practices had disappeared as the pits became progressively more mechanized in relation to the introduction of longwall working. This had enlarged the scale of operations and led to aggregates of men of considerable size having their jobs broken down into-one-man-one-task roles, while coordination and control had been externalized in supervision, which had become coercive. Now they had found a way at a higher level of mechanization of recovering the group cohesion and self-regulation they had lost and of advancing their power to participate in decisions concerning their work arrangements.

The lesson of the composite systems was that it was not necessary to increasingly bureaucratize an organization each time that it adopted new technology. Successful organizational alternatives could be found. "Organizational choice" existed (Trist et al. 1963).

STS theory evolved from the research studies undertaken in the United Kingdom to other initiatives in India and Norway (Trist 1981, pp. 12–28).[1] However, most of the undertakings were isolated ones or restricted to selected "experiments" and a limited number of applications. The researchers evolved their concepts from both observations of managerial innovations in situations of changing technologies (Trist and Bamforth 1951) and from action research[2] to change organizational arrangements to develop more democracy in the workplace (Emery and Thorsrud 1976, Rice 1958).

Sociotechnical systems theory is established at three levels in society: the primary work system, whole organization systems and macrosocial systems (Trist 1981). Our interest in AMT environments will focus our attention to the primary work system level, particularly in factory environments where the concepts, the design principles, and the applications have been most fully developed.

Many variations on the theory, principles and design approaches of sociotechnical systems (Emery and Trist 1960, Trist 1981) exist, sometimes under quite different names (e.g., high commitment work systems (Davis 1983, Walton 1980, 1985), high involvement systems (Lawler 1986), and a new paradigm of work (Emery 1978b)). Another range of names refers to organization designs that are sometimes similar to STS and sometimes quite different (e.g., participative management, employee involvement, workplace democracy, high performance management, new human resource management, work restructuring, workplace reform). The term "quality of working life" (Davis and Cherns 1975, Kolodny and van Beinum 1983) is often used in conjunction with STS, sometimes to mean an outcome of an STS

---

[1] Some of the working papers of this period are reproduced in Emery (1978b) and the conceptual explanations are best expounded in Emery and Thorsrud (1976) and in a condensed way in Trist (1981). A trilogy of books are currently in production that will provide a comprehensive view of the antecedent developments and the major influences on the evolution of sociotechnical systems theory. The first of these has been released (Trist and Murray 1990).

[2] Action research describes a form of research in which interventions by the research team are studied to observe their impact and corrective actions taken according to the results of the observations.

designed organization, sometimes as an overarching field under which STS falls as a specific organizational design approach, and sometimes as a way of identifying an organizational process (Mansell and Rankin 1983). Finally, there are organizational arrangements that relate to only one or a few aspects of STS (e.g., self-managing work groups or multiskilling or skill-based pay systems or search processes). Unfortunately, some of these arrangements, particularly self-managing or self-directed work teams, are sometimes misconstrued to be self-sufficient changes in themselves rather than just components of a changed paradigm of work.

THE DESIGN OF SOCIOTECHNICAL SYSTEMS

Organizational theory informs organizational design and sometimes the concepts that underlie one or the other are difficult to disentangle. Various lists of characteristics (Emery 1978a), attributes (Davis 1977), principles (Cherns 1976; Trist 1981), and work-force strategies (Walton 1985) have been developed, many of which are redundant, but all of which contribute to a better understanding of STS and assist in transforming the theory into practice.[3]

A sociotechnical systems approach to the design of organizations is based on several broad themes, each of which is further elaborated as follows: (1) the organization is an open system in continuous interaction with its environment, (2) the design of organizations is based on joint optimization of the social and technical systems, and (3) there is choice in organizational design. The latter theme rejects the classical idea that there is a technological imperative that determines the design of work organizations.

The concept of organization as an open system has been adopted from biology (von Bertalanffy 1968) to recognize that organizations are complete undertakings with inputs, outputs, throughputs, and feedback mechanisms to monitor and adapt their performance. Unlike biological entities, organizations are capable of changing their state (Buckley 1967) in a relatively short amount of time. Organizations are embedded in an external environment that shapes the values of the members, particularly those of the dominant coalition (Thompson 1967) and may expose them to high degrees of uncertainty (Lawrence and Lorsch 1967) and turbulence (Emery and Trist 1965). They are collections of individuals formed into organizational units that can be viewed as "minisocieties" (Davis 1982) with highly differentiated roles, structures, rewards, and processes for interacting, developing and sustaining themselves. Organizations are purposeful (Ackoff and Emery 1972) and capable of

---

[3] Davis's (1982) list, which incorporates that of Cherns (1976), is probably the longest with twenty-one "principles," but within each many more attributes or subprinciples are subsumed (e.g., sixteen criteria are listed under the principle of "Human Values-Quality of Work Life"). Emery's 1978a criteria entitled "Characteristics of Sociotechnical Systems," Walton's (1980) extensive treatise on "Establishing and Maintaining High Commitment Work Systems," Trist's (1981) comprehensive paper entitled "The Evolution of Sociotechnical Systems," and Davis's (1982) article on "Organization Design" are the most detailed conceptual discussions of STS currently available.

following alternative paths (the principle of equifinality) to the achievement of their objectives.

Organizations are social inventions or socially constructed realities (Berger and Luckmann 1966). They are learning entities, sometimes learning to maintain the status quo, sometimes learning to change it continuously or drastically in the light of changing contexts, values and objectives. Open systems theory is a broad general theory, and STS theory, as an organizational theory, is a subset of open systems theory.

Joint optimization is a second basis of a sociotechnical systems approach. Joint optimization puts the social and technical systems on a more equal footing in the design process. It has always been accepted that the first step in the design of, say, a production process, is to lay out the logical flow of the technology to achieve the desired final product, as in the input-throughput-output sequence previously mentioned. Workers in the process are then added in as needed, usually when it is less expensive to add a worker than to add an additional technological component (e.g., installing an operator to monitor a machine because automatic monitoring is too expensive or technically not feasible). At this point, the technical system is optimized for efficiency, for the cost of the investment in technology, for the quality of the output, and so on. The workers and the social systems are seen as extensions of the technical system, as necessary appendages to make the system function properly.

Under joint optimization, there is an understanding that there may be trade-offs whereby giving up some efficiencies on the technical side to realize social system benefits may yield a better joint outcome. For example, physically arranging a manufacturing cell in a U-shaped sequence rather than placing the machines in a straight line may make it more technically difficult to lay out a tow chain to move programmable carts. However, the physical proximity of operators and loaders and setters to each other should allow them to cover for each other during absences, advise each other, learn from each other, and generally improve their social system functioning as well as the overall effectiveness of the cell.

Joint optimization may be an optimistic objective, since optimizing any system or subsystem, technical or social, is difficult. However, achieving a "good fit" between the technical and social subsystems is a feasible objective and within the belief system of STS design it will result in a more effective organization than an approach that optimizes only the technical or the social subsystem or an approach that makes one subsystem subordinate to the other.

Joint optimization has acquired new significance under computer-based technology. With the programmability and the flexibility responsiveness that is now being approached in the technology, it is conceivable to design to achieve a desired social system outcome (e.g., an organization structured around teams, without incurring extra cost for the technology or without sacrificing technical objectives (Walton 1983)). The number of alternative organizational arrangements is expanding (Kolodny 1990). The challenge for designers will be how to optimize two subsystems that both have flexibility in the range of alternatives they can assume.

Organizational choice is the cornerstone of STS design. An organization exposed to a changing and turbulent external environment must constantly adapt to fit the externally driven changes, amongst which will be found changing societal and managerial values. Those values inform choices about the direction and kind of adaption. Earlier organizational theory positions maintained that there was one best way of organizing, a perspective that is still widely held. The concept of organizational choice is in sharp contrast to that position. Even after value positions are established and organizational philosophies are identified, organizational designers still have choices about the technical and social subsystem parameters that will accomplish the desired objectives. The types of technology, its first and second order consequences (Walton 1983), the assumptions held about people's learning abilities, the degree of autonomy granted to operators and supervisors (Klein 1991), are among many decisions that can affect the sociotechnical design. Organizational choice is the guiding design philosophy. This is pursued further in the next chapter.

Sociotechnical systems has a well developed design methodology (Davis 1982, Emery and Thorsrud 1969, 1976, Taylor and Asadorian 1985) that has had most of its application in production areas where the work tends to flow in a linear sequence from operation to operation. Some of the tools and techniques of this design methodology (e.g., variance analysis, described shortly) are most applicable when the flow of the work is linear. Nevertheless, STS approaches have evolved from factory settings to office and white collar settings (Mumford and Weir 1979; Pava 1983; Ranney 1986) and, of late, to some of the less linear operations in the organization, such as research and development (Pasmore 1988) and computer operations (Taylor 1986).

The basis of STS design is described as follows and illustrated with current practice in AMT settings. The discussion is selective rather than comprehensive, with only the major design concepts described.

The Primary Work System

The unit of analysis for STS design is a bounded subsystem (Trist 1981) that transforms an input into an output. The boundaries are drawn where there is a natural break between the interdependencies of one subsystem and those of another. Although there will always be some interactions between subsystems, the internal interdependencies will be greater than those external to the subsystem. This subsystem is referred to as a *primary work system*; it is a sociotechnical system comprising a technical subsystem and a social subsystem.

Small and large primary work systems may exist side by side in the same plant or organization. For example, within a CAM environment, a manufacturing cell could have two to three members while another, perhaps because it was handling larger part families, might have four to five persons and require up to three times as much physical area. An adjacent FMS line could have an area twice as large again as the largest manufacturing cell and have seven to eight operators, loaders, and programmers associated with it. Each cell and the FMS line would constitute a primary work system. Alternatively, the boundary of a primary work system might be drawn to incorporate several cells if their interdependencies were high.

Cummins Engine Company designed and built a high commitment plant in Jamestown, N.Y. in 1974, where diesel engines were assembled and engine parts machined (Spector and Beer 1985). JIT and SPC systems were added to and integrated into the plant of 1000 employees in 1986. The plant was subdivided into businesses that were further subdivided into teams organized around engine parts machining lines and an engine assembly line. Each illustrates the concept of a primary work system in the descriptions that follow.

Engine parts were machined for the engines assembled in the Jamestown plant, for engines assembled in other Cummins plants and for the parts distribution network. Machining lines had automated materials handling and automated gauging linked to the machine tools. There were several flexible machining centers. Each parts machining area (e.g., flywheel, piston, cambox, crank, camshaft, etc.) was a primary work system dealing with a family of similar parts and with its social system organized as a team. Each team handled setup and maintenance in addition to operations. There were about twenty persons per machining team, though they occasionally expanded to fifty to eighty people over three shifts. However, the complications of holding meetings and coordinating between shifts led to poor communication and dissatisfaction on some of the larger teams. The cambox team, in response, subdivided its eighty plus team into three smaller teams (casting and drilling, assembly, variable machine timing and inventory control) (Spector and Beer 1985). Although each team could be considered a primary work system, the entire area across all shifts would more likely be the relevant unit of analysis. Hence, there is some discretion in choosing the unit of analysis/design.

In the Jamestown plant, engines were assembled on a nonpaced assembly line with computer-assisted assembly and test operations. Assembly is a sequential process of continually adding parts onto an engine block, but subtasks that could reasonably have bounds drawn around them were broken out (e.g., piston assembly, head assembly, manifold assembly). Seven teams of five to seven workers were associated with each of the subtasks. Quality was monitored at the output of each subtask (although operational tests had to wait until all assembly was complete). Each team ordered its own materials (within blanket purchase agreements negotiated by purchasing with suppliers), received them at the team location, checked them on receipt when necessary, and negotiated with the supplier when problems arose. Teams received daily reports on costs of their operations.

Buffers decoupled one team and its subtask from another, giving a team autonomy to meet and plan and coordinate its activities (Kolodny and Dresner 1986). With the buffering and the physical proximity and team specific data, one could argue that the subtasks were sufficiently autonomous to each constitute a primary work system. Yet the sequential dependence of each team on the previous team was sufficiently high that, under our definition, the entire engine assembly line would probably have constituted a primary work system. Again, there is some choice here as to what constitutes the basic unit of analysis or primary work system.[4]

---

[4] A subsequent JIT redesign removed the buffer areas as part of the process of removing excess in-process inventory (Klein 1989, 1991), but the subtasks and associated teams remained with approximately the same divisions.

Within manufacturing organizations, then, a particular production unit has generally been considered as a primary work system, subject to the discretion mentioned above in selecting the boundaries of the system. However, Susman and Chase (1986) have noted that CAM requires more "upstream" analysis than has been traditionally considered, with the result that integrated CAD/CAM activity or design-to-manufacture units (Dean and Susman 1989) may be entering the STS calculus of what constitutes a primary work system.

The Work Group

The work group associated with a primary work system constitutes the social subsystem part of the particular sociotechnical system. As an organizational entity, it is a sharp contrast with the single job and the superior-subordinate dyad that lies at the base of traditional organizational design. Rice (1958, p. 36) defined a work group as "the smallest number that can perform a "whole" task and can satisfy the social and psychological needs of its members." He viewed it as "the most satisfactory and efficient group" from the viewpoint of both the task to be performed and those performing it.

It was from the field-based development of STS in the U.K., India, and Norway (Emery and Thorsrud 1969, Rice 1958, Trist and Bamforth 1951) that the concept of "semiautonomous" work teams evolved (Gulowson 1979, Herbst 1962, Susman 1976).[5] In time, good descriptive examples of such teams were published. Trist (1981, p. 34) described the characteristics of such groups as follows:

> Autonomous groups are learning systems. As their capabilities increase, they extend their decision space. In production units they tend to absorb certain maintenance and control functions. They become able to set their own machines. The problem-solving capability increases on day-to-day issues. They negotiate for their special needs with their supply and user departments. As time goes on, more of their members acquire more of the relevant skills. Yet most such groups allow a considerable range of preferences as regards multi-skilling and job interchange. The less venturesome and more modestly endowed can find suitable niches. The overall gain in flexibility can become very considerable, and this can be used to enhance performance and also to accommodate personal needs as regards time off, shifts, vacations, etc.

The size and tasks of a primary work system dictate the size of the organizational unit or work group. Wide variations in size can be expected. Furthermore, assumptions about the roles of team members, their training and cross-training, the levels of multiskilling sought, and the role of the team supervisor can all moderate the size of the team. A highly automated machining cell can have a two-person team—an operator and a loader—with occasional support from maintenance, setup, and

---

[5] Semiautonomous work teams have sufficient autonomy to regulate their own affairs while remaining connected to the organization, sometimes through a supervisor, sometimes through planning and objective setting systems, and always through the interdependencies they have with other units around the task itself so that they are never completely autonomous. This is discussed in detail in the next section around self-regulation of work groups.

tooling, whereas an alternative configuration for the same cell might include the tasks of minor maintenance and setup within the skill set of the cell team and may or may not require an additional cell team member. A hybrid cell of about the same size, comprising conventional as well as CNC machines and more manual than automated parts transfer and loading, might easily have a team three times the size.

In smaller plants, even with several distinct production lines, all the members of a single shift often comprise a single team. For example, Crown Cork and Seal's aluminum can production plant in Toronto is a sociotechnically designed plant with several production lines operating in parallel. However, the plant is automated enough and small enough that all the operators on one shift, approximately ten on the day shift, comprise a single work team. Plant designers have choices about where the boundaries will be drawn, although there are design principles that inform those choices (see following).

Self-Regulation of Work Groups

A fundamental belief in STS design is that people are capable of regulating themselves, that they do not have to be externally controlled by supervisors or control mechanisms to accomplish their tasks or to do good work. Work groups in STS designs have been referred to as semiautonomous or self-regulating to reflect the responsibilities they assumed for various aspects of their own performance. The term "semiautonomous" has always been a cumbersome and uncomfortable one—perhaps because it made the issue of "How much autonomy?" all too obvious (Klein 1991). Alternative names evolved: self-regulating work groups, self-managing or self-directed work teams, and self-maintaining organizational units (Davis 1982).

Susman (1976) developed three categories of decisions for determining autonomy: task interdependence, self-governance, and self-regulation. (Autonomy as a design variable at the level of the individual is discussed in Chapter 8.) Other variations on these basic formulations of autonomy have been developed (Cummings 1978, Cummings and Molloy 1977, Wall et al. 1986). For instance, Gulowson (1979) proposed seven criteria of autonomy for a work group from a study of eight different applications (Table 7.1). Gulowson ordered them to reflect a Guttman scale of increasing autonomy (i.e., each level of autonomy achieved included the previous levels).

At the Cummins Jamestown plant, each of the seven engine assembly teams described earlier functions as a self-regulating work team. Each team member has both a horizontal and a vertical task. Horizontal tasks are directly involved with the team's primary work system (e.g., assembling the pistons into the engine block) and team members may choose to allocate and accomplish the work in whatever way they choose, as long as their choice does not impede the work of adjacent teams. Vertical tasks are the kind of support tasks normally carried out by a supervisor. The Cummins teams still have supervisors (at one time called team advisers and now called team managers), but their activity is more directed at managing the boundary between their team and other parts of the organization, including other

TABLE 7.1 Criteria of Autonomy

1. The group can influence the formulation of its goals, including (a) qualitative aspects and (b) quantitative aspects.
2. Provided that established goals governing relationships to the superordinate system are satisfied, the group can govern its own performance in the following way.
   (a) The group can decide where to work.
   (b) The group can decide when to work.
   (c) The group can decide which other activities it wishes to engage in.
3. The group makes the necessary decisions in connection with the choice of the production method.
4. The group makes its own internal distribution of tasks.
5. The group decides on its own membership.
6. The group makes its own decision with respect to two crucial matters of leadership:
   (a) The group decides whether it wants to have a leader with respect to internal questions, and—if it does—who this leader shall be.
   (b) The group decides whether it wants a leader for the purpose of regulating boundary conditions, and—if it does—who this leader shall be.
7. The group members decide how the work operations shall be performed.

Source: Davis and Taylor (1979).

engine assembly teams. One team adviser of a machining group listed his team's vertical tasks as:

> ... vacation scheduling, tracking illness, budgeting and forecasting, quality planning, preventive maintenance, material management, participating in the Board of Representatives, and affirmative action task force. People get these vertical tasks assigned first by volunteering and then by a lottery or by alphabetical order. People perform one vertical task usually for one year, but for those of budgeting and forecasting and preventive maintenance, people perform them for two years.[6]

Wall, Kemp, Jackson, and Clegg (1986, p. 283) conducted a three-year study of autonomous work groups at both a "greenfield" or new plant site and an existing site of a British confectionary manufacturing company. They found that production employees worked in groups of eight to twelve people, all of whom were expected to carry out each of eight types of jobs involved in the production process. Their list of team member duties included collective responsibility for allocating jobs among themselves, reaching production targets and meeting quality and hygiene standards, solving local production problems, recording production data for information systems, organizing breaks, ordering and collecting raw materials and delivering finished goods to stores, calling for engineering support, training new recruits, and participating in selecting new employees.

---

[6] From an interview conducted in 1984. The Board of Representatives had twenty elected representatives from the business and functional areas throughout the plant. It served as a forum for new ideas, complaints, and issues of concern such as compensation.

The autonomy of teams in STS designs is highly conditioned by the supervisor's role with respect to the team, with design alternatives ranging from traditional supervisory roles to no supervisors at all and with a wide range of in-between positions (team advisers, coordinators, team leaders, team managers, supervisors). The historical conflict felt by the supervisor as the "man in the middle" has not been better addressed in STS designs than in traditional organizations, with the exception that some STS designs have resolved the issue by totally eliminating the role. The spread of interest in employee involvement has heightened the concern with the supervisory role, a phenomenon that seems to reappear at intervals in the literature of organizations, probably because it has never been successfully addressed or resolved at any of the prior periods of concern (Rosow and Zager 1989). The changing roles of supervisors in the settings that are the specific focus of this book—AMT environments—are discussed in Chapter 8.

Teams in STS designs can have a wide variety of structural configurations with issues such as the degree of autonomy, the amount of multiskilling (discussed as follows), the assumptions about supervision, and the characteristics of the task itself serving as key design variables.

STS Design Principles and Processes

The theoretical underpinnings of sociotechnical systems and the design concepts that inform organizational arrangements based on this alternative paradigm of work and organizations were described and illustrated above. In the sections that follow, specific *design principles* to guide organizational designers are described. They are significantly different to the work and organization design principles that inform traditional scientific management (Davis, Canter and Hoffman, 1972).

The *design processes* used within this alternative paradigm are also significantly different. They, too, are based on the theory and concepts identified above but also on processes of participation that include those affected by the design decisions. This is in contrast to the traditional design process in which "expert" designers determine the organizational arrangements on behalf of the organization's employees.

Design Principles. Cherns (1976) developed a list of several principles of organization design based on STS theory. The discussion that follows uses AMT applications to illustrate Cherns' principles.

Boundary Management and Variance Control. Controlling key variances as close to the source as possible is one of the basic design principles of STS. Variances are deviations from goals or objectives. They require corrective action. They are the exceptions, out-of-spec conditions, errors, and many other ways in which things can go wrong at any point in an operation or activity. They generally reflect the uncertainty associated with process operations, but they can also result from unpredictable problems with the product itself (e.g., in performance criteria).

If variances are controlled where they arise, quality can be considerably improved because deviations are not transmitted beyond the boundaries of the unit in question.

Workers who create the variances or who are the first to encounter them should correct them. The high costs of having supervisors coordinate corrective action because the variances are detected in places where they do not originate are eliminated or considerably reduced. Organizational units or primary work systems should be designed so that those closest to where the variances arise have the necessary information, skills, knowledge, authority, and commitment to detect them and control them. The boundaries of a primary work system should be drawn to minimize the export of variances across that system's boundary. This complements the earlier stated concept of drawing a primary work system's boundaries where there is a natural break between the interdependencies of one system and those of another or, restated, drawing the boundary to contain the system's interdependencies to the extent possible.

For example, at a Westinghouse switchgear control assembly plant, in preparation for redesigning the plant to reduce cycle time, groups of workers were asked to describe in minute detail the process of fabricating and assembling the control panels manufactured in their plant and to lay out the flow of the activities on paper (Kirby 1985). From this assessment, the workers and consultants then determined where the initial boundaries had to be drawn to divide the work into primary work systems (Taylor and Asadorian 1985). Then the workers were asked to identify in detail everything that could go wrong in the processes of production (i.e., the variances). From these they selected the key variances, and from the combined analysis they were able to reconfigure the different primary work systems so each would be capable of controlling variances within its own bounded unit.

Wall et al. (1986) conducted the first empirical test of this principle in a CNC component insertion machine assembly area. Two different organizational design arrangements were installed: one in which the CNC machine operators controlled and repaired machine breakdowns, when they were able to, and one in which they deferred to specialists to correct breakdowns when they occurred. The experiment demonstrated unequivocally that control of variances closest to where they originated, in this case by the CNC operators, was the most effective way to minimize machine downtime. (This study is described in detail in Chapter 8.)

Controlling variances close to their source of origin usually requires the judicious selection of boundaries in the design of the primary work systems. The same boundaries are the ones that determine the team arrangements and team size, which reinforces the systemic nature of STS designs. However, the systemic nature cuts both ways. When Cummins Engines removed the buffers between its assembly teams, it also reduced the autonomy that the teams had previously enjoyed. Slack was removed and, with it, the pace and the stress in the situation quickened (Klein 1989). New ways of working were required because, essentially, alternative design philosophies had changed the characteristics of the boundaries of the different primary work systems (Klein 1991).

Organizational Philosophy and Values. STS designs are based on values that are explicit and often reflected in organizational philosophy statements or charters (Poza 1983). Every organizational design has inherent assumptions about people.

More often than not they have not been made explicit for the members of the organization. STS designs, in recognition of the fact that their open systems character makes them a product of their environment, usually have embedded into their design process a process of exploration of the changing values of their environment and of its impact on their organization and members. This "search process" (Clarkson 1981, Emery and Emery 1973, Weisbord 1987) confronts the organization with current environmental trends and future directions that are then compared with current and sometimes historical practice. The identified "gap" then serves as impetus to address the question of whether current practice will allow the organization to proceed effectively toward the mapped out future or whether changes are in order. This process exposes the organization to the value positions of the key decision makers and to the compatibility of those values with environmental trends as well as with the environmentally influenced and changing positions of the organization's members.

Cummins Engine, for example, espoused a "factory culture" based on growth, trust, equity, and excellence (Klein 1989). At GE Canada's Bromont plant, the philosophy statement is more detailed, and the subheadings illustrate its thrust: Mission, Participation, Personal Development, Social Climate and Structure of the Plant, Justice and Dignity, Family and Community Involvement, Organizational Renewal (Arnopoulos 1985).

Compatibility. This principle means that the design outcome and the design process used to achieve the outcome should be compatible. It is difficult to achieve a participative outcome with a nonparticipative design process ("How you start is how you finish"). Many plants interested in engaging in innovative redesign now take teams of employees on visits to other innovative plants and to conferences describing such innovations to change their expectations and expand their conceptual understanding about what is possible both in structural arrangements and in the processes of how they will operate. After extensive education, these "design teams" take a key role in the design of a new work structure capable of building in the very kind of participative behavior in which the teams were engaged during their educational process.

Multiskilling. Multiskilling is intended to develop more functional competencies in a person than they might utilize at any one time. This allows them to adapt to uncertainties and unanticipated changes. Within a team of multiskilled members, a high degree of team adaptability can be realized. (Multiskilling which focuses on the role of the individual in AMT environments, particularly the changing role of machine operators in the direction of becoming more multiskilled, is discussed in Chapter 8.)

Three illustrations from three different STS designed plants illustrate the range that multiskilling can cover. In one Procter and Gamble plant, line operators also take turns as forklift truck drivers transporting the team's consumer products to the warehouse, whereas in the warehouse, a second skill for the warehouse workers is accounting, as it is their responsibility to maintain the cost reporting systems for the plant. The latter role also serves to increase the status of their positions and reduce perceived inequalities in worker roles. In a Shell Canada polypropylene plant, process operators on teams all learn second skills as instrument or laboratory technicians (Halpern 1984). In a Westinghouse switchgear control panel fabricating plant, team members acquire administrative and group process skills to complement their operating skills (Kirby 1985).

In most AMT environments, the use of multiskilled workers is the dominant organizational approach to achieving flexibility. In a survey of users of cellular manufacturing in the United States, Wemmerlov and Hyer (1989) found that of the thirty-one companies with manned cells, twenty-seven or 87 percent had multifunctional or multiskilled operators. Furthermore, "The extent of intracell mobility was fairly extensive. Using a 5 point scale, where '1' represents 'very little operator movement and '5' represents 'a great deal of operator movement' within a given cell, the average response was 3.6" (Wemmerlov and Hyer 1989, p. 1522). The researchers also found that 39 percent of the companies claimed to move operators between different cells (with an average extent of movement of 2.5 on the same five-point scale) in order to achieve more flexibility in the manufacturing system. The result is interesting in that such movement would complicate the creation of self-regulating work teams.

Multiskilling is usually supported by a skill-based pay system or a pay for knowledge system (Gupta, Jenkins, and Curington 1986, Lawler and Ledford 1985) that rewards people for acquiring more skills than they might use at any one time. Skill acquisition under these systems consists of several types: horizontal skills have ladders of increasing knowledge and skill associated with a person's primary specialization; secondary skills can be support skills that contribute to the work groups' task in its area of responsibility; and administrative or vertical or coordinating skills are those associated with the day-to-day management of the team, usually roles assumed by most team members.

Minimal Critical Specifications and the Principle of Incompletion. No more design detail should be specified than is necessary so as not to constrain subsequent changes and to provide those involved in the design with "space" to take ownership for the aspects of the design they are able to influence or affect. This principle is in direct contrast to a traditional engineering one that aims to specify every possible details in advance.

STS designs are never complete; they evolve as the environment changes and as members change as a result of multiskilling and team membership. If STS oganizations are learning systems, then freedom to incorporate learning must be allowed. Klein (1991) studied several manufacturing plants that changed their manufacturing systems to incorporate JIT. When the plants implemented JIT and participative or team-based management at the same time, implementation was relatively trouble free. However, in the Cummins Jamestown plant, where a high commitment environment had existed before JIT was implemented, the tightly coupled nature of JIT operations constrained the autonomy the teams had previously enjoyed and implementation was very stressful. Although the principle of minimal critical specifications would appear to ease the change process from a more authoritarian system towards a more participative one, it is not evident that it is as helpful moving from a more autonomous organization to a more controlled one.

Support Congruence. The support systems of STS designs should be compatible with the design objectives of the different features of the design. For instance, if people are to be granted local autonomy, they must be given the information and the authority to make local decisions. If equity and open communications is a desired attribute of the organization, then status differentials, which normally get in the way of achieving these ends, must be lowered. If work is organized by way of teams, some component of the reward system should recognize team achievements to reinforce such arrangements.

Design Processes. Within STS approaches, one particular design process has achieved a measure of commonality bordering on standardization—a tiered approach with a steering committee at the top that gives sanction, support, direction, protection and resources to a design committee (or several design committees), considers the recommendations that emanate from that committee (or those committees), and monitors the design process to ensure consistency with corporate goals and mission statements (Davis 1982; Sherwood 1988). The design committee, in turn, identifies particular needs and opportunities within the design process, delegates specific aspects to different technical study teams, usually composed of people working in the particular area where the activity is centered, or, if the function being studied is broader based (e.g., a skill-based pay system), drawing its membership more widely, monitors the progress of the study teams and coordinates their activities. The study teams carry out organizational analyses consistent with the organization's philosophy, do cost/benefit/risk/feasibility analyses, develop implementation strategies, and participate in communications to the total organization.[7]

When the design or redesign undertaking is a substantial one (e.g., the redesign of a large factory), the entire design process just described may be replicated within each department or area of the factory (Davis 1982 p. 2.1.19). The overall factory redesign committee may create a temporary design process that comprises a steering committee, design team, technical subteams, consultants, and process managers within each department or area. It is not unusual to find several different consultants, both external and internal, working within the same plant at the same time but on different redesign processes within different departments or areas of the plant. The leadership of some of these temporary design processes may be the very people tapped to become the managers of redesigned departments or areas.

The composition of the steering committee is usually a balance between a small number of people with authority to provide the sanction[8] and support and resources required (e.g., top level managers and union officials) and a larger body of people that by their composition would take into consideration the subject of diffusion (i.e., members would be drawn from adjacent organizational units to which the

---

[7] From presentation materials by N. Halpern at the University of Toronto's workshops on Organization Design for a Changing Workplace (Toronto, November 1990).

[8] The "sanctioning" process has been brought to a high level of prescription and sophistication in the many STS designs within the General Motors Corporation. This was described by William Duffy at a University of Toronto workshop on Advanced STS Design (Atlanta, March 1991).

changes might subsequently diffuse or which are important support for the changing organization's ability to sustain its innovative activity). Design committees tend to be structured as a vertical slice of people with wide representation and legitimacy in the eyes of the organization's members.

Early STS implementation concentrated on the shop floor work organization areas and these areas continue to be the dominant focus of such change activity. Design issues have become more dominant in recent years as organizational issues have moved up and out from the shop floor units to encompass managerial and support areas and customers and suppliers. More recently, the focus of attention has shifted to top management and the areas of "legitimization" for innovative work restructuring (Kolodny and Stjernberg 1986). In practice, then, the change process has evolved from the bottom of the organization up toward the top levels.

Despite this, the design process that has been described (steering/design/study committees) is top down. It is so because it has evolved from the experience of early organizational innovators who struggled, often unsuccessfully, to keep their innovations alive without the kind of top-level support that steering committees provide. The diagonal slice basis for membership on the design committee and the composition of the study teams are intended as mechanisms to incorporate the widespread participation of employees in the design process.

Davis (1982 p. 2.1.21) has identified four phases of a "comprehensive, integrated organization design process" as follows:

TABLE 7.2 A Comprehensive, Integrated Organization Design Process

Development of preliminary data
    Assay of organization
    Scan of locality, labor market
    Community analysis

Generation of criteria for design decisions
    Analysis of organization environment set
    Development of organization philosophy

Integrated design of organization
    Technical variance analysis
    Technical boundaries of subunits
    Design of self-maintaining teams
    Preliminary organization design
    Sociotechnical design of organization including job design
    Design of social support systems

Implementation of organization design
    Organization design
    Design of transitional organization
    Implementation
    Evaluation
    Redesign

Source: Adapted from Davis (1982).

The design process described above has organization wide applicability. In practice, however, the unit under design has tended to be a manufacturing organization or a contained manufacturing plant. Most applications have been in "greenfield" or new designs (Arnopoulos 1985, Halpern 1984) but "brownfields" or redesign situations have also made use of the process (Kirby, 1985). While the design process is best sustained through explicit changes to the culture and value basis of the organization, often the change philosophy is a less ambitious one in which pilot projects are undertaken or parallel structures are developed (Bushe and Shani 1991). The danger of each of the latter is that they require no substantial change in orientation from top management and, as such, are difficult to sustain and diffuse when they encounter difficulties because the commitment of management is often suspect when their own investment has been low.[9]

The design process within the language of STS approaches refers to steering committees and design committees that are based on the involvement and widespread participation of the organization's members from workers to top management. The design process within the language of AMT is a narrower one and a lower-level one because it engages in a dialogue between the equipment designers and the equipment users (Corbett 1990). However, it is based on similar philosophies of involving those most affected by the decisions made and a recognition that human considerations have as significant a role to play in the design of AMT as social systems have in the design of STS work organizations.

Human-centered design criteria and philosophies directed at end-users are the basis of such design. "The basic principles are that people should not be subordinated to machines, and that the human skill, ingenuity, flexibility, creativity, and knowledge that comes from experience should be appreciated, and enhanced (rather than suppressed) in the technological system" (Symon 1990 p. 223). Symon also notes that the criteria must also consider the endusers as the "owners" of the technology, as opposed to the technologists, and requires educating potential direct and indirect users about the technology and informing them about its design and implementation, as they progress. She suggests that human-centered working practices entail a management approach in which local knowledge is assumed to be more accurate than programmed knowledge.[10]

The design process advocated by Symon for AMT design is similar to the STS design process described in that interdisciplinary design teams are proposed, teams that include both technical and organizational perspectives are advocated, and the involvement of users is mandated. The language of "user involvement" often replaces that of "worker participation" because the users of the equipment are usually or-

---

[9] From observations by Carl Bramlette at a University of Toronto workshop on Advanced STS Design (Atlanta, March 1991). He advocates "vertical slice" structures to ensure higher involvement and commitment from the top and has expressed reservations about the steering/design/study committee structure because it tends to create its own parallel hierarchy within the organization.

[10] Symon's comments invite a revisiting of Noble's (1979) arguments about the misdirection management took when it opted for programmed knowledge in place of playback methods because the former retained management control, whereas the latter placed control in the workers' hands.

ganizationally distinct from the designers. However, some organizational designers have long advocated processes that integrate user and worker perspectives (Mumford and Weir 1979).

The practical manifestation of the interface between the designer and the user is often the on-screen representation of the work being performed because, in many cases, the operator may have no direct access to the manufactured object. "It is the interface that defines the way the users can handle a problem and the alternatives they have" (Havn 1990 p. 231). This approach places the equipment designer in a dialogue with an unknown operator who is organizationally located within a system that can range from being traditionally bureaucratic to one based on a sociotechnical or high commitment work culture. The assumption increasingly in the mind of the designer must be that operators can interact with the production process in a meaningful way, that they have control over their daily work and that there are "opportunities to exercise their engineering skills and (tacit) knowledge" (Symon 1990 p. 226). Symon described one such experimental situation that was designed to provide graded levels of interaction from fully automatic to fully manual.[11] In the fully automatic case, the operator could allow the system to suggest appropriate speeds, feeds, tool heads, and cuts to give the most efficient performance and automatically generate the necessary parts program. Alternatively, the operator could override the system by choosing other tool heads or cuts than those recommended by the system according to his or her knowledge and experience (Symon 1990 p. 226).

STS Critiques

Pasmore and Tolchinsky (1989) have noted that STS design approaches have tended to focus within a single unit or department such as manufacturing, often overlooking the interdependencies with other units such as marketing, research and development, and design engineering. This latter critique is especially crucial for the design-to-manufacture boundary and its increasing importance for CAM environments. STS methodology developed within the confines of relatively linear production processes, and although nonlinear alternatives have been proposed (Pava 1983), particularly for white-collar areas, success with these alternative methodologies has been limited. Pasmore and Tolchinsky (1989) also critique the reluctance of STS designers to ask whether the product itself, not just the organization, might not merit redesign. They also note that STS processes have not taken up the call of continuous improvement that marks so much of current quality management approaches.

Hirschhorn has noted another problem with STS designs based on work teams (in Adler and Howard 1990). The assumption of the work team is that member roles are similar, so the cross-training and multiskilling that provides the team with its flexibility is possible. Hirschhorn notes that technological change forces more

---

[11] The Kraus Manufacturing Company in West Germany offers a production version of such a machine tool controller that, although 10 percent more expensive than a comparable conventional control, has gained a high level of acceptance among small and medium sized manufacturers (as reported by Dr. T. Martin at a conference at Monash University, Melbourne, Australia, March 1, 1991).

specialized roles on people, roles that STS is not designed to incorporate easily. "As we get more and more technical depth, people will occupy more and more dissimilar roles" (Adler and Howard 1990 p. 12). This works against the creation of the more holistic organization designs espoused by STS principles. Symon (1990 p. 228) cited such experience with respect to working across disciplines: "The system designers and the social scientists could not understand each other's technical language and expressions, and neither could appreciate the practical problems faced by the engineers in the user site."

STS has faced criticism because it is difficult to understand and is couched in jargon (e.g., semiautonomous work groups, joint optimization, variance control). The design processes are slow because they are broad based and attempt to involve as many people as possible to gain their commitment. This problem was also experienced by Symon (1990) and compounded by the realization that the design process is an iterative one and if empirical and local knowledge were to have primacy (i.e., in "learning by doing"), then patience was required to let the consultative process succeed despite the additional time required. In the turbulence of a rapidly changing external environment, some argue that there is not sufficient time to wait for commitment to build.

STS theory is strongly driven by explicit values positions that translate into an ideology that some find unacceptable. The values drive the theory toward democratization, and this generates resistance from both managers committed to hierarchical models of control and many traditional union leaders who have been elected within representative systems that would be challenged by advocacy of direct participation. For example, some team-based designs seek representation from each team in organization wide councils and committees, which flies in the face of a parallel representation system that exists through the union hierarchy but is not necessarily distributed according to the STS or team-based structure. Similarly, there is resistance in the managerial ranks because the "parallel" nature of the STS design process (e.g., diagonal slices through the organization) is a challenge to middle managers' hard-won traditional authority, but this resistance tends more often to be covert and difficult to identify because overt resistance is eliminated by "delayering" the middle ranks of managers and supervisors.

Another critique of STS theory is that it accepts technology as it is (Maton 1988), even facilitates its application, rather than criticize its deskilling potential, ". . . the STS priority is mainly towards adjustments by people to technology rather than the opposite" (Maton 1988 p. 880). As such it creates a division between the "haves" (those with technological skills) and the "have nots" (those unable to use the technology).

The strongest conceptual critique of STS was advanced by Kelly (1978) in a careful synthesis of the early conceptual and empirical work of the principal STS researchers. Kelly raised many issues that troubled him about the development of STS theory, several of which follow. He detected two trends in the theory that he felt were important to differentiate: one that focused on the group and down played the role of individual job design and a second that concentrated more on job design and individual job characteristics. He questioned whether joint optimization had been observed by the researchers or whether they had ot seen an "intensificiation of labor" on the social side that contributed more to the joint outcomes than any change in the technology. Despite the advocacy of organizational choice, so many STS designs ended up with group arrangements that Kelly wondered whether STS did not really advocate group design as its own "one best way" of designing work organizations. He also suggested that STS theory, rather than transcend the limitations of one person/one job designs, had not rather identified the limiting conditions of product and process uncertainty under which one person/one job arrangements were less effective than group arrangements.

Kelly's critique is written in the context of someone who was supportive of the paradigm but who also asked enough penetrating questions about the basis of the conceptualizations that, at a time when the STS paradigm of work is confronting many new domains that must be addressed (e.g., nonlinear applications, applications in the professional ranks, linking strategic and top management concerns to shop floor innovations, incorporating new manufacturing technologies and methods such as total quality management into the STS philosophy), his comments merit revisiting.

ALTERNATIVE FACTORY ORGANIZATION DESIGNS

Scientific management theory, as originally articulated by F. W. Taylor (1911) and as practiced in most of the organizations of the Western world over the last fifty years has informed and dominated modern organization designs. Other "paradigms" of work have arisen of late that have had significant impact on the design of manufacturing organizations. These alternative approaches to the design of factory work are compared and contrasted in this section and some of their implications for AMT are drawn out.

As the stability of their external environments disappeared, managers found that the closed system logic of scientific management—the division of labor, hierarchical control and work specialization—had ignored too many other variables that had become important in their rapidly changing situations, in particular, technology and environmental uncertainty. Increasingly, they realized that the structures of their organizations were *contingent* on variables such as the kind of technology they used and the uncertainty, variety and complexity in their environments and less on the internal organizational variables they had used to rationalize the design of their organizations. The theory that explained this perspective is referred to as contingency theory (Lawrence and Lorsch 1967). Contingency theories were a challenge to the orthodoxy of scientific management's internally driven logic because environmental uncertainty and technology were viewed as variables external to the organization. Within the manufacturing parts of the organization, technology was considered by contingency theorists as the dominant explanation for the social structure of the organization (Thompson 1967, Woodward 1965). By the 1960s, contingency theories of organization design had taken their place alongside scientific management theories and became alternative explanations of societal organizational structures.

During the same time, sociotechnical systems theory also arose to challenge the traditional "one best way" of organizing. Although STS theorists accepted the dominance of technology as a primary design variable, they questioned whether technology had to be viewed as deterministically as it had always been. They made the strong argument that there were always alternative technological arrangements that, when combined with appropriate social system designs, could yield joint sociotechnical designs that would be more effective than those that were driven by technological considerations alone.

The STS innovations were at the factory floor level, but the organization designers recognized that the social system could be more influential if the value system that drove the social structure was directly confronted. To do so, STS concepts addressed the design of the total work process.

Engineered Organizational Design

The STS innovations that were undertaken in a small number of sites in Norway were hardly taken up in the rest of the country. In Sweden, however, they diffused quite widely in a range of organizational settings (Sandberg 1982). Most were fabrication and assembly areas. As with contingency theory and perhaps because they were oriented towards the manufacturing floor, the Swedish work innovations were also dominated by the role of technology. The designers of the organizations, however, sought explicitly to increase the satisfaction of the workers by modifying the technical systems to create more desirable structural arrangements for workers. It had become increasingly difficult to attract people to shop floor jobs and it was believed that better working arrangements, particularly work group configurations, would overcome some of the negative attitudes towards factory work. Hence the Swedish stream of work innovations concentrated primarily on the design of the factory and, sometimes, in white collar settings with linear work flows.

This Swedish variation of STS design attended to the technical and social systems more in a parallel fashion than in an integrated way. As a design approach, it was very conscious of the impact of the technical system on the work organization and technical systems were deliberately designed to take advantage of that understanding. However, the approach also tended to underemphasize how the social system might in turn influence the technical system design. Industrial engineers and organizational designers viewed the psychology and sociology of the workplaces as engineering problems to be solved through traditional engineering rationality, sometimes in concert with social scientists, but almost never in consultation with those affected by the design decisions. Although there were noble exceptions, such as Volvo's Kalmar plant (Aguren, Hansson, and Karlsson 1976; Gyllenhammer 1977), for the most part the Swedish approaches failed to confront the value systems underlying the organization of work precisely because so much of the design was initiated in industrial engineering departments.

Jenkins (1982), referred to this approach, which he suggested as being German in origin, as *engineered organizational design*. The majority of the Swedish work innovations could also be so classified. Jenkins' description of engineered organizational design follows, though he gives more credit for the possibility of initiatives in the social system than observation would uphold (Kolodny 1985).

> This approach . . . aims to shape the social (and psychological) system through the physical design of the technology. In general, the psychological guidelines are the same as those included in the (sociotechnical) theories, and to an extent are built into the technology. For example, certain widely applied principles, e.g., group work, a high degree of employee control over operations, and task variety, may be introduced through supervisory practices, but they may also be tied into the technology in such a way that the system cannot be operated without following the principles. . . . The thinking is similar to that of the sociotechnical systems theories, but taken a step further. Instead of looking at an organization consisting of social and technical systems, we think of a single unified system incorporating two highly independent and interactive aspects. A particular organizational objective may be attained by changing one or the other (p. 11).

For scientific management, the organizational emphasis was the control structure. For the Swedish stream, referred to here as an "engineered organizational design" (EOD) approach, the emphasis was on the work organizational arrangements at the bottom of the organization. For the STS approach, the emphasis was on the total organization design process, including its value basis. EOD, then, is an intermediate step somewhere between scientific management and the STS approach described here. EOD approaches diffused widely in Sweden. In contrast, STS approaches, which were initiated primarily in North America, tended to be relatively isolated "experiments," particularly during the period when the Swedish work innovations were arising in large numbers.

The forces that drove American companies to adopt focused factories (described shortly) drove Swedish ones as well. However, unlike the Americans, Swedish volumes were not large enough to dedicate factories to single product lines. Early on, Swedes had to learn to organize focused factories to economically handle several product lines, with the result that they adopted flexible focussed forms or product shops widely throughout the country. Much of this was well documented in a series of publications produced by the former Technical Department of the Swedish Employers' Confederation.[12] Unlike the ways in which STS and high commitment evolved in North America, much of Swedish work innovation was driven by industrial engineers and has a clear engineering perspective. There has been criticism that too much of it was also driven by social scientists (Sandberg 1982), which may also be valid. The point of raising this is to note that there has been a qualitative difference in Swedish work innovation to that of North America. That difference has mostly revolved around a few issues: participation and direct involvement in workplace decisions, participation in design processes, and the place of values and corporate culture in the organizational change processes.

---

[12] There are many reports from the Technical Department of the Swedish Employers' Confederation (e.g., Norstedt and Aguren 1973); however, most of the conclusions have been incorporated in two summary books by Lindholm (1975) and Aguren and Edgren (1980).

The codetermination agreement of 1977 developed framework legislation that ensured trade union representation on the boards of companies and opened up their books to their unions. However, more often than not, it also served as an upper limit on how much cooperation would take place. Representative involvement in Sweden was widespread because it was so legislated. For the most part, however, it stopped at the legislated involvement of the union representatives. It did not directly involve the workers on the floor (Haas 1983, Rubenowitz, Norggren, and Tannenbaum 1983).

Sweden, despite the highest unionization percentage in the world and a strong history of social democratic governments, does not have any more of a tradition of democratic participation in its workplaces than do any other countries. A debate took place in the mid 1970s in Sweden within the trade union movement as to whether the union's orientation should be toward shopfloor democratization and participation or for legislation to guarantee representation. The latter won out.

In essence, the forces that drove Sweden's work innovations abandoned the technological determinism that continues to drive so much North American new technology, but they were also different to the sociotechnical systems that have developed in North America. A word of caution is in order here. The North American examples, while many in number, remain as special cases. They still do not represent practice in the mainstream. Furthermore, recent trends in Sweden (Gustavsen and Engelstad 1980) have increased the emphasis on direct participation such that the gap between EOD and STS approaches is narrowing.

The International Motor Vehicle Program

Information about Swedish experiences and STS designs have been published extensively, yet they have had little impact on the design of most manufacturing organizations. Technological determinism has continued to dominate the design process. Recently, however, a massive new study has developed a set of data-based conclusions that reinforce the concept of joint sociotechnical design very strongly and severely challenge the rationality of technological determinism. That evidence is arising from the International Motor Vehicle Program (Krafcik and MacDuffie 1989, Womack, Jones, and Roos 1990) a research project that comprises eighty assembly plants of twenty-four companies doing automobile assembly work.[13]

In a brief extract of that research, Kochan (1988) examined six sites from the data set and placed them on a grid drawn between innovation in human resource management practices (e.g., few job classifications, flexible work organization, and extensive communication) and amount of technology (including information and manufacturing technologies) (see Figure 7.1). For his sample he chose two

---

[13] The IMVP research comprises eighty assembly plants of twenty-four companies doing automobile assembly work in fifteen countries. Twenty-two are in the United States (seventeen) and Canada (five); thirty-one are in Europe (of which eight are in the United Kingdom, six in West Germany, six in France, and four in Belgium); eight are in Japan; six are in Australia; and thirteen are in countries new to the auto assembly business—Brazil (five), Mexico (five), Korea (two), Taiwan).

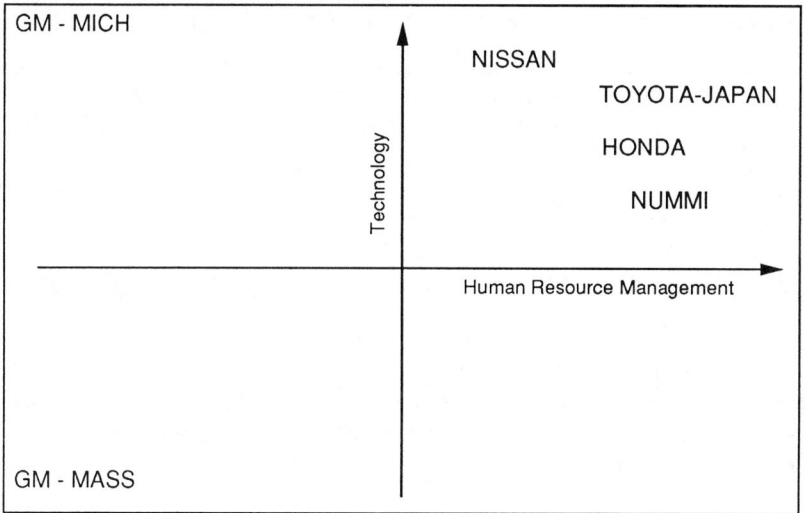

FIGURE 7.1. Human resource management vs. technology in a selected sample of automobile assembly plants. From Kochan 1988. Used with permission.

General Motors sites that were low on the human resource management dimension but were at opposite ends on the technology dimension (the high technology plant had invested $650 million in new information and manufacturing technologies); three Japanese owned plants, one in Japan and two in North America—all three medium to medium-high on human resource management practices and on amount of technology; and a joint venture between General Motors and Toyota—the NUMMI plant in Fremont, California that is medium in its amount of technology and medium-high in human resource management practice.

Using data from the IMVP database, Kochan ranked the six plants in terms of productivity and quality (Table 7.3). The plants with the poorest productivity and quality performance were the plants lowest on the human resource management dimension. Interestingly, the plant with the highest level of automation and the plant with the lowest level of automation were similarly poor on these dimensions. The plant with the highest technology (automation level) had the poorest quality. Conversely, plants with only moderate levels of technology had the best quality and productivity performance.

Although the sample size in Table 7.3 is small, it is drawn from the larger database to illustrate the trends in that data set. Clearly, in automobile assembly operations, *high productivity and high quality cannot be achieved by way of technology alone*. Table 7.3 illustrates that even among Japanese managed plants, the relatively automated Nissan plant in Tennessee has productivity well below Toyota in Japan (although quality is comparable).

Those plants that paid no attention to human resource management practices were poor performers, on both quality and productivity dimensions, regardless of

TABLE 7.3 Quality and Productivity Comparisons of Major Automobile Plants

	Productivity (hr/unit)	Quality (defects/100 units)	Automation Level (0 = none)
Honda, Ohio	19.2	72.0	77.0
Nissan, Tenn.	24.5	70.0	89.2
NUMMI, Calif.	19.0	69.0	62.8
Toyota, Japan	15.6	63.0	79.6
GM, Mich.	33.7	137.4	100.0
GM, Mass.	34.2	116.5	7.3

whether or not they invested in high technology. The best performers had moderate to moderately high levels of technology and were high in the application of human resource management practices.

The Integration Hypothesis. One of the studies in the IMVP centered on "volume" products (automobiles selling for less than $19,000 U.S.) to prove this latter point which the researchers refer to as the "Integration Hypothesis" (Krafcik and MacDuffie 1989 p. 7). It states that:

> High performance manufacturing results from the integration of technology and management policies that govern production practices, work organization and human resource management.

To evaluate the hypothesis, the IMVP created a "Management Index" that comprised the "management policies that govern production work practice, work organization and human resource mangement." This index distinguished between two types of management.

- *Robust/buffered.* Minimizes the role of human resources (e.g., with narrowly defined jobs and inflexible work organization) and uses "buffer" inventories and large repair areas.
- *Fragile/lean.*[14] Depends on a skilled, flexible, motivated workforce (for problem solving and for continuous improvement) and is "lean" in the avoidance of problem-hiding buffers.

Krafcik and MacDuffie hypothesized that it is the fragile/lean management style that leads to high performance manufacturing, or, more precisely, automation and fragile/lean production systems, are important separately but together yield high productivity and quality. Their conclusions with respect to technology places a strong contingent condition on technology as a design imperative and an even

---

[14] In a later book describing the IMVP, Womack (1990) the "fragile/lean" term was replaced by the term "lean" manufacturing.

stronger one on the management style. Although technology plays a strong role in boosting productivity as plants move from very low levels of automation to moderate levels, "there is little gain in moving from moderate to high levels of technology for plants operating a robust/buffered production system" (Krafcik and MacDuffie 1989 p. 19).

Technology, then, is essential for the achievement of productivity where technology has been underemployed. Too much technology, however, is a squandered investment, particularly if human resource practices are of the kind found in traditional or "scientific management" organizations and if manufacturing practices also follow the traditional ones of using buffer inventories to hide poor practices. High productivity and quality will only result from a management-centered approach that depends on people, on good manufacturing practices and on an appropriate level of technology. Krafcik and MacDuffie also found that technology deferred to management practices in predicting quality improvements.

Designing AMT Environments. This study has several particularly important messages for AMT designers. In the continuing dialogue about the merits of highly integrated installations versus islands of automation, these studies would (1) caution the designer about investing too quickly in too much integrated technology and (2) insist that the technical design only proceed in concert with the design of the organizational and managerial variables.

Martin (1989) recently concluded that cells, in straddling the middle ground between stand-alone machine tools and full-blown manufacturing systems "have proved to be the compromise of choice." He argued that cells can systematize and integrate the manufacturing process in a way that stand-alone machining cannot, "while simultaneously remaining a reasonably priced and very manageable production improvement project that doesn't overload the capital and engineering resources of a company the way flexible manufacturing systems can" (p. 49).

Martin's comments underscore part of the message of the IMVP—that it is important to take advantage of new manufacturing technology but not desirable to go too far down the technology road. However, Martin's comments only reflect part of the IMVP conclusions in that he understands the significance of the work organization in making the "middle ground" effective. The cells he refers to have realized their results because they are joint social and technical subsystems. Martin's concern is that some flexible manufacturing systems can demand a level of integration that will make them resemble the situations of too much technology that the IMVP results caution against and not pay enough attention to management and work organization factors (i.e., FMS with traditional or robust/buffered organization is a recipe for poor performance). The unanswered question is "Is the combination of FMS and the sociotechnical design principles described earlier a formula for high performance?"

The IMVP data provide evidence that supports the outcome of joint optimization of the technical and the social subsystems. The data do not necessarily support STS design principles and processes as a way to achieve the outcome. It may well be that one can accomplish the kind of results experienced by the higher performing plants in the IMVP sample through a less comprehensive process of design than that described earlier in this chapter. Some would argue that the best of the Japanese plants, while definitely based on "lean" manufacturing, have not proceeded as far along the road of participation and involvement as STS design advocates (Womack, Jones, and Roos 1990). Others would agree, but would point to high social costs for not doing so (Nohara 1991).

PRODUCT-FOCUSED FORMS

Engineered organizational design was earlier identified as an approach that incorporates some of the concepts of STS theory and many sociotechnical design principles. EOD is increasingly expanding to embrace more and more of the STS paradigm. Sweden has a reputation for being a world leader in the design, production, and export of computer-aided manufacturing equipment, particularly machine tools and robots. EOD arose in Sweden in concert with innovations in production processes in this sector as well as in the automotive sector in order to focus factories and reduce cycle times. EOD and product focussed forms of organization (described shortly) went hand in hand to provide Swedish industry with an early lead in work restructuring and alternative production approaches, particularly in the mechanical industrial goods sectors.[15]

This sector describes the product-focused forms (PFF) that weaned Swedish mechanical industry away from traditional functional forms on the shop floor, a trend that later spread to North America.

Forsaking the Functional Hierarchy

Manufacturing management's increasing inability to buffer product-market uncertainties from its domain, the high interdependencies associated with increasing integration and coordination requirements, and the unpredictability of process uncertainties have forced an unfamiliar set of demands on the production environment (Figure 6.1). Management has turned to computer-based technologies and organizational arrangements to restructure manufacturing to adapt to these events more effectively. The primary outcomes have been an increase in manufacturing flexibility as well as a more rapid rate of response resulting from reductions in the time aspect of most events (e.g., lead time, cycle time, startup time, and so on (see Table 2.2)). In the past these two outcomes would have been considered incompatible. Faster cycle times would have been the result of greater scale economies from increasingly dedicated equipment offering less and less flexibility. In the current environment

---

[15] Japanese industry, during a slightly later time period, evolved the "lean" manufacturing approaches described earlier, particularly in the automotive sectors. The Japanese dominance in the consumer electronics sector during this time period seemed more a consequence of an emphasis on quality. Fabrication and assembly within this sector appear to have been carried out with more traditional structural arrangements.

of manufacturing, both are accomplished by combining AMT and work organization arrangements into better sociotechnical designs.

Flexibility increases are accomplished by linking programmable technology and systems with innovative organizational arrangements (e.g., multiskilled operators, self-managing teams) that allow the flexibility features to be realized. Faster startup times, lead times cycle times, and other time dependent parameters are achieved through changing from functional organizational arrangements to product-focused or task-oriented forms of organization and by supportive manufacturing techniques (e.g., faster die changes) and AMT (e.g., automatic tool changers).

In addition to speed of response, there are several other factors influencing manufacturing organizations to adopt alternative structural arrangements, particularly product-focused forms (PFF) of organization (Kolodny 1986). For example, there is a need for the following:

- Good project or task visibility through clearly designated responsibility for the task and equally clear points of contact for customers, suppliers, and managers in the organization
- A full-time orientation toward all the aspects of the task (lowering inventory, materials and direct labor costs, maintaining production schedules, improving quality)
- Decentralization to encourage more commitment and innovation by providing local autonomy and control (which also decouples the organization from excessive coordination complexity and unnecessary control costs)
- The ability to cross functional lines more easily
- The ability to process several tasks in parallel
- A goal orientation that is more measurable and more understandable in terms of the actual products produced

An accompanying set of factors have been driving manufacturing organizations away from functionally organized structures including:

- Division of labor and specialization that leads to fractionated jobs and worker alienation
- Overspecialization that favors professionalism (e.g., computer science, engineering), at the expense of more generalist roles such as administration and management
- Poor interunit coordinating
- Long lines of communication because of the constraints of hierarchical structures.

An Hierarchy of Product-focused Forms

Product-focused forms fit within the context of a range of innovative organizational structures that have been arising since the early 1970s as alternatives to traditional functional organizational forms. The different variations of the form are all characterized by considerable use of computerization and automation, by work units organized in teams and by cross-training of individuals in the work units. As Table 6.3 illustrated, the features of product-focused forms are flexibility, small size, relative autonomy, and organization with the flow of production. As with all organizational forms, these forms also share some problems:

- It is difficult to make cost/budget allocation decisions for resources that are shared between product-focused units (e.g., common inspection or transportation equipment).
- Some physical resources do not divide up easily (e.g., tool cribs, specialized support services such as centralized programming).
- A tendency exists in product oriented forms to confine innovative and other activities that could be shared to one's own product line, particularly if the measurement system differentiates the products.
- Scale issues—resources are duplicated within product-focused units and sometimes utilized inefficiently.

Within metal working operations, PFFs usually take the form of manufacturing cells (Wemmerlov and Hyer 1989) and flexible manufacturing systems (Figure 7.2). Within assembly operations, PFFs are usually found as assembly cells or flexible assembly systems (Edgren 1981, Kolodny 1986b). When a sufficient level of support activities and authority are added to machining and/or assembly operations to make them relatively autonomous, the form is referred to as flexible focused factories

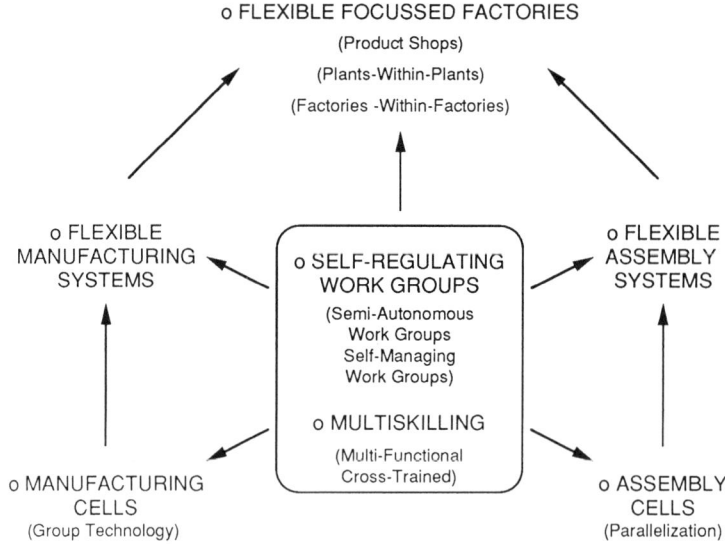

FIGURE 7.2. A hierarchy of product-focused forms.

(Rohan 1981, Skinner 1974) or product shops (Aguren and Edgren 1980) or factories within factories (Lindholm 1975) or plants within plants (Skinner 1974). As Figure 7.2 illustrates, all the structural arrangements build on a base of self-regulating work groups with multi-skilled team members.

The hierarchical layout of Figure 7.2 is intended to connote increasingly integrated activities. Manufacturing and assembly cells normally integrate only the programmable (and manual) equipment of the cell, which may also include both computer-based and manual material-handling equipment, plus the accompanying personnel that constitute the permanent (operators, loaders) or partial (toolsetters, supervisors, etc.) members of the cell. FMSs and FASs are generally more integrated in terms of information technology and computer-based fabrication or assembly technology and broader in scope so as to incorporate some of the support functions associated with the particular system (e.g., mechanics, forklift drivers, tool setters, supervisors, industrial engineering, and programming). However, the boundary between the categories is an overlapping one. The range of system sizes in the cell and system categories is so wide that the required support for a cell could be more extensive than the support for a system. For example, Wemmerlov and Hyer (1989) surveyed one manufacturing cell with forty machines, whereas Adler (1989a) studied an FMS with four machines (plus a measuring machine and some materials handling equipment).

The forms illustrated in Figure 7.2 are all oriented "with the grain" (Lindholm 1975) (i.e., with the flow of production or in the primary direction of the task to be accomplished). This task orientation has drastically reduced cycle time, often as much as eighty percent or more (e.g., from ten weeks to two weeks) (Rohan 1982). Other reasons for product-oriented organization include setup time reduction and WIP inventory reduction (Wemmerlov and Hyer 1987). The forms identified also have in common changed roles for the cell members, either through increases in multiskilling or through membership in work teams, or both.

Work Groups

Semiautonomous work groups (Gulowson 1979, Trist and Bamforth 1951) or self-maintaining work groups (Davis 1977) or self-regulating work groups (Susman 1976) or self-managing work groups (Hackman 1987) lie at the basis of product-focused forms of organization. Work groups do not have to be product oriented. Such groups could form within a functional unit, for example. The many versions of quality circles represent just such entities. However, unless such groups or teams are task oriented rather than problem oriented, as is the case with quality circle groups, they do not assume the characteristics of the product-focused forms described here.

For instance, functional groups or quality circles do not develop and evolve their structural arrangements; the members seldom pursue multiskilling paths; support systems to facilitate and encourage learning behavior—such as skill-based pay systems—are rarely implemented; they are not granted authority to select replacement members for their unit; and they do not assume responsibility for coordination of intragroup activities or for the maintenance of the work group. The type of work groups referred to here carry out some or all of these activities and are organized around a relatively meaningful or whole task that lies in the direction of the output. Their task is clearly bounded, but within that boundary they have autonomy in how they accomplish their work.

Vast numbers of self-regulating work groups are arising now in industry and in a very wide variety of situations (Hackman 1990). Self-regulating work groups may well be the Western industrial world equivalent of what quality circles have been in Japan and their impact on the organization of work in Europe and North America is likely to be as dramatic as quality circles have been in Japan (Cole 1985). Applications in manufacturing have been numerous, both in continuous process (Halpern 1984; Walton 1972) and in discrete manufacturing technologies and, in the case of the latter, in both consumer-oriented products (Carnall 1982, Wall et al. 1986) and industrial products (Aguren an Edgren 1980, Kirby 1986). However, this discussion will be confined to applications in AMT environments (Arnopoulos 1985, Betcherman, Newton, and Godin 1990, Dean and Susman 1989, Klein 1989, Mishne 1988, Spector and Beer 1985).

One recent case study illustrates just such an operation (Betcherman, Newton, and Godin 1990). In 1982 Pratt and Whitney Canada (PWC), a subsidiary of the United Technologies Corporation, established a Manufacturing Modernization Program to "plan and implement leading edge manufacturing technology at PWC." Although some computerization already existed in the company (e.g., manufacturing planning and control, some CAD applications in product design and some CAD-CAM activity, process planning, CNC), there were major gaps in the use of new technologies and there was no integration of the various technologies.

> To achieve major gains in flexibility, productivity and predictability, a flexible manufacturing system was needed that would integrate the process and material handling equipment with the planning and control systems into a single, synergistic system known generically as computer-integrated manufacturing. (Betcherman et al. 1990 p. 28)[16]

In 1985–1986, PWC began the planning for a new manufacturing plant in Halifax, Nova Scotia that would machine castings for aircraft engines. The plant was to take account of integrated strategies for manufacturing-improvement, human-resources, and market/business. The organization design was to be based on sociotechnical principles, and a design process was instituted that engaged a steering committee of top managers and a design committee comprising the plant manager, external consultants, representatives from all key departments involved in the design, and plant employee representatives as they were hired.

The design committee soon established the need for the people in the plant and the decision-making systems to have a perspective that was as holistic as the computer-integrated system they were attempting to put in place. Hence they looked

---

[16] Material from Betcherman et al. 1990, *Two Steps Forward: Human-Resouce Management in a High-Tech World*, Economic Council of Canada is reproduced with the permission of the Chairman of the ECC, 1991.

for "an integrated work system built around teams and roles, rather than a fragmented work system built around functions and jobs" (Betcherman et al., 1990 p. 33). Their analysis led to the design of three types of teams: management, support, and production. Initially, the committee designed a single production team for the entire plant, but as the plant became fully operational and employment rose to 200, they subdivided it into teams organized around different aspects of the transformation process.

Production team responsibilities included:

1. Monitoring their own performance in terms of lead times, quality, and cost and making suggestions for improvement
2. Understanding the key variances in the product resulting from the process and knowing how these variances affect subsequent operations
3. Inspecting their production to ensure conformance to product requirements
4. Doing routine maintenance on their equipment
5. Alerting the management team if unexpected events put their production plan in jeopardy
6. Ensuring inter- and intrateam coordination (Betcherman et al. 1990 p. 33)

Support teams were to provide specialized technical advice to the production and management teams with regard to planning, computer systems, electrochemical metal cutting technology, special maintenance, and so on. Workers rotated from duties on production teams to support teams to enhance their understanding of the relation of production and support and to help them become multiskilled. Both production and support teams became involved in many of traditional management's functions, which included:

1. Control of absenteeism, attendance, and vacation
2. Selection of new employees and performance improvement and dismissal of unsatisfactory employees
3. Performance evaluation and the reward system, including a pay-for-skill program
4. Training opportunities and nonfinancial rewards
5. Descriptions of roles internal to the teams (Betcherman et al. 1990 p. 33).

The management team was accountable for the performance of the factory, its long-term evolution, its interactions with its environment, corporate headquarters, local government bodies, and the local community. The team was also responsible for the management of the sociotechnical design of the plant and ongoing redesign.

Group Technology/Manufacturing Cells

Group technology (GT) is a manufacturing management philosophy to group together and standardize families of parts having similar design characteristics and requiring similar machining operations (Burbridge 1975, Hyer 1984) so they can efficiently be stored and retrieved (Hyer and Wemmerlov 1989). Manufacturing cells or cellular manufacturing arrangements are an application of GT. The traditional functional organization of machines is replaced by product focused manufacturing cells that are reorganized according to a GT philosophy.

The composition of such manufacturing cells can vary widely. Wemmerlov and Hyer (1989) report an average cell size of 6.2 machines for manned cells (4.7 for unmanned cells), with a range from 2 to 40 in their survey of 32 companies with cells. The average number of product lines produced in a cell was 2.8, with a range of from four components per product to 5000 and an average of 956. Even within a GT philosophy of grouping families of parts with similar characteristics and operations, the products produced in a cell can themselves vary widely.

Flexible manufacturing cells may be fully automated with CNC machines, automatic tool changers, programmable robots, automated materials handling equipment, and remote DNC. However, Wemmerlov and Hyer (1989) reported that only one company of the thirty-two in their survey that claimed to have manufacturing cells in operation had only unmanned cells, whereas six others had both unmanned and manned cells. Some cells may be hybrid in composition (Rahimi, Hancock, and Majchrzak 1988), with combinations of manual and programmable controllers, manual loading of parts and tool changing, and tedious and slow setups. The number of operators associated with a cell, and hence the size of the team, will vary with such factors as the size of the cell, the level of automation in the cell, how activities are divided between machines in the cell, the layout of the machines in the cell, the relative utilization of the different machines, the type of product produced, the cycle time for the operations in the cell and at each machine, the work arrangements, the labor agreement, the number of machines a single operator can control, the proximity of the machines to each other, and so on.

Manufacturing cells can provide many of the conditions necessary for the formation of a cohesive, integrated working group. They can offer opportunities for social interaction, greater job variety, greater independence, and participation in decision making that may provide workers with greater job satisfaction than traditional functionally organized job shops (Kaimann and Bechler 1983). Organizing teams around cells provides team members with job characteristics generally considered conducive to motivation (Hackman and Oldham 1980, Kaimann and Bechler 1983). For example, operators assigned to several different machines in a cell have the opportunity for variety and job rotation that entails cross-training and an opportunity to increase multiskilling.

At a feeder plant for turbines and generators, Westinghouse Canada had constructed five group technology cells. The machines were purchased from a wide range of countries (United States, West Germany, Japan, Sweden). However, Westinghouse insisted that all the machines use a standard Westinghouse-designed controller with each machine. The control panel was general enough to allow every different machine's control requirements to be represented on the panel, even if many of the functions were not used on different machines. The result of this decision was that cross-training on different machines within a cell was made easier because the operators had ready familiarity with the controls.

Manufacturing management in the plant had not explicitly intended to organize the cells into self-regulating teams. However, with cross-training made easy and because of the proximity afforded by the physically contained configuration of each cell, operators could rotate jobs through the different machines in a cell, and cover for each other when necessary. They did so and several self-regulating teams soon evolved in the plant.

Running smaller batches through a cell means more diversity in the work of the team and of individuals. Cells are relatively independent entities, with relatively complete tasks to accomplish within their physical boundaries,[17] which can provide teams with the satisfaction of seeing a task through to completion. Team allocation of tasks, schedules, batch sizes, training, days off, vacations, shift work, and the like results in work that is high on autonomy and offers team members opportunities to participate actively in the decisions that affect them.

The composition of cell teams can differ between ongoing operations and startup situations. One plant manager, reporting on his company's successful implementation of cellular manufacturing, attributed the success to the fact that "we had only manufacturing engineers on the cells at first, working with the machine tools, running parts, getting solutions" (Martin 1989). Jaikumar (1986) has reported that similar practices of using engineers as cell operators have differentiated the Japanese successes from the less successful American installations he studied.

Despite cellular manufacturing's conduciveness to work team organizational designs, widespread adoption is still at an early stage in the process of implementing cellular manufacturing. This is partly because work team implementation in manufacturing in general has proceeded at a slow pace overall and partly because a technocentric mentality still dominates much AMT design. In the earlier mentioned survey of users of cellular manufacturing in the United States, Wemmerlov and Hyer (1989) found that only two of the thirty-two companies that claimed to have manufacturing cells in operation had explicit autonomous work team structures. Good work team design follows from redesign of the work flow itself in order to identify the primary work systems. That design process only takes place when the manufacturing environment, and particularly the AMT environment, are viewed and understood as sociotechnical systems.

Assembly Cells

The term "assembly cells" refers to parallel operations that replace a single, long-linked, balanced assembly line with several shorter lines that each carry out the same total set of operations as the long line. Each operator on a parallel line has a cycle time for activities that is considerably longer than the short cycle times that balanced the long line. This can considerably expand the required skills of the assemblers if they learn to test, inspect, adjust, and repair as well as assemble products. Parallelization achieves the desired output volume by adding enough parallel lines.

---

[17] Wemmerlov and Hyer (1989) report that of thirty-two companies with manned cells only three had parts machined completely within their cells. For the remaining twenty-nine, however, an average of 78.3 percent of the parts were processed inside a single cell.

Manufacturing cells achieve product focusing by dedicating a cell to a product line or a family of products in place of having the products produced in a traditional functionally organized factory. Traditional assembly lines are generally dedicated to a particular product line or product family a priori. The rationale for categorizing assembly cells under the general heading of product-focused forms is only that they divide up longer assembly lines into several different and smaller lines, or cells, which may then be differentiated so each becomes even more dedicated to a particular product or product line.

Manufacturing cells achieve their flexibility from computer control of the metal working and material-handling processes and from the responsiveness of the organizational arrangements, the basis of which may be multiskilled operators or cell teams or both. Assembly cells generally have a lower degree of mechanization and computerization than manufacturing cells. From the point of view of jointly optimizing the social and technical subsystems, assembly cells achieve their flexibility goals via a stronger emphasis on the organizational arrangements. As with manufacturing cells, assembly cells are natural organizations for team-based structures. Many cells are so organized, though they do not have to be. Some are organized around individuals, although generally with expanded sets of skills.

One of the first parallel assembly arrangements was established in Saab's automobile engine assembly operation in Sodertalje, Sweden (Norstedt and Aguren 1973). This particular arrangement had six parallel lines drawn off the main moving belt assembly line, each dedicated to the assembly of a complete engine, with teams of three people assembling each engine. Target cycle time for assembling each engine was set at thirty minutes. Teams were free to arrange the work according to their preferences, as long as they achieved the required engine output at the end of the day. Hence teams could differentiate their roles—one person might pull material while the other two were assembling, they could rotate jobs, or they could work more quickly during one part of the day and slow down or meet to plan activities at another time.

General Motors Corporation of Canada's truck plant in Oshawa, Ontario provides a recent example of parallel cell assembly within an AMT installation. A $600 million modernization completed in 1987 made it one of the largest technological change projects in North America. The plant has 3300 unionized workers and a production capacity of 250,000 trucks per year. Extensive use was made of synchronized manufacturing to allow inventory to arrive hourly at the receiving docks, located in a 360-degree perimeter of the plant to facilitate JIT point-of-use receiving.

The moving assembly system was replaced by AGV technology to provide "stop-and-go" transport of parts to the workers who halted the AGVs to carry out their assembly activities. Cycle time at each stop was about five minutes. "Operator-release of the AGV was a conscious decision (and a change from conventional North American auto-plant use of AGV technology), in order to place more quality control and responsibility in the hands of the assembly worker, who is expected to release the work when it has been done right the first time" (Painter 1991 p. 10).

Under centralized computer control, the AGVs delivered the parts and major units (e.g., cab, motor, chassis) to "assembly islands" or assembly cells. Each island had five parallel spurs, each with an input and output buffer, and two operators at each work station (Figure 7.3). The "left-hand" and "right-hand" workers at each spur made up a miniteam. The structure was also designed for the development of assembly island teams comprising the five miniteams and a supervisor. The teams would learn to rotate, regulate their own activities, and develop into self-regulating or "natural" work groups. However, at the time of writing, this was more an intention of the system than a realization because both supervisory management and the plant's union found the technical system changes easier to accept than those that a redesigned social system demanded.

The spur design at General Motors Oshawa resembles the former "dock assembly" layout at Volvo's well-known Kalmar plant in which carriers were moved off the main assembly line to a dock assembly area where teams of two or three people carried out assembly on a stationary carrier (Aguren, Hansson, and Karlsson 1976 p. 11). Several carriers were worked on in parallel. The entire work cycle was carried out at a single work station. The dock assembly method at Kalmar has since been abandoned, "primarily because the number of model variants has increased, and it would now be virually impossible to store all the needed components conveniently in the limited area available" (Aguren, Bredbacka, Hansson, Ihregren, and Karlsson 1984 p. 41). Kalmar's operations have continued with "straight-line assembly" where input and output buffers precede and follow several work stations that are in sequence. Two-person teams move with the carrier through the several stations until their cycle is complete, about twenty-five to thirty minutes, then return to start assembly of the next car.

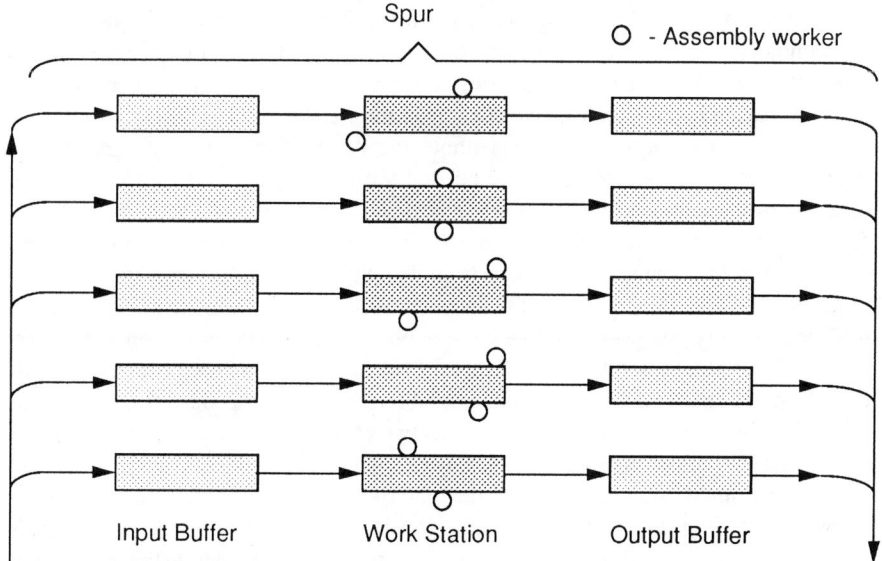

FIGURE 7.3. Assembly island at General Motors' Oshawa Truck Plant.

The Kalmar plant, with approximately 750 employees, was designed around teamwork (Gyllenhammar 1977) with approximately thirty work groups of 15 to 20 workers each and was constructed to provide each team with its own physical area (entrance, changing room, coffee room, and assembly area). The teams were the strongest social groups in the plant. "The cohesiveness in the team is so strong that pool workers, who are temporarily assigned to a different team, usually come back to their home team areas for coffee breaks, even though they must use part of the time to travel from one part of the plant to another" (Aguren et al. 1984, p. 45). In an evaluation conducted after ten years of operation, fifty-four out of sixty-seven workers surveyed (80 percent) rated teamwork as good or very good (the question was not applicable to 13 percent and only 6 percent rated teamwork "rather poor") (Aguren et al 1984).

Job rotation was practiced within the miniteams and between tasks such as assembly and preassembly. Because of the buffers, operators were able to "work ahead" and gain minibreaks of between four and eight minutes (Aguren, Hansson, and Karlsson 1976). Breakdowns were handled by changing the production pace and enlisting the help of others in the production area. Absenteeism and disturbances were managed by creating an "absentee pool" of operators who were willing to learn several jobs and act as replacements wherever they were needed (Aguren et al 1984).

Reorganizing Material Handling. Table 7.4 lists some of the gains and losses from parallel arrangements of work. The table indicates that material management and handling is one of the more significant disadvantages of parallel operations. As noted above, this was the problem that forced abandonment of the dock assembly system at Kalmar's assembly plant. In assembly environments, changing the arrangements to be flexible enough to strive for efficient assembly of a batch of one is essentially a material-handling problem. Gathering together the parts needed to assemble a product is a time consuming process and arranging them for convenient operator (assembler) access is a space consuming task.

To handle a range of products without the delay of trips to a distant storage requires local level access to all the needed variety of parts and equipment. This means more space for materials, more sophisticated materials racks with, for instance, slide-out drawers to accommodate much more material, and layouts for loading material from one side and consuming from the other. It may also entail modifications to the design of centralized testing or inspection devices to allow them to be inserted into the assembly process to eliminate time-consuming detours away from the natural flow of the product assembly line (Agervald 1981). It means assembly arrangements that allow one assembler or a team of assemblers to not be impeded by an assembler who might be slower or might be experiencing a problem (i.e., parallel arrangements or designs that allow passing).

Some computer-aided solutions to the materials problem of parallel arrangements allow operators on assembly lines to call for the delivery of their needed materials via a local terminal. Programmable, self-propelled carriers fetch the parts from high-stacked programmed storage areas. Such systems also allow unmanned or

TABLE 7.4 Gains and Losses in Parallel Arrangements

Gains

Lower production losses
  Increased resistance to disturbances and less downtime
Better balancing and use of time
  Better and more even workloads at different stations through less rigid tying of operators to specific tasks, longer task cycles, and more tolerance for different product models
Less need for replacement personnel
  Manning can be varied and rest periods can be taken without the need for replacements
Lower costs for checking and adjustments
  Increased control done by operators themselves, better follow-up through direct referral of errors to the operator concerned
Faster change-overs
  Less need for balancing and running-in in connection with changes in production volumes or models

Losses

Lower methods level
  Reduced advantages of long series, somewhat less systemic methods planning, increased handling of components at work stations
Higher tooling costs
  Need for parallel equiping of special tools
More costly material-handling systems
  Investment in more complex system alternatives and advanced technology, increased maintenance
Increased costs in materials supply
  Increased need of materials at the work stations, transport, and handling work in stockrooms
Larger premises
  System layouts spread over larger areas

Source: Adapted from Aguren and Edgren (1980) p. 29. Used with permission.

lightly manned production through second and third shifts. Semiautomated responses have materials stored in programmable racks that operators can access via a computer terminal, but they must physically walk to the stores area to use the terminal and take possesion of the parts. More local level control results as operators plan and control their work. At the same time, more line control of material means fewer materials clerks, forklift drivers, and materals handlers.

Stiga AB, a Swedish manufacturer of power lawn mowers, reorganized their assembly area to reduce cycle time and improve quality of working life (Agervald 1981). In one organizational arrangement, they confronted the issue of materials management by combining the locations for assembly and storage in order to reduce transportation and inventory administration costs (alternate aisles between storage racks were widened to create room for assembly benches). In a second product line, they formed people into two-person, parallel work team cells, located the materials stores adjacent to the assembly area, and provided the teams with wheeled carts to collect their own parts. Teams could then see a total assembly operation through from start to completion, including having control over the material management aspects.

Flexible Manufacturing Systems

The organizational arrangements used to date with FMS have tended to maintain traditional functional structures. However, as the earlier-cited IMVP studies suggest (Krafcik and MacDuffie 1989, Womack, et al. 1990), there may be great peril in continuing to do so. If the findings from these assembly plant studies are transferable to FMS, technological integration beyond some as yet undefined level can lead to less efficiency rather than more, particularly if the organizational arrangements do not lead to changes in the roles of operators and/or teams and support workers. Because FMS tends to connote integrated and complex technology, often in its infancy with respect to its integrative and systemic characteristics, the resolution of the problems likely to arise during its implementation and operation will almost certainly call on several different kinds of skills. Traditional functional structures have been weak in fostering cross-functional skills. Organizational arrangements that encourage multiskilled operators and team operations are likely to be far more successful in supporting FMS.

Adler (1989a p. 8) cited one FMS he studied as "the only teamwork FMS installation in the U.S." Adler's comparative analysis of the team-based FMS with a comparable one organized functionally and against other functionally organized FMS studied by Blumberg and Gerwin (1984) suggested that there were enough desirable characteristics associated with team based organization to expect further adoption of these forms in FMS installations, a position Blumberg and Gerwin also advocated. At the same time, however, Adler pointed out that FMS worker satisfaction can also be achieved without necessarily adopting team based structures. Multiskilled roles would probably lead to this result (see also Chapter 8).

FMS can be viewed as a larger, more complex, and more integrated manufacturing cell that has evolved out of group technology (Majchrzak 1988). Yet the implementation problems associated with FMS (Ettlie 1988b) and the high interdependencies it invokes with other parts of the manufacturing and engineering organizations (Bomba 1989), suggest that it is much more than an extended cell. Graham and Rosenthal (1986) cited FMS technology's characteristics of integration, intelligence, and immediacy as factors that make it qualitatively different than conventional machining technologies. They noted that FMSs required planning of a scope and comprehensiveness seldom in evidence for regular facilities programs.

FMS is a relatively contained manufacturing process. Oney (1989) described some of the organizational difficulties Apple Computer encountered because so much of FMS philosophy is based on a "process-oriented" focus. All Apple Computer manufacturing had been based on product line parameters. As a result, products selected for manufacture in apple Computer's new FMS tended to be accompanied by their product-specific organizational arrangements, that is, dedicated operational teams for each product line were assigned to manufacture their product in the FMS. For example, the measurement systems emphasized quality and schedules achievements specific to each product line. This strong product culture and its reinforcing set of support systems worked against the integration of product functions. It caused people to avoid coordinating shared tasks. In fact, they competed for common resources, which detracted from the overall efficiency and acceptance of the FMS.

Once the project team and upper management recognized the problem, the operations teams' measures were altered to promote cooperation rather than focus exclusively on their own product line. Production scheduling across products was better coordinated, and process and quality engineers were reorganized to concentrate across a broader base of products. Operators were cross-trained to encourage a process-orientation (Oney 1989 p. 21).

Flexible Assembly Systems

At the General Motors' Oshawa plant and Volvo's Kalmar plant, each of the assembly cells is imbedded in a larger assembly system. Given the centralized computer control, each plant could be considered a flexible assembly system, though there are constraints on the degrees of flexibility. Kalmar's rerouting possibilities are limited, but there is considerably more at Oshawa because the AGVs carry their own materials. Volvo has taken this one step further in its new plant in Uddevalla in Sweden (Eurojobs 1990, Kapstein and Hoerr 1989). Kits of materials are delivered by AGVs to each of forty-eight different locations, but it is the work teams who decide when they want kits delivered, and they can revise the planned delivery schedule with as little as two hours' notice.

A different kind of assembly system can be found in CNC-controlled printed circuit board assembly where several or many parallel, single operator-single machine combinations are set up. In such situations, mix, changeover, and modification flexibility are achievable because of the programmability of the CNC equipment. However, given that the work is usually organized in single operator-single machine combinations, other kinds of flexibility—volume flexibility or rerouting flexibility—would be very limited.

Assembly systems comprising several or many assembly cells enjoy several flexibility attributes. Organizing units in parallel reduces the vulnerability of the overall assembly process to many unanticipated disturbances because the effects are not generalized to all the assembly cells. For example, material or labor shortages may only impact one or several cells at a time or it may be possible to shift work to other cells to achieve rerouting flexibility. When product adjustments are necessary, line balancing can be simplified if only one or two lines need changing (Aguren and Edgren 1980). In effect, modification flexibility is facilitated in parallel operations. Parallel cell organization also allows new product variations to be introduced into one cell without upsetting ongoing production in other cells (Butera 1975), a form of changeover flexibility.

If volume variability is anticipated, limited volume flexibility is possible and achievable in assembly systems with parallel cell organization by having different production volumes in different cells. Mix flexibility in FAS can be achieved at both micro and macro levels. In the former, multiskilling of individual cell members and team-based attributes of the cell can offer considerable adaptability if the product mix varies. In automobile assembly operations, of the kind described above, macro-level mix flexibility can be achieved through the combination of parallelization and programmable transportation systems. Less flexible technical configurations in automobile assembly, such as the Toyota Production System, have also achieved mix flexibility through a combination of efficient computer control, high responsiveness on the floor, and multiskilled workers.

At General Motors' Oshawa truck plant, the computer-integrated assembly system has increased the interdependence between assembly workers and tradespeople and other support people, adding a new level of coordination to the supervisor's role. At the same time, the supervisors' work in proximity with the teams has physically distanced them from other supervisors, a change they have found dissatisfying (Painter 1991). Loss of control has also become an issue for supervisors (Blumberg and Gerwin 1984), brought about by the technology's control over system integration, worker control over AGV release, and management pressure to operate with a more humanistic style than they have utilized in the past.

Flexible Focused Factories

Product shops is a literal translation of the Swedish term *produkt verkstaden*. The closest English language term is "focused factory" (Skinner 1974). North America experienced a flurry of focused factory activity in the mid 1970s, particularly in the then booming oil rig equipment supply business (Rohan 1981). However, when the recession of the early 1980s began to be felt, a large number of the focused factories or "satellite" plants found themselves without enough work to keep their dedicated factories viable. Many were closed and the equipment was withdrawn to the "mother" plant. Comprised of group technology cells and/or assembly cells, the dedicated focused factories were too inflexible to accommodate the highly uncertain environment. There was a need for all the benefits of focusing, but an equally strong need for flexibility to adapt to the uncertainties of a quite unpredictable environment. Flexible focused factories was the resultant response.

The discussion on product shops and focused factories has been towards product focusing—concentrating on a limited number of products. However, manufacturing involves a range of characteristics to produce a product and they must be handled consistently with the philosophy of focused manufacturing. Skinner (1974) has suggested that these different characteristics (process technologies, market demands, product volumes, quality levels, and manufacturing tools) must be attended to in ways that make them all manageable and controllable. It is by limiting the concentration on each of these characteristics and by learning through repetition that the focused factory can learn to do better.

Skinner has also noted that it is not necessary to think of focused factories as stand alone entities. Within a plant, focused manufacturing can be achieved by organizationally and physically dividing up production into "plants within plants (Skinner 1974, p. 121) or "the small factory" (Lindholm and Flykt 1981 p. 72) or "small workshops in large workshops" (Aguren and Edgren 1980 p. 90), each of which has their own workforce management approaches, production control, organization structure, and the like to accomplish a particular manufacturing task.

Volvo's large truck assembly plant in Tuve, Sweden was designed as a flexible focused factory or "product shop." It employed 225 people at the time of a plant visit in 1983. Assembly was carried out in two parallel lines. Teams completed a cycle of operations of approximately thirty minutes at each assembly station. Buffers separated the teams from each other and allowed some variability in each work team's pace. Parts inventories were delivered directly to the location on the line where they are used. Technical support for the production lines was located in offices directly adjacent to the plant making the plant relatively self-sufficient. The teams were self-regulating and multiskilling was encouraged through a skill-based pay system. Despite its small size and self-contained nature, Tuve was considered a very efficient assembly operation.

Product shops tend to be small. They promise simplified planning and control: there are fewer planning points required since production is organized "with the grain" and the flow of production itself informs the planning process. Cycle times and throughput times are reduced, which improves delivery and offers better responsiveness for custom-made parts or special orders. Table 7.5 lists further reasons for adopting product shops or flexible focused factories.

For a period of time, in the 1980s, the small size of the Tuve plant was considered to be the prototype of all future production facilities Volvo would build. In what at first appeared to be a turnabout in production philosophy, in 1988, Volvo opened up its large assembly plant in Uddevalla designed for eventual employment of 1000. However, the plant is structured around six distinct product workshops or product plants of approximately 80 to 100 persons each. Each plant is independent, with its own manager, although all the plants share a common materials center that has the ability to ship parts to each of forty-eight different locations on a pull basis or JIT basis via an elaborate AGV system.

Within each product workshop, eight teams of eight to ten persons each carries out the total assembly of an entire automobile, managing itself, handling schedules, quality control, hiring and other duties normally assumed by supervisors (Kapstein and Hoerr 1989). The product workshops have no first-line foreperson, but each team has a spokesperson who reports to the plant manager. All managers and all white-collar workers in the plant have themselves built a car on the floor of Uddevalla.

The plant is designed to be highly flexible and capable of dealing with a large number of variants in the Volvo 740 car it produces. The level of mechanization in assembly is relatively low. The product workshops are a series of plants within plants and the complex encompasses most of the structural attributes illustrated in Figure 7.2.

Tuve and Uddevalla are examples of flexible focused factories dedicated to assembly work. There are also many examples of flexible focused factories dedicated to machining and fabrication (Rohan 1981, 1982), but fewer that have combined both FAS and FMS. More often, the two streams are separated rather than integrated

TABLE 7.5 Reasons for Forming Product Shops/Flexible Focused Factories

Easier to manage a bounded unit
  Produces a limited variety
  Is controlled independent of the rest of the factory
  Relatively small number of people
    Everyone knows everyone else and what each is doing
    Contacts can be quick and informal
    Problems are identified early and measures arranged quickly
More opportunities to use self-regulating work teams
Reduced cycle time, better customer delivery, reduced capital investment
  Smaller size and informal contacts
    Removes waiting, transportation, and storage delays
    Reduces work-in-process inventories
    More responsive to customer orders
Simplified planning processes and production follow-up
  Easy to have a direct overview of materials flow and production processes
  Flow of production guides much of the process
Good work environment, good work motivation
  Broader work tasks and longer task cycles
  Increased skill variety
  May assume planning, coordinating, training, quality, materials roles
  Better connections between own task and outcomes
  Easier to understand the significance of quality, product design, etc.

Source: Adapted from Grondahl (1979).

in a single factory. This may be a consequence of the differences in orientation between assembly and fabrication people or it may reflect manufacturing strategies that separate parts production plants from final assembly operations. Cummins Engines Company's Jamestown plant, however, does combine the two. Although large in total size (about 1000 employees), it is subdivided into businesses of up to about 150 people, three machining businesses, one assembly business, and one assembly and test business. Figure 7.2 depicts these manufacturing organization variations by presenting an hierarchy of product-focused forms with flexible focused factories at the top of the hierarchy integrating both streams of manufacturing.

SUMMARY AND CONCLUSIONS

This chapter has focused on the introduction of a new paradigm of work—sociotechnical systems design—that has changed and has the potential to change the "middle" of the organization where a particularly large amount of pressure is being exerted on the organizational and managerial systems. STS and product-focused forms, as alternative ways of organizing for AMT, are ech solutions to the pressure on the middle. STS designs focus explicitly on the vertical dimension of that pressure. They decentralized responsibility and delegate traditional middle management tasks to teams of people capable of self-regulating their activities. Product-focused forms of organizing AMT environments shorten response times and attack the horizontal pressures on the middle of the organization. Together, STS and PFF provide an alternative to traditional ways of organizing manufacturing that hold the promise of higher peforming manufacturing entities.

Product-focused forms of organizing were introduced under the rubric of "engineered organizational design," a perspective that lies between the technological determinism of traditional manufacturing organization and the joint optimization of sociotechnical systems theory. Where an organizational culture exists that reflect less willingness to devolve decision making and authority downward to those who are most affected by the decisions, engineered organizational design may be an appropriate work restructuring perspective. In time, however, as lower organizational members acquire the knowledge and the competence to assume responsibility for the design decisions that affect them, a realignment of the authority structure is inevitable.

Per Gyllenhammar, the chairman of the Volvo Corporation expressed this viewpoint (Gyllenhammar 1977 p. 68):[18]

> People often ask why we concentrated so much on the technical systems and so little on the people for whose convenience they were designed, especially at first. The answer is fairly simple. If we failed with these technical systems, the chances of changing the work organization would drop drastically. If we succeed with the technical systems, we gain a visible, economic success, which is the prerequiste for acceptable overall performance. We could not "succeed" with people themselves unless we succeeded with the technology for people.
>
> The best we can hope for, in Kalmar and our other plants, is to achieve a situation in which technology does not limit the freedom of the men and women who work there. Then we may experience a dynamic kind of organization development that comes not from management but from the work force itself. An organization that develops and changes at the instigation of its members rather than its managers has a better chance of renewing itself all the time, evolving to fit the true situation of its people.

Commentators on the design of AMT and AMT environments, particularly those approaching the subject from the technological side, have increasingly mentioned the necessity of paying attention to the sociotechnical dimension. For the most part, they mean the "people" aspect and usually refer to the issue in terms of resistance to change and implementation and sometimes recognize the importance of participation and involvement to obtain commitment. What they are less aware of, and what this chapter has emphasized, is that paying attention to the sociotechnical dimension means appreciating that a real revolution in the ways of working with new technology has to be entertained or a sociotechnical outcome will not be achieved. The paradigm change required is as significant for the organizational and managerial systems as the technological changes taking place in computer-based manufacturing systems are for the operating processes in AMT environments.

---

[18] Used with permission.

8
AMT AT THE LEVEL OF THE INDIVIDUAL

Many of the traditional and routine functions carried out by operators and assemblers in their interactions with conventional machines and equipment are absorbed into the software and hardware of computer-based technologies. For operators who had experience with conventional technology, the embedded operations are still remembered and inculcated in their experience. For newer and/or younger operators, familiar only with the new technology, the embedded routines are elements in a flow diagram or blocks of software code in a program or concepts in their minds or, sometimes, completely invisible.

For example, tools that cut metal need incremental adjustment by operators to compensate for the wear on the tool bit itself. In a CAM environment, the skill and experience to make appropriate adjustments to the cutting tool may be incorporated into a program block in the central controller of a DNC system that records the usage of each tool bit by the amount of time it has been actively cutting metal since the last time it was sharpened, even if the last time was months earlier. The information on the initial state of the tool bit was entered into the system from the tool crib by the person who ground it according to its specification. The particular location of the tool bit in one machine's automatic tool changer was entered by a setup person who installed it in accordance with a prescribed setup procedure or by an operator who passed a hand wand over its ID (identifier) chip to verify placement (Swenson 1990). Each time the cutting tool was called upon to cut metal, the central controller automatically and continuously compensated its cutting position according to a calculation that recorded actual cutting time, the pressure exerted during the cut, and the estimated wear based on a formula that considered usage and pressure and the characteristics of both the tool metal and the metal being machined. Swenson (1990 p. 33) described just such an application:

Piezoelectric transducers mounted in each tooling position measure the cutting force and feed data to a Krupp Widia tool monitor database. The transducer and monitor can learn a force envelope for each cut of each tool and can build a histogram of cutting force. If that force is exceeded, the system locks the tool by returning it to the magazine at the end of its cut and holding it there until replaced. Meanwhile, the system pulls a sister tool from the magazine, reads the ID chip, loads appropriate parameters, and continues operating. If a tool breaks during use, the cell goes into a "feed hold" state until the machinist decides whether the job can use a sister tool.

THE CHANGING RELATIONSHIP OF WORK AND TECHNOLOGY

The knowledge and experience of the operator who had previously adjusted the cutting tool manually has essentially been embedded in the software routine of the machine's controller. Before computer control, the operator made compensating adjustments to the tool's position incrementally, between carrying out many of the other required operations and routines associated with his or her role. Under computer control, the tool is adjusted continuously, smoothing the cut and easing the wear on the tool bit and on the metal being cut.

When the tool breaks, when it has not been properly ground, when the tool steel does not wear according to the calculation, or when the metal being cut has an imperfection, the operator has to take actions that were not prearranged in the design of the control software. With appropriate skills and proper training, the operator's actions may quickly correct the problem. Alternatively, if the problem that arises is an unusual one, and the skills to correct the situation have not been taught or acquired, or if the experience or intuition to deal with the uncertain situation does not exist or is inappropriate, the tool or the part or the machine may be damaged.

Operator tasks in advanced manufacturing technologies are intermittent machine feeding or machine loading rather than repetitive tasks and include considerable monitoring of machines, controls, and manufacturing processes. Indirect labor and some maintenance and inspection tasks are additional new skills. In some AMT situations, programming has become an integral part of the operator's role. In other situations, all programming is considered the job of a specialist. The skills of maintenance people have similarly changed, in many cases to include programming and electronics.

The roles of operators and maintenance workers change, then, as different skill sets are needed to interface with computer-based technologies. In turn, this changes the relationship between these roles and the new technology. With changed roles and relationships, the social systems differ from those found in traditional manufacturing plants (Davis and Taylor 1976), as do the premises that guide the design of jobs and organizational arrangements.

This chapter addresses the relationships between individuals and computer-based automated manufacturing technologies. The characteristics of AMT that affect individuals and their social systems are first identified. Advanced manufacturing technologies are integrative in nature and demand a corresponding level of integration

188 AMT AT THE LEVEL OF THE INDIVIDUAL

and coordination from the social and support systems that accompany them if a good sociotechnical fit is to be realized. Hence, issues of coordination and integration are discussed early in the chapter. The nature of AMT also gives rise to a series of concerns about the effects of the technologies on the skills of those who operate them and whether they deskill and displace people or provide them with opportunities to enhance their skills.

Computer-based technologies change the nature of the skills required to operate, supervise, and design them. These changes are described in this chapter and are followed by a discussion of how specific roles change to accommodate the skill changes. Operator roles are those most affected, but supervisors, maintenance workers, and design engineers are not exempted from the impact of AMT. The relationship between technology and humans, at the operator level and at the designer level is also brought out in this chapter. The chapter ends with a discussion about the changes in human resource practices needed to successfully support the AMT technologies, particularly training and education.

THE INDIVIDUAL AND THE CHARACTERISTICS OF AMT

Older, simpler, deterministic technologies break down in expected ways. Computer-based technologies, with their complexity, their interdependence, and their as yet poorly understood characteristics (Goodman, Griffith, and Fenner 1990, Patterson 1983), and in which "a network of electrical, electronic, hydraulic, pneumatic, and mechanical devices is subject to the stresses of vibration, corrosion, and electrical failure" (Hirschhorn 1984 p. 72), behave unpredictably. These automated, new technologies are stochastic rather than deterministic (Davis and Taylor 1976). They inject uncertainty into the situations where they are applied. The uncertainty is experienced both with respect to the *nature* of the stochastic events and their *timing*.

Turning first to timing, operators must respond quickly, which means granting them greater discretion over work decisions and more autonomy in their roles if they are to respond appropriately (Majchrzak 1988). They cannot wait for supervisory instructions and this changes the supervisors' roles.

The unpredictable nature of the stochastic event demands a wide repertoire of responses from operators (Davis and Taylor 1976). This affects their training, their skill levels, their styles of learning, and their problem-solving approaches. The uncertainty of unpredictable breakdowns can also lead to restructuring around operators' roles to include team arrangements that can benefit from the experience and ability of several different people, decentralized decision making to provide more local discretion, and flattened organizational hierarchies because vertical control chains cannot contribute to the response rate needed.

The stochastic nature of technology imposes a complex role set on the human and machine or system combination. Routine operations are interspersed with periods of high attentiveness and activity. During routine operations, typical activities may include returning dull tools to tool cribs, setting up parts in fixtures for the equipment's next operation, carrying out minor maintenance on nonoperating machines in the cell or system, and making notes to manufacturing or design engineers for suggested improvements to drawings or procedures. Routine tasks are similar to the kinds that have been associated with manual and deterministic technologies. When a nonroutine variance arises, the operator's mode of functioning changes significantly. He or she must understand how the technology functions in order to isolate the source of deviance. The requirement for attentional commitment suddenly becomes very high. Data-based reasoning is needed to diagnose the error information on the controller screen and abstract reasoning skills are called into play to integrate that diagnosis with both visual data about the part and cutting tool and with a conceptual appreciation of the technology's functioning (Zuboff 1985).

For computer-based manufacturing technologies where the investments are relatively small (e.g., CNC equipment), the consequence of incorrect diagnoses of problems because the technology's ways of failing are not well understood may not be large. For direct numerical control (DNC) installations and FMS, the consequences can be significant. If variances arise in unpredictable ways, operators have to do more than understand and conceptualize the technological process to respond appropriately. They may have to call on their experience and intuition because the effects of the variances were not predicted or planned for, nor allowed for in their diagnostic or problem-solving procedures or their training. If there have been no opportunities for operators to acquire the experience to accompany their conceptual appreciation, there will be no guidance for their intuition when the variance is one never before encountered.

> The legacy of technology-centered design can be stated as follows. Having fragmented the job of the system operator and removed a large part of his/her skilled control of the system, the designer then relies on him or her to deal with all unforseen (unprogrammed) disturbances. However, a technical system that does not provide the experience out of which operating skills can develop will be vulnerable in those circumstances where human intervention becomes necessary. (Corbett 1988 p. 76)

Weick (1990) extended Davis' and Taylor's (1976) observations about the stochastic nature of new technologies by identifying two additional characteristics of new technology: abstract events and continuous events. Abstract events refers to the abstract nature of computer based manufacturing technology, with most of its features buried in "black boxes" beyond the observation of the operator. Continuous events refer to those periods when discrete CAM technology resembles continuous process technology, as when the programmed nature of CAM technologies results in repetitive operations with few interventions. When this is combined with instruction sets that are increasingly embedded in the software and hardware, CAM technologies may resemble some aspects of continuous processes, particularly in the demands they place on the operators of the equipment or systems (Adler 1986, 1988, Davis and Taylor 1976).

The demands abstract events place on the operator and on maintenance personnel are for conceptualization skills that were not required in traditional manufacturing technology. New technology exists as much in the imagination of the operators as it does on the plant floor. ". . . (M)anagers and operators experience increased cognitive demands for inference, imagination, integration, problem-solving, and mental maps to monitor and understand what is going on out of sight" (Weick 1990 p. 14). Operators in proximity to operations can observe metal being cut or plastic being formed or parts being assembled and call on their experience and judgment to intervene to change or adjust tooling or setups or fixtures when they see the need to do so. However, whatever interactions the operator may have had with the control aspects of the technology are increasingly diminished as they become imbedded first in local controllers, then deeper inside local computer controls, which may or may not allow for operator intervention, and then still deeper within the system when, for example, DNC places the computerized control completely outside the operator's access.

For continuous events, especially those occurring in concert with stochastic ones, reliability has replaced the efficiency objective that dominates in deterministic technology. "This shift from efficiency to reliability may constitute the single most important change associated with new technologies" (Weick 1990 p. 11). For example, JIT systems, which are increasingly integrated into AMT environments, limit the availability of backup or alternative parts. They place very tight forward demands on production units for the delivery of quality products on time, that is, for reliability. As the number of different technologies involved in AMT increases—and it is not uncommon to find systems simultaneously utilizing mechanical, pneumatic, hydraulic, electrical, and electronic technologies—and as the number of internal interdependencies also increase, the probabilities of failure increase rapidly. Reliability of each component becomes very important. This is another aspect of AMT that has parallels with continuous process technology.

Weick (1990) has analyzed the arousal state of operators when stochastic events occur to illustrate how the diagnostic process they use makes sense of the symptoms emanating from the technology. The high mental workload on operators forces them to use more of their own interpretations to understand what is occurring, interpretations that may be quite incomplete and so result in severe consequences: "When people use fallible models to cope with stochastic, continuous, abstract events under conditions of excessive arousal, interactive complexity is one consequence" (p. 33).

Integration and Coordination

Advanced manufacturing technologies such as DNC and FMS have highly integrative characteristics that reverse the trend of continuously subdividing tasks that has characterized most of the technology introduced into production domains to date (Lund and Hansen 1986 p. 141). For example, General Motors' Vanguard plant, with twenty-three manufacturing cells and twelve assembly cells, has a four-level Factory Control System (FCS). The first level is the machine controller or process controller for each computer-controlled machine in the cell. The second level is a cell controller (and cell composition can very significantly, given the range of machine tools, material-handling robots, and quality control guages and equipment assembled for each family of parts). At the third level, each cell controller is hooked into a communication network to the FCS as well as the material handling AGV system. At the fourth level, an interface exists between the FCS and the divisional information system (for finance, payroll, etc.) (Painter 1990).

This level of integration changes the roles of operators. Majchrzak (1988 p. 12) reported that in CIM plants, operators spend 12 percent of their time identifying process improvements. In CAM equipment plants, operators spend only 4 percent of their time doing so. She suggests that the level of *integration of the manufacturing technology* will be an important parameter in assessing the impact of technology on human resources (i.e., the integrative nature of AMT is likely to be as disruptive for the roles of skilled trades, engineers, and managers as they appear to be for the machine operators).

Organizational coordination increases per the sequence of coordinating mechanisms listed in Table 6.2. The cost to the organization increases as more coordinating mechanisms are utilized. If the analogy is applied to AMT and is appropriate, the cost of integrating manufacturing technology should increase as the more integrated technologies (e.g., DNC, FMS) are adopted. In fact, ". . . the greatest difficulties have been experienced in development of total factory control systems which efficiently integrate all the various levels of machinery control as well as the ancillary plant functions such as materials handling" (Painter 1990 p. 24).

However, it may also be that such actions, embedding some of the coordination in information technology, merely substitutes one kind of coordination for another. As such, it might be argued that total real coordination costs do not increase. This notion is represented in Figure 8.1, which illustrates a constant amount of coordination but with the mix between organizational type coordinating activities and computer integrated coodination changing as the integrative nature of the manufacturing technology increases from hybrid cells, comprising a mixture of manual and CNC equipment, through to CIM. The assumption in Figure 8.1 is that interdependencies within the manufacturing processes are constant (Klein 1991). In fact, they more likely increase; in which case the right-hand ordinate of Figure 8.1 would be larger than the left, making the figure more of trapezoid than a rectangle.

For those who design organizational structures to organize the work, the stochastic, continuous, and abstract nature of new technology injects uncertainty into the design process in addition to the issue of the appropriate mix of organizational coordination and information technology integration. This is evident from the frequent instances

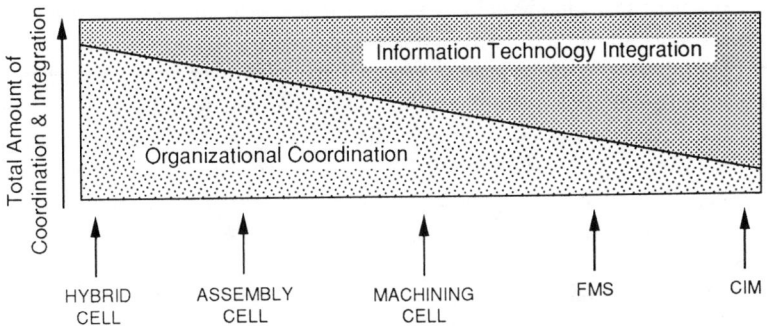

FIGURE 8.1. Organizational coordination and information technology integration.

of companies using different organizational arrangements when installing almost identical equipment (Gerwin and Tarondeau 1982, Goodman, Griffith, and Fenner 1990, Majchrzak 1988). Wilkinson (1983) has documented the situation particularly well in describing the case of two operations that installed the same CNC machine using almost opposite assumptions about the capabilities of the operators who would be tending them.

Deskilling or Reskilling

AMT has led to high labor displacement; reputed to be as high as 30:1 in FMS, though 7:1 to 10:1 is the range more often cited (Lund and Hansen 1986 pp. 40, 41). In CNC-based machine shops in engineering, reductions of 30 to 50 percent in the number of operators and setters has been reported (Bessant and Senker 1987).

In addition to labor displacement, these new manufacturing technologies have also created high controversy with respect to their impact on the design of jobs and on the nature and distribution of skills. The widely held deskilling criticism of new technology is that human know-how is increasingly incorporated into technology (Braverman 1974, Wilkinson 1983) and this deprives workers, draftspersons, and other specialists of their power (Cooley 1984, Noble 1979) and leads to either a deskilling of them by confiscation of their knowledge (Cavestro 1990) or to a polarization of their skills into some that are diminished and a smaller set that are increased. For example, Blumberg and Gerwin (1984) found that direct workers in an FMS establishment suffered from loss of control and reduced discretion. The radical criticisms contend that new technologies, by confiscating knowledge, aim to extend to these areas the dehumanizing principles of the assembly line. Unfortunately, they offer as alternatives only a return to systems that limit the excesses of new technology by restricting its application.

The deskilling argument is based on two factors (Wall and Kemp 1987): the first is that new technology increasingly provides opportunity to separate control from execution (e.g., as in DNC). The second is that management will take advantage of this opportunity to reduce costs and exercise more control.

This deskilling perspective has itself been criticized on the grounds that it is a poor representation of the reality of work and that it is only a representation of prescribed work (Cavestro 1990). The critics point out that in reality, workers do not carry out their tasks in the numbing way that scientific management prescribes. Some reskilling is, in fact, currently taking place, with machine operators' situations cited as examples in which support skills such as maintenance have been added to their tasks. Rothwell (1987) also notes that deskilling and reskilling can take place simultaneously, as when an operator loses direct machining skills but acquires CNC and programming skills. Furthermore, ample evidence exists to counter the belief that managers actively pursue opportunities to deskill the workforce in the existence of many new technology implementations that have not done so (Wall, Clegg, and Kemp 1987).

Another challenge to the deskilling criticism questions the linearity of technology's advance and suggests that its rate of change in manufacturing may be curvilinear with respect to its effects on skill requirements and the quality of jobs (Adler 1989a). That is, only the early phases of mechanization are deskilling and degrading with respect to jobs and subsequent technological advancement will upgrade skills and the quality of jobs.

Adler (1989a) designed a study to explore the deskilling controversy, to replicate some of Blumberg and Gerwin's findings, and also to test the above-mentioned curvilinear model. By self-report, the FMS operators in Adler's study reported a significantly increased level of skills and experience in their jobs over previous jobs, which included working with conventional machine tools and NC machine tools. Furthermore, the FMS operators found their jobs to be improved over both conventional and NC equipment (i.e., the curvilinear model, like the deskilling criticism, was not supported in this study).

Neither the Adler nor Blumberg and Gerwin studies are very satisfactory tests of the deskilling controversy. Each study raises too many methodological questions, particulary because the questions asked to explore the operator-technology interaction may be too simplistic for the complexity of the situation, complexity that is a product of the training supplied, the changed relationships with supervisors and specialist staff, the high interdependencies experienced (Adler 1989a), the differential work demands in different sites studied—including the approach to the site as a work-producing entity or as an experimental location, different resulting pressures on operators and other FMS line and staff personnel, the noncomparability of product lines, and so on.

The deskilling-reskilling controversy, at it has been framed here and is usually presented in the AMT literature (Wall, Clegg, and Kemp 1987), may becoming obsolete. It is probably one of the false dichotomies that Roethlisberger (1977) has warned against. One can use technology to deskill workers and others and diminish the quality of their jobs and their careers. Equally, technology can be used to upgrade people's skills and provide new career opportunities for them.

What the controversy makes clear is that there are choices in the design and implementation of new technology that have implications for both productivity and the quality of working life, choices that can only be made if the range of alternatives is understood (Susman and Chase 1986, Wall, Clegg, and Kemp 1987). Technology is not neutral (Patterson 1983); how it is configured dictates particular outcomes that influence the deskilling vs. enhanced skilling or the autonomy vs. control arguments toward one orientation more than the other. But it is not the technology doing the dictating and it is nothing inherent in the technology that is doing so. The design of technology is *always* informed by humans with assumptions about the behavioral relationships between machines and people, even if those assumptions are imbedded deep within their psyches and out of the range of their consciousness. It informs the shape of the line they draw on paper or the direction of the cursor they move on the CAD screen.

The debate over this issue will likely continue. Some of it is driven by ideological positions; for example, managers who believe AMT can reduce their dependence on unions who are unwilling to be as accommodating as they might want. Some of the controversy has been stimulated by a poor understanding of these technologies;

for instance, the desire on the part of some managers to use technology to deskill their work forces so they can operate with a lower level of and less expensive skills. As Adler (1986) has noted, this "wishful thinking" outcome seldom occurs.

The deskilling-reskilling debate has centered on the factory floor. With increased use of expert systems, there will likely be a resurgence of the controversy at the level of professionals and managers. The more expert systems are able to incorporate the knowledge of people at the workplace, the more they will intrude onto domains that will again raise questions as to what should be the work of people and what should be the work of machines (Jordan 1979). The issue is likely to be strongly revisited first in the CAD/CAM areas, focusing on the roles of engineers—one step up from the already mentioned concerns for draftspersons and designers (Cooley 1984), but expert systems to replace the functions of management will not be far behind. At that point, the controversy will move to a different plane because until now much of its ideological basis has been pitted as a technology vs. workers issue. At the level of technology vs. managers, the controversy may actually fade. Managers do not have strong unions to defend their interests.

THE NATURE OF SKILLS

Stochastic events, continuous events with reliability requirements, and abstract events require skill sets quite different to those used in situations where the technology is deterministic and predictable (Davis and Taylor 1976, Weick 1990). There appears to be general agreement that the direction in which skills evolve under these technologies is from "hands-on" or motor skills or physically involved and tactile skills to those that are conceptual, cognitive, and based on abstract understanding (Cummings and Blumberg 1987, Lund and Hansen 1986, Wilkinson 1983). Davis and Taylor (1976), reported on a study by Hazelhurst, Bradbury, and Corlett (1969) that compared the skill requirements of NC and conventional machine tools for eight pairs of jobs in four companies. Although NC tools reduced physical effort, demand for motor skills, and the number of decisions an operator was required to make, they also involve an appreciable increase in demand for perceptual skills (machine monitoring and controlling) and conceptual skills (e.g., interpreting drawings and instructions and doing calculations).

Conceptual skills and cognitive understanding increasingly characterize the skill requirements of automated technologies. However, they do not completely replace the skills that operators' acquired in working with manually controlled machines and traditional technology. Despite the embedded nature of automated technologies, operators can still directly observe deviations or variances in metal cutting or plastic molding or component assembly. The appropriate actions to take when such variances are observed are informed by previous experience or intuition as much as by conceptualization and/or abstract understanding. Furthermore, the nature of stochastic events in AMT strongly underlines the need to act quickly, an act that experience and intuition are sometimes best equipped to inform.

The different theorists that have approached the understanding of the required skills needed to cope with computer-based technologies appear to have converged on a similar set of the skills. Lund and Hansen (1986 pp. 92, 93) developed a list of the minimum set of skills required by factory workers to cope with computer/telecommunications technology that appears to subsume most of the recommendations. The list, in no particular order, is a comprehensive way to describe the nature of the skills of operators and others, such as engineers and maintenance personnel, close to the factory floor in AMT environments. Lund and Hansen drew from experiences in continuous process plants because the computerization and closed-loop automation that began in the 1950s in that industry had become a mature technology with characteristics resembling the computer-based technologies of AMT environments. Although others have put forward a similar view about the similarity of AMT to continuous process operations (Bessant and Senker 1987), the question of how similar and how different they are has been speculated on but not studied. This is explored in more detail in Chapter 13.

Visualization

The embeddedness of so much of computer-based technology into the equipment and beyond the concrete view of the operator or maintenance person requires an ability to manipulate mental patterns. The operator may observe actions at the point where operations are physically occurring (e.g., the point at which an electronic component is inserted by a CNC machine or the tip of a tool that is cutting metal). However, the connection of those events to their controls requires an ability to visualize the relationships in the operator's mind, in effect to construct a mental flow chart or block diagram.

Some AMT operations must be monitored at a distance from the actual equipment, sometimes for safety reasons and sometimes because of the large physical area covered by a CAM technology, such as FMS. Visualization in these kinds of situations is initiated from data on a distant terminal. Such visualization has some parallels with *data-based reasoning* (Zuboff 1988), a skill in which only key pieces of relevant information are extracted from large amounts of data presented in order to arrive at a reasonable interpretation of the phenomena being examined. In the words of one operator:

> You must keep important data in your mind as you continue to scan. People learn to organize data in their minds. They build models in their heads about what is really happening, and they build on the model with data until they have a complete picture. (Zuboff 1985 p. 11).

Information system designers can aid the operators and others who must diagnose from the screen data by providing explanation and elaboration. For instance, in one petrochemical refinery, the information system was designed to advise the operator of the meaning of an out of specification condition, ". . . our target temperature is 250 C, we are now actually at 225 C, the incentive to attain target is an improvement in yield outturn to the extent of $550/day" (Halpern 1984 p. 50). However, this requires an ability to simulate the system reasonably accurately and prepare *a priori* responses to problems that can be anticipated. Stochastic technology with unpredictable variances makes the accomplishment of this unlikely despite the increased usage of simulation and expert systems in complex new technologies (Liu et al. 1990).

Conceptual Thinking (or Abstract Reasoning)

This skill category arises from the need to conceptually think through (Adler 1986, Zuboff 1988) and analyze a technological system that is increasingly integrative in its nature. For design engineers, software engineers, or systems analysts, design in this arena requires an appreciation of production processes, business attributes (e.g., accounting, inventory management), ergonomics, and psychology apart from the basic requirements of metal cutting, computer engineering and systems architecture (Lund and Hansen 1986). If the setting is an assembly rather than a machining operation, the sociology of work teams may replace some of the above listed knowledge categories. Support staff and engineers and managers with an integrative orientation are clearly required, but realistically, the diversity of knowledge and skill called for will likely demand teams drawn from engineering, manufacturing and human resources.

Mokray, in Adler and Howard (1990), suggests that skills cannot be treated in isolation of each other and their use follows an integration from lower order to higher order in keeping with the needs of the situation. She proposed four layers of increasing skills that must be learned as the complexity and integration of new technology increases:

- Doing skills—the actual performance of the task
- Controlling skills—based on techniques such as JIT or SPC
- Learning skills—to reflect on ones work and improve on it
- Creating or discovering skills—the process of innovation to find new patterns or new contexts

Zuboff (1988) compared the work organization to a database and suggested that the quality of skills that people bring to the new information is a determinant of whether the emerging database will be experienced as "overload or as an opportunity to reach for a new level of comprehension and innovation" (p. 10). Mastery of this information is accomplished through the *intellectual skills* of abstract thinking, data-based reasoning, and theoretical understanding.

As the complexity that requires conceptual clarity increases, there is a serious danger of overload and degradation of human performance. Weick (1990) has suggested that the performance of complex tasks is more vulnerable to disruption from excessive arousal than simple tasks. "Those disruptions that do occur (for example, perception narrows, and dominant responses dominate) tend to reduce learning and induce error" (p. 10).

Understanding of Process Phenomena

A process orientation is sometimes used to contrast with a technological approach.[1] The latter would attempt to automate as much as possible. However, using people who have existing process related skills and qualifications (Schultz-Wild 1990) to bridge the gaps in process sequences may be considerably less costly than attempting to fully automate a process sequence via hardware and/or software. Zuboff (1988) focused on information technology and drew a parallel distinction between its use as an *automating* technology that can be used to displace human effort and to substitute for human skill and as an *informating* technology whose implications, although still not well understood, can generate information "about the underlying processes through which an organization accomplishes its work" (Zuboff 1985 p. 8).

Schultz-Wild (1988 p. 82) has referred to these process-related skills as "taking actions and decision-making close to the manufacturing process on the shop floor." He suggests that doing so can:

- Save planning and investment costs in complex and expensive automation technologies by using workers who are able to bridge gaps in the process sequence
- Save implementation costs
- Offer rapid introduction of technological and organizational innovations
- Reduce training expenditure
- Decrease running-in time
- Reduce the risks and duration of systems failures
- Make better use of manufacturing equipment
- Reduce labour costs, because of the broad training of operators

Others have referred to this differentiation between integrative[2] and technological approaches as the difference between a philosophy of operator control vs. specialist control and have empirically shown operator control to be more effective when variances are high in CNC assembly operations (Wall et al. 1990). The success of process-related skills or operator control is dependent on the availability of operators with the requisite skills. Schultz-Wild (1990 p. 89) has identified the training of the German skilled worker ". . . having practical and theoretical production knowledge and experience and having had further training in the field of information systems and control technology . . ." as the type of skills needed.

The problems associated with continuity and reliability are different than those associated with discreteness and efficiency (Weick 1990). The technologies in CAM

---

[1] Corbett (1988) and others use the term "technology-centered approach."
[2] In the literature there have been several variations on the term "integrative" (e.g., "human-centered" (Corbett 1988), process-oriented (Schultz-Wild 1988), and computer/human integrated manufacturing (C/HIM) (Haywood and Bessant 1990)).

environments can be highly integrated and workers and specialists who deal with them may have to reassess their ways of operating and thinking. Thinking in terms of discrete products usually dominates discrete technology, but computer control invites process thinking. The next *user* of the product is not an anonymous inventory clerk but rather a known colleague, perhaps a member of one's own work team or of a subsequent team, depending on where the boundaries are drawn. There is a human interaction incentive to be concerned about the quality and timeliness of what one produces, an incentive that is often emphasized by conceiving of the next user as one's "customer" (Schonberger 1990). With a flow orientation, there is a holistic perspective about the product that is not captured in discrete manufacturing.

Statistical Inference

High quality standards have become a requirement rather than an option in manufacturing firms. One of the key attributes of success in increasing quality levels has come from decentralizing responsibility for quality measurement and control to the level of the operator who produces the product. This has meant teaching operators to appreciate trends, limits, and the meaning of data through charts, graphs, and other tools of statistical process control. Statistical inference is a learned way of thinking and a necessary skill for those closest to the production of goods. More significantly, it is a required way of thinking when technology is stochastic and variances arise in a random and unpredictable way. A mind set geared to deterministic technology would have difficulty comprehending the nature of the variances encountered in AMT equipment.

Verbal Communication

Effective verbal communication has become a required skill for operators who must communicate with engineers and maintenance personnel to correct problems and foster process improvements. An encountered variance may not be present to be objectively displayed. Instead, it must be explained from an interpretation or an inference drawn from a piece of data, by describing an event that has already occurred, by calling on the type of conceptual understanding cited earlier, or from some combination of these. The skill to communicate effectively can be an even greater requirement in assembly operations where parallelization results in team organizations that must manage themselves autonomously or in cell manufacturing that has similar organizational arrangements.

The demand for skill in the area of verbal communication is also high in engineering and design areas because almost all CAD/CAM organizational arrangements involve teams drawn from people with different orientations (Dean and Susman 1989). The long time orientation of engineers and the short time perspective of manufacturing personnel can be the seed-bed of misunderstanding (Lawrence and Lorsch 1967). An inability to communicate effectively will likely result in friction and conflict.

Verbal communication also serves as a surrogate phrase for a range of behavioral skills associated with the increasingly interdependent organizational arrangements found in CAM environments. A range of interpersonal skills are needed as responsibility is decentralized to the operator and team level, skills such as conflict management, consensus decision making and listening (Adler and Howard 1990). Team-based structures that accompany integrated technologies need to match the technologies' integration with comparable coordination and communication within the team. In Thompson's (1967) framework, reciprocal interdependence requires mutual adjustment to achieve required coordination requirements. Mutual adjustment, in turn, requires good verbal negotiating skills.

Attentiveness

The paradox of operating with stochastic technologies is that some systems can function for hours in relative stability during which staying alert and attentive is difficult to do (Lund and Hansen 1986) and then be punctuated by frantic periods of extreme arousal (Weick 1990) when the system deviates from plan and, sometimes, in perilous ways. Zuboff (1986), referring to this phenomena as "attentional commitment" quoted one operator:

> Sometimes hard work is the easier kind of work to do. I find that now I have to pay more attention. I have to watch things closer. . . . There is less manual work and a lot more strain. You can't break your concentration from what you are doing in the middle of a sequence. You have to concentrate; you have to pay attention. (p. 123)

Individual Responsibility

Uncertainty can arrive at the level of the operator from the hierarchy, because of product-market changes or from process uncertainties such as material defects or machinery breakdowns. The assumption of responsibility for the events that occur becomes one of the most important of the new skill requirements (Lund and Hansen 1986). Bright (1958, as referenced in Davis and Taylor 1976) referred to responsibility as the only job demand that consistently increased with rising technological levels. Speed of response alone would dictate the need to assume local responsibility, but so too do current concerns for excellent quality and for customer service. Flattened organizations with supervisory roles that are changing to devolve more autonomy, discretion, and decision-making authority onto team members are putting pressure on people to assume more responsibility for results at the level of their respective roles.

At the same time, as the locus of control is shifting from that of external direction towards internal responsibility, information technology is entering the picture in ways that can lead to alternative outcomes. The information available at the level of engineer or operator is often also available at the level of supervisor or manager. It can be used to monitor individual actions in a policing way, or it can be used to intervene in a helping way. Errors made can be treated as learning opportunities or in a punitive fashion that can build a climate in which experimentation is avoided because the cost of failure is high. How the information technology is used can have significant impact on how much individual responsibility is assumed.

ROLE CHANGES IN AMT ENVIRONMENTS

Changes in skill requirements because of new computer-based technologies have had and will have profound effects on those who work with the technologies. One view of these changes is that they will lead to greater centralization and control, create more bureaucratic structures, and offer less discretion to those who operate and maintain the technology. The view taken here is the opposite: that computer-based technologies will foster decentralized decision making, delegation of authority and local control (Burnes and Fitter 1987). This dichotomous position runs throughout the conceptualizations of advanced manufacturing technology. It appears in the deskilling vs. enhanced skilling debate, in the centralized decision making vs. flattened structure perspectives, and in the autonomy and discretion vs. coordination and control arguments.

It is clear that the choices that are made affect the roles of those who interact with AMT, some roles much more than others. Operators, maintenance workers and skilled tradespersons, and supervisors are the roles most directly affected. The roles of technicians and professionals are also affected, although less so at the moment. However, as the boundaries between manufacturing and the design areas continue to disintegrate, it is likely that these roles will also undergo as significant a degree of change as those currently closer to the physical technology and perhaps an even greater amount of change.

Operators

There is increasing overlap between operator and maintenance functions in AMT environments, even to where "technician" and "maintainer" terms are sometimes used as operator titles (Painter 1991). Operators may clean sensors, make minor adjustments to equipment, and carry out other routine maintenance functions depending on how much of the maintenance skills they have acquired and how much organizational change has taken place to reallocate roles. Operator roles have also changed to include a greater mix of direct labor (materials loading, process monitoring) and indirect labor (troubleshooting, scheduling) as operators assume some of the activities previously carried out by other manufacturing support personnel (e.g., setup specialists, inspectors).

> The job of Technician at Tektronix Forest Grove involves mental as opposed to manual dexterity; namely judgments about the proper temperature and speed at which to run the soldermasking coating process, and organizational skills to reduce changeover time at the drilling machines, and continuous learning of computer programs and their many quirks. (Painter 1990 p. 30)

The demands of AMT for new skills have expanded the operator's role set to include portions of adjacent and supporting roles. The effect is to develop a more comprehensive (i.e., more integrated) operator role, one more able to adapt to the extended integrative requirements of the AMT that the operator must confront. The decentralization of more authority for decision making and more responsibility to the level of the operator, the provision of the information and the training needed to make the appropriate decisions, the reduction of job classifications to facilitate the creation of holistic roles, cross-training, and job rotation opportunities to practice the expanded roles are all part of the support required for restructuring these critical roles in AMT environments.

In effect, the collapsing of organizational boundaries that is being experienced throughout manufacturing organizations is also being replicated at the operator level as a new configuration of roles has evolved to reflect the characteristics of computer-based technologies and their different skill requirements. The constraints on further integration are the specialized nature of the skills and the difficulty in learning them, and traditional job demarcations that exist as a result of job classification systems and/or because of collective bargaining agreements.

In one plant, operators were able to learn and apply maintenance skills continuously and were reinforced with the incentive that they could progress to the level of competence of an electrician or a process mechanic (Painter 1990). A continuous process plant structured shift arrangements so teams of operators have up to twenty days on "day assignment" out of every sixty-three days of work, during which they join the journeymen craftsworkers as their assistants and in the process acquire maintenance skills as second skills (Halpern 1984).

Cross-training on each other's machines is relatively easy to accomplish when machines are organized into cells. Job rotation augments this process if appropriate training is carried out.

Maintenance Workers / Skilled Trades / Craftsworkers

Electrical work has changed the most among skilled trades as a consequence of computer-based technologies. Because of the high information technology content, electricians have become the skilled workers most involved with the new technology. Computer programming and electronics have become integral parts of their jobs and many electricians have had to acquire knowledge of their systems' architectures (Painter 1990 p. 45). With multiple levels of communications and controls it is sometimes difficult to determine at which level a fault lies. Problem diagnosis has always been a major part of an electrician's job, and computer-aided and expert systems will increasingly ease their diagnostic work. With computer-based technologies throughout the manufacturing environment, "the diagnostic power" of problem solving has shifted to them (Painter 1990). With it, electricians have acquired considerable influence amongst skilled workers because their actions have an important effect on downtime.

Mechanial work has remained relatively stable in nature, although there is more work with pneumatics and hydraulics than in the past. Computer interfaces have been developed to help mechanics with adjusting and repairing tasks (and as in the case of electricians, there will probably be an increase in the use of expert systems for these purposes). Predictive maintenance is the new dimension of mechanical work and total process maintenance, with its preventive orientation—to minimize downtime, will become an increasingly important aspect of mechanical work. Ergonomics have improved for most maintenance workers (e.g., platforms on AGVs that tilt to ease the physical arrangements for assemblers also ease access for mechanics), but the small size of so much of the microelectronic component-based controls and equipment require "baby fingers" in order to work on repairs (Painter 1990 p. 44). Concerns are also high about hazardous chemicals and security of lockout that is remotely controlled (with at least one plant implementing a warning light system on the robot to indicate whether it is under remote or direct control (Painer 1990 p. 44).

Computerized technology has also changed the nature of the diagnosis-repair cycle:

> Unlike work with the electro-mechanical technology where the source of problems was much more apparent while repair was usually time-consuming, the relationship has been reversed for maintenance of computerized technology. Identification of problems has become more difficult, while repair can often be done very quickly. (Painter 1990 p. 44)

Although there has been a reorienting of influence within skilled trades in manufacturing toward the electrical side, the skilled trades have remained the high status and highest paid of the hourly work categories. Skilled tradespeople have generally not supported multiskilling efforts and skill-based pay system. The acquisition of journeyman status takes many years, and the skilled trades have been reluctant to enter into programs that invite them to acquire second skills or to share their acquired skills with others. They have usually been the group that has fought against pressure to reduce job classifications and undermine carefully won protections under collective agreements.

Supervisors

There are opposing views as to whether AMT will lead to a reduction in the need for supervisors or whether it will increase the demands on supervisors. Burnes and Fitter (1987) have advanced a series of arguments on both sides of the issue as follows.

AMT requires fewer operators, which in turn requires less supervision; semi-autonomous work groups have proliferated and assumed many of the tasks of supervisors; CAPP and information and control systems preplanning remove many of the control tasks that supervisors would otherwise carry out; and the more differentiated specialist roles that evolve in AMT environments assume some of the functions traditionally carried out by supervisors.

The arguments on the side of an increased need for supervision as a result of AMT are that mix, changeover, modification, and volume flexibilities are important and impossible to fully predetermine; supervisors are needed to handle these different manufacturing aspects; breakdowns due to process uncertainties will also arise and for reliable operations and effective equipment utilization, supervisors, ". . . must have a good working knowledge of the equipment so they can make decisions such as when it is necessary to call a maintenance person, and what kind of maintenance person is needed" (Blumberg and Gerwin 1984 p. 125); and supervisory control in addition to automated processes may be the most effective way to manage integrated manufacturing systems.

A wide variety of roles is evolving for supervisors in advanced manufacturing technology situations. For example, some supervisory roles are changed rather than reduced or increased, as when supervisors buffer a team from uncertainty by managing the team boundary and/or acquire needed resources for team members. Variety in the way AMT is diffused and even more variety in the beliefs managers bring to the design of supervisory roles are primary reasons for the differentiation but so too are the issues previously mentioned as affecting the roles of operators, such as the flattening of organizational structures, the decentralizing of autonomy and discretion to workers and teams, and information technology that makes information available simultaneously to both those who report to supervisors and those to whom they report. Although it is difficult to generalize about patterns in the range of roles of supervisors in AMT environments, Painter (1990) has identified four different supervisory role variations in a study of jobs in fifteen situations with new, computerized technology (ten were manufacturing sites, of which six were greenfield sites, six were unionized, and seven were in discrete production technology).

The *shift supervisor/shift superintendent* role is closest to the traditional supervisory role, but the new roles tend to have wider spans of control (typically twenty to twenty-five reports) because layers have been removed and more responsibility is delegated downward. Some supervisory roles have begun to acquire managerial dimensions and some supervisors work as members of "management or leadership teams" to oversee all operations and because teams of supervisors may be the only way to respond to the complexity of integrated manufacturing systems (Burnes and Fitter 1987).

The *process/shift team coordinator* role, more often found in continuous process technology, has a span that covers the entire plant process. It is a role that tends to share substantial authority with team members—and it is often a single team that oversees an entire process plant—and is expected to get very involved with the team in problem solving.

The *unit/zone manager or coordinator* is responsible for multiple shift crews or several cells in physical proximity producing similar families of parts.

In the *first-level management team*, members have no direct supervisory role, but have substantial managerial responsibilities.

> Each group of "A" Team members at CGE Bromont are accountable for a small business (e.g., small rotor production) within the larger business of the plant. "A" Team members are responsible for their own budgeting, hiring, payroll, production

scheduling, quality, and technical development. Only 10–20% of the time are "A" Team members on the shopfloor. (Painter 1990 p. 52)

Most supervisors gained their positions because of craft or technical skills, but increasingly they do not maintain the technical expert role as more specialized knowledge devolves to operators and maintainers or because the increased sophistication of the technology resides with staff specialists. In addition, their enlarged spans of control and an increasing load of administrative tasks do not allow them time to perform specialist roles. In the process, they may experience a loss of control to specialists such as electronics maintenance and quality control when they have to call on their expertise to handle problems with the technology.

The supervisory position is becoming an entry level one to management for graduate engineers, partly because of the challenge of its increasing technological complexity, partly because the role retains a high accountability for performance, but especially as manufacturing has again become a viable career path to the top.

Supervisors have acquired increasing responsibility for people development. As such, they need to learn interpersonal and team-building skills. Supervisors' roles as coaches and boundary managers have become more prevalent. For example, they have to interface more with technical staff because of the technological complexity of AMT. They need excellent communication skills, both for their various boundary roles and for their relationships with subordinates.

With the integrated nature of advanced manufacturing technologies, coordination is a major challenge for supervisors. An additional new dimension of their roles is that they are expected to be involved with innovations to manufacturing processes, often as members of CAD/CAM design teams, and they are expected to manage the change process as new technology and its accompanying organizational arrangements are implemented.

Opportunities to manage in participative styles can reduce the tension of their traditionally adversarial interactions with hourly and union workers. It can also reduce the stress of a continuously increasing and highly variable workload (Blumberg and Gerwin 1984). For some, the opportunity to work in a participative style makes supervisory work rewarding (Painter 1991); but it can have the opposite effect if supervisors have no intellectual or personal commitment to a participative philosophy.

Technicians and Computer Specialists

The dividing line between maintenance/craft and technician skills is a difficult one. Sorge, Hartman, Warner, and Nicholas (1983) noted a tendency to use tradespersons or more skilled people as CNC operators because of the need to integrate operations and machine setting. German technician level training follows after craft training and builds on it, giving these workers a combination of the skills needed to operate the more complex computer-based technologies. Sorge et al. (1983) contrasted this with the less formalized British training in which people were either trained as craftsmen or as technicians but not both.

CAD has changed draftspersons jobs substantially. Cooley (1984) has noted their loss of drawing and detailing skills as they have acquired facility with the CAD equipment. Furthermore, the desire on the parts of managers to minimize their investment in relatively expensive CAD equipment, coupled with a need to maximize utilization of the equipment that has been purchased has led to shift work for draftspersons who never had to work off day shift. Similar situations have been described for engineers using CAD equipment, although there has tended to be more discretion for them in the choice of such work. Shift work unquestionably connotes a deterioration in the quality of white-collar working life.

Computer specialists have become critical for AMT, but they have also been difficult to find and retain. For some, their limited business and manufacturing and factory experience constrains their usefulness. In other cases, the remote locations of many factories do not have the attraction of big cities and with the general shortage of computer specialists, the factories do not provide enough incentive to attract or retain them.

Consequently, some innovative responses have arisen. One remotely located "feeder plant" that machined parts for turbines and generators in group technology cells with DNC found itself unable to retain computer programmers for precisely the reasons previously mentioned. It was quite successful, however, in retraining several of the local machinists to become programmers and received the additional bonus of having programmers who understood metal cutting, machine tools, and factory life. Perhaps the most innovative of these responses was that of Teli in Sweden (Stymne 1989). It was faced with a surplus of blue collar workers in a country where layoffs were not acceptable and with a principal supplier (Ericsson) in need of considerable computer programming help as it upgraded its switching products from mechanical to electronic switching. Teli responded by successfully teaching a team of eighteen blue collar workers how to program.

Professional Engineers and Managers

Engineering roles have changed considerably as a result of AMT. Designers working with CAD/CAM have had to learn to appreciate manufacturing, work in teams comprised of people of nonprofessional status—draftspersons, manufacturing technicians and managers, machine operators, and the like—and most of all compromise on "good engineering" designs in favor of designs that emphasized producibility and reproducibility. In effect, designers have had to acquire both engineering and management skills in order to design for manufacture and/or assembly.

One type of manufacturing engineer has also had to change roles but in an opposite direction. With the integrated characteristics of AMT, manufacturing engineers who previously spent their time as troubleshooters on the machine shop or assembly shop floor have been delegating those tasks to operators and maintenance workers and taking more time with "systems design" matters (Painter 1990).

As the complexity of AMT increases, some manufacturing managers do not have the technical expertise to challenge the engineers and specialist staff. Others have had primary roles as information communicators and these have become unnecessary as information technology has made the same information known and available to all simultaneously. Coupled with pressure to flatten organizations, this has placed enormous stress on the ranks of middle managers, a stress that is largely being removed by extensively thinning these ranks.

At the same time, there has been a new vitalization in the career trajectories of manufacturing managers. Those with qualifications have attractive opportunities as AMT spreads; however, for the most part, the base of those qualifications is an engineering degree. The base, however, is only that. Individuals are needed who can design integrated systems; have a technical understanding of production processes; can grasp marketing, inventory, and accounting; and have an understanding of psychology and physiology and an appreciation for the politics of their institutional structures (Lund and Hansen 1986).

WORK DESIGN

Unpredictable and unanticipated variances are inevitable at the interface of the operator and the technology and emanate from a range of causes. Some are direct effects of product-market uncertainties created in the external environment and are manifested at the operator-technology interface as variety in the parts handled, as major and minor design changes, and as changes in production levels. Other variances are the result of process breakdowns, some due to machine failures and to uncertain rerouting alternatives that must be selected to compensate for the preferred equipment not being available, and some due to variations in the composition and dimensions of the input materials.

Systems-induced variances result when the complexity inherent in factory systems with a range of programmable controllers, computers, and communication systems (e.g., direct numerical control, local area networks, and wide area networks) cannot anticipate all system errors. Corbett (1990 p. 115) noted that ". . . when software is developed to replace certain operator functions, this software tends to create uncertainties that are more difficult to control because they are less 'visible.'" System-induced variances also result when several different manufacturing systems are linked together (e.g., MRP and automated material handling through programmable robots, high-stack retrieval equipment and AGVs, automated inspection, and gauging systems). A third source of system-induced variances are the collapsing organizational boundaries that draw different organizations and organizational units into manufacturing's immediate orbit each of which injects variety from their different operating philosophies, cultures, systems, and procedures (e.g., just-in-time systems, computer-aided process planning, computer-aided design, design-to-manufacture teams, and so on).

AMT system designers have three options to control problems and unpredictable disturbances (Corbett 1988). They can attempt to eliminate all material imperfections entering the system and ensure high quality and reliability of all tooling, maintenance, and other processes that support operations. In our conceptual model (Figure 2.1), this is a strategy to reduce uncertainties. However, despite total quality management and statistical process control and total process maintenance and "zero defects," perfection is still an ideal, not a reality. Second, AMT can "predict all possible disturbances and develop software and hardware to control them" (Corbett 1988 p. 76)—the technological approach. This approach includes predicting the behaviors of people involved in the production process, a somewhat daunting task unless minimum levels of behavior are established for each task and position. If this is done, this then becomes the traditional hierarchical and bureaucratic model with its associated dysfunctions. Third, they can explicitly plan to depend on the skills of a trained operator with a wide repertoire of responses to be applied against a range of possible situations that can arise—the integrative approach.

Referring again to the conceptual model, the second and third approaches are adaptive strategies, however, the manufacturing philosophy that informs the design of the interface between the operator and the technology will differ significantly for the integrative approach from that which informs a technological approach.

> The interaction between human and computer may thus be viewed as a social interaction between operator and designer, in which the designer predefines the situation through the type and scope of the information given to the operator. (Corbett 1988 p. 78)

Complementarities between Technology and Workers

Industrial engineering has conducted a long and still unsuccessful search for complementary relationships between the things that machines do well and the things that people do well (Jordan 1979). In AMT, this search is thrown into high relief with the changed relationship between technology and the operators and maintenance workers who interface with it directly. Hirschhorn (1984) pursued this theme by noting that the control of expected errors or steady-state functions have been built into the feedback-based machine controls, but the control of unpredictable events depend on workers' knowledge, attention, and watchfulness, whether the event is due to machine breakdown or material defects or variances arising from the design of the machines and equipment. Hirschhorn suggested that there is a *complementary* relationship between automated manufacturing systems and workers with respect to a typology of errors: "machines control expected or "first-order" errors, while workers control unanticipated or "second-order" errors" (p. 72).

Corbett (1990 p. 115) noted that one of "great ironies of technological design" is that it is the operator's ability to cope with uncertainties that ensures system reliability, whereas at the same time system designers with a technological orientation attempt to design out the human element in order to eliminate it as "the major source of uncertainty." Nevertheless, it may well be desirable to design AMT systems to remove as much of the routine work as possible by giving it over to the systems' controls. Combining automation and computer control and expert systems can remove large amounts of activity from what is normally the domain of human workers. In many cases, the eliminated tasks can be performed more effectively by the technology and such activity, which is essentially machine extension activity, is better allocated to the technology in the long run. Although this frees operators and maintenance workers from carrying out routine activity during normal operations and does give them time to carry out support and preparation and administrative functions, it also offers them limited opportunity to apply their conceptual understanding of the technology and integrated manufacturing systems or to hone their concentration skills or to gain experience with the technologies and systems.

Errors and breakdowns and unpredictable events are inevitable in automated manufacturing environments. They cannot all be planned for. They cannot all be simulated beforehand in a training environment so adaptive routines can be practiced. There is an inherent uncertainty in complex human-technology interactions that cannot all be thought through in advance. The design of the technological system must take account of the need of those close to the source of variances to possess the requisite understanding (e.g., visualization, conceptualization) and to be able to practice the requisite skills (e.g., attentiveness, analysis, coordination, communication).

Appropriate design for such conditions must see the operator's role as primary in the control of the work, or in Hirschhorn's (1984 p. 73) words, ". . . we would then see the worker moving from being the controlled element in the production process to operating the controls to controlling the controls." The technological response views the operator's role as secondary to the technological design. If the manufacturing system is to be one that evolves, if the system is to be a learning system, if continuous improvement is an objective, if joint optimization of the technical and social systems is to be achieved, then the roles of human operators in the system must be complementary to the technological system (Trist 1981) and not secondary to it. This cannot be accomplished without putting discretion into the operator's role.

Choice in the Control of Work

Wall et al. (1990) conducted a longitudinal change study in a department assembling printed circuit boards with seven CNC insertion machines. Three of the machines were categorized as high variance systems because they were particularly prone to operational problems. The other four low-variance machines required considerably less human support (downtime averaged between 25 to 35% of that of the high variance machines). The work was organized initially according to *specialist control*, in which each operator's role was to load, monitor, and unload the machine and if operating problems occurred, they informed the supervisor who, in turn, called a specialist engineer or programmer to rectify the fault. The work was subsequently redesigned to transfer tasks from the specialists to the operators to give them greater control over key variances that affected performance, a condition the researchers referred to as *operator control*.

Machine downtime, as a measure of AMT utilization, was used as the measure of performance and recorded during a fifty-day period of specialist control and a subsequent fifty-day period of operator control. Between the two periods, operators had a twenty-day period of off-the-job and on-the-job training for the new work practices. The researchers hypothesized that increased operator control would reduce downtime for high variance systems but not for low-variance systems. The data supported their hypothesis. "The change to operator control is clearly associated with a reduction in downtime for high variance systems . . ." and ". . . these results suggest that the primary effect of the introduction of operator control was a reduction in time taken to deal with operational problems . . ." (Wall et al. 1990 p. 694).

Their research design demonstrated the choice that exists between *specialist control* that restricts the range of operator activities at the interface of the operator and the CAM technology and *operator control* that expands the operator's discretion at that interface. Their results also showed unequivocally that the latter choice was a more effective one for the particular manufacturing organization studied.

AMT designers can opt for the equivalent of specialist control (i.e., assume limited capacity on part of the operator to manage uncertainties that arise) and design the information and control technology so that it limits the information available to the operator. The requirement then will be to turn to a specialist who has detailed knowledge of the system to diagnose errors and prescribe corrective action. Alternatively, they could make a different set of assumptions about the capacity of those who will operate and maintain the technology and design with assumptions about their skills, knowledge, understanding, experience, and capacity to learn. AMT designers have such choices in theory. In practice, they have had no training to inform them that there is more than a single technological design philosophy (Liu et al. 1990, Stymne 1989). Although there are design principles that could guide them (e.g., designing for minimal critical specifications (see Chapter 7)), their training and experience probably did not expose them to such design principles or to the philosophies that inform them.

There are always constraints on designers. Most new technology will not be installed in greenfield sites but rather in plants where many of the existing systems, procedures, and people will have to be adapted to or accommodated. This may mean having hybrid systems and having to take account of the skills of the existing workforce, the mix of semiskilled and unskilled workers, the existing shift work patterns, different pay systems, and so on (Wilkinson 1983). Nevertheless, even within constraints, there is choice.

Job Design and New Skills

Jobs to accommodate the new skills required by AMT must be designed into viable work roles that can provide opportunities for the required skills to evolve and develop. Current knowledge of the changing nature of AMT, the skills required to engage with it, and the needs of the human workers involved should inform those who design work sufficiently to have them at least put aside traditional notions about how jobs should be designed (e.g., to minimize learning, training, and skills and to achieve repetitive job content (Davis, Canter, and Hoffman 1972). Bessant and Senker (1987) developed a list of how work patterns have changed from traditional manufacturing technologies to advanced ones (Table 8.1), patterns that could inform industrial engineers and work designers.

TABLE 8.1 Patterns of Work in Present and Future Factories

Present Patterns	Factory of the Future
Single skills	Multiple skills
Demarcation	Blurring of boundaries
Rigid working practices	Flexible working practices
Operation mainly by direct intervention	Mainly supervision of advanced operations
High division of labour	Moves towards teamwork
Low local autonomy	High local autonomy and devolution of responsibility
Training given low priority	Training and organizational development given high priority

Source: Bessant and Senker (1987), with permission.

Lund and Hansen (1986) are skeptical that engineers will change their job design philosophies toward the "Factory of the Future" without explicit direction from management. If that direction is given, it will likely take the form of new work philosophies. Such philosophies are rarely promulgated without some prior examination of basic values about work and workers held by the key actors and usually made explicit in mission and vision statements that then become the basis for the new work philosophies (Kolodny and Stjernberg 1986). Engineers may adopt some of these characteristics of the Factory of the Future on their own, but the type of comprehensive change effort that will ensure success will require the commitment of the total organization, particularly those at the top.

Unfortunately, too many managers have failed to appreciate that radical changes in thinking are required to successfully implement new technologies. Cummings and Blumberg (1987) studied three different case studies of AMT implementation in which new technologies resulted in increased interdependence between the organization's production system and various support, supplier, and user groups. They concluded that work design appeared to receive only scant attention in all three and the organizations also underestimated the degree to which the work context had to be modified to support the new tasks.

Job Design Models. Job design models are applied theories of motivation. Several job design models have been advanced to improve the quality of the jobs held by individuals. A model developed by Hackman and Oldham (1976, 1980) has been the most widely cited and tested. It proposed designing jobs to allow job holders to experience the "psychological states" of *responsibility, meaningfulness*, and *knowledge of the results* of their performance as ways of enhancing several different outcomes: work motivation, work satisfaction, quality of work performance, absenteeism, and turnover. To achieve these psychological states, five core job characteristics have been identified that should receive explicit design attention. Three of them (*task variety, task identity* (whole tasks), and *task significance*) have their strongest effect on meaningfulness; *autonomy* loads onto responsibility, and *feedback* directly affects knowledge of results.³

The Hackman-Oldham model has had extensive development of the instruments used to measure the core job characteristics, the psychological states and the outcomes (Hackman and Lawler 1971, Hackman and Oldham 1976). Standards or normative results for each variable and for an overall measure called the MPS (motivating potential score) have been compiled for a wide variety of occupations. Subsequent empirical testing of the model has resulted in reasonable levels of support for it in particular settings (generally, well-defined jobs in white collar settings), although the support has been weaker in many others. There have also been some concerns about whether the model actually evaluates what it purports to (Roberts and Glick 1981). However, the model has strong face validity and intuitive appeal and, as such, has had extensive application, including in CAM environments.

Blumberg and Gerwin (1984) used the model to explore the job characteristics of FMS workers (supervisors, mechanics, operators, tool setters, and loaders) in a traditionally organized U.S. plant that machined major housing for tractors. They found that most workers scored low on the job characteristics and psychological states compared to a somewhat comparable normative sample of U.S. workers. They interpreted the low scores of their data to mean that most of the workers who responded felt they had low levels of control over their jobs and little discretion in their work.

Adler (1989a) compared their results with two similar FMS situations he studied, "both having been built by the same vendor in the same time-frame to very similar specifications" (p. 12), one organized traditionally and one as a team-based installation. His data, in contrast to Blumberg and Gerwin, showed high self-ratings by the FMS operators on most of the job characteristics, psychological states, and outcomes— ratings significantly higher than the same normative sample. FMS operators found the quality of their jobs enhanced as a result of the new FMS technology in both the traditionally organized and team-based installations.

The sharply contrasting results of these studies led to a questioning of the methodology and the appropriateness of the model used to evaluate jobs in AMT settings. As previously discussed, the roles operators have in AMT are complex sociotechnical arrangements with complementary relationships between the technology and the human operator that are not well captured by models such as the job characteristics one described above. STS designs are systems and systems must take account of the context (external and internal environments), the interdependencies of their situation and interaction effects of people working in concert, such as in teams. The job characteristics model was not developed for such situations.

Hackman suggested that a more comprehensive sociotechnical approach might be a more appropriate way to design work. To this end he proposed enriching jobs by combining tasks to form natural work groups, establishing 'client' relationships

---

[3] Variations of the model considered different job characteristics such as distinguishing between initiated feedback and received feedback (Kiggundu 1983) or considered the psychological effects of job strain (Karasek 1979).

upstream and downstream of the work operations and giving job incumbents access to data and information to make decisions (Hackman 1977). This would situate the job characteristics previously cited into a larger framework and take account of some of the work system and work group characteristics described in Chapter 7.

Other job design proposals overlap the characteristics in the Hackman-Oldham model and some have extended the domains included so considerations are given to much more than just the job itself. In this respect, they come closer to capturing the reality of the complexity of the technical-social system interface that is the hallmark of work roles in AMT. For example, Emery and Thorsrud (1969) proposed a set of general psychological requirements that job design should address that are somewhat broader than the Hackman-Oldham job characteristics:

A minimum of variety and reasonable challenge
Continuous learning
Minimal areas of decision making
Recognition and social support
Meaningful social contribution
A desirable future

Emery (1978c) has also suggested how tasks should be combined to make up more meaningful and holistic jobs (Table 8.2).

Multiskilling to provide employees with the knowledge and/or skills to do more than the minimal requirements of their jobs is basic to achieving flexibility in AMT environments. Emery has identified this as the difference between a "redundancy of functions" and a "redundancy of parts" philosophy (Emery 1976). In the latter, workers are viewed as replaceable parts. Their work characteristics can be specified in a job description and if there is a need to replace a worker for any reason, it is only necessary to find another with the same detailed characteristics. Such characteristics tend to be the ones that make workers easily interchangeable; therefore, characteristics such as learning ability, experience, and breadth of knowledge are deemphasized. Redundancy of function design, in contrast, attempts to create jobs where workers are employed for what they know and are capable of learning rather than what they can do at any specific time. Value is placed on the capacity to adapt to situations that are not always predictable, on the discretionary aspect of jobs (Trist 1981).

Work Teams. Perhaps because of the varying kinds of manufacturing technologies, researchers into AMT have sometimes arrived at opposite predictions of the impact of the technology on work design. For instance, one prediction suggests that operators in AMT will work in teams (Bessant and Senker 1987), whereas another sees technology disconnecting operators from the process, physically isolating them from others, and having them work alone (Lund and Hansen 1986). Both scenarios seem plausible. Many CAM environment situations are appropriate for individual job design (e.g., stand-alone CNC machines), but some CAM technologies such as

TABLE 8.2 Combining Tasks into Jobs

1. Optimum variety within the job
2. A meaningful pattern of tasks that gives to each job the semblance of a single overall task
3. Optimum length of work cycle
4. Some scope for setting standards of quantity and quality of production and a suitable feedback of knowledge of results
5. The inclusion in the job of some of the auxiliary and preparatory tasks
6. The tasks included in the job should entail some degree of care, skill, knowledge, or effort that is worthy of respect in the community
7. The job should make some perceivable contribution to the utility of the product for the consumer
8. Provision for "interlocking" tasks, job rotation, or physical proximity where there is a necessary interdependence of jobs
9. Provision for interlocking tasks, job rotation, or physical proximity where the individual jobs entail a relatively high degree of stress
10. Provision for interlocking tasks, job rotation, or physical proximity where the individual jobs do not make an obvious perceivable contribution to the utility of the end product
11. Where a number of jobs are linked together by interlocking tasks or job rotation they should as a group
    - Have some semblance of an overall task that makes a contribution to the utility of the product
    - Have some scope for setting standards and receiving knowledge of results
    - Have some control over the "boundary tasks"
12. Provision of channels of communication so that the minimum requirements of the workers can be fed into the design of the jobs at an early stage
13. Provision of channels of promotion to foreman ranks, which are sanctioned by the workers

Source: Emery (1978c). Used with permission.

cells and FMS appear to lend themselves naturally to the organization of work into teams.

Cummings and Blumberg (1987) suggested that self-regulating work teams were the most appropriate choice for organizing the work when technical interdependence and uncertainty were both high, when people have both social and growth needs, and when the task environment is dynamic. They suggested that the conditions that led to high levels of technical interdependence, technical uncertainty, and environmental dynamics were to be found in advanced manufacturing technologies, an argument also put forward by others (Blumberg and Gerwin 1984, Susman and Chase 1986).

In their 1984 study of an FMS system, Blumberg and Gerwin recommended team organization as an antidote for the felt loss of control and the low levels of satisfaction the workers had identified. Adler (1989), in his replication study of the Blumberg and Gerwin research, selected one team-based FMS organization as one of his sites—to evaluate the many proposals for team organization—and a comparable traditionally organized FMS. He found that the nonteam organized FMS scored higher on many job characteristics and the employees were more satisfied than the one that was team based. He concluded that it was not necessary to organize on a team basis; if workers felt effective, this would swamp any advantages of autonomy that might accrue to team organization.

Autonomy. Autonomy on the job can have a range of meanings. It can be manifested in physical actions, such as not being tied to the job. It can mean discretion to influence the pace of work (e.g., pulling the cord that stops operations when quality is unsatisfactory). It can mean designing technology to provide freedom and/or choice (e.g., switching the machine to automatic control when appropriate and, conversely, having a manual override to return control of a programmable operation to the operator). Autonomy can be cognitive—having the ability to make decisions about matters related to work and the context of work (work and vacation scheduling, training, selection of new work group members) and legal—having the authority to do so. Autonomy can be viewed as discretion at the level of the individual or it can be a shared concept, as when a work group has sufficient autonomy to become self-regulating (Susman 1976).

> Comparisons between machine operators' jobs on conventional stand-alone machines and those on the automated machining systems demonstrated substantial changes in the nature of the jobs. Despite the fact that the workpieces and the machining operations performed were technically almost identical in nature to that which had been performed in the building for years, the change in the nature of the machines toward much higher levels of automation significantly altered the job content and skill requirements of the operators. The older jobs on conventional machines had a degree of autonomy, discretion, freedom of movement, responsibility, group interaction and control. The new jobs on the automated machines required, in contrast, the performance of prescribed functions involving limited judgement, some conceptual skills and abstract reasoning, carried out over long periods of monitoring vigilance in isolation. The key job characteristics in the eyes of management was "responsible behavior" that safeguarded the investment in the machines and the product. Relative to the comparable older jobs, these newer jobs had less autonomy and control. (Lund and Hansen 1986 p. 99)

Of the different job and work attributes, autonomy merits special attention. It loads directly onto the important psychological state of *responsibility* in the Hackman-Oldham model. The granting of autonomy is considered one of the prime ways of facilitating self-regulation in work groups. For example, buffers were utilized in many early sociotechnical designs to give work groups autonomy from the control of the production pace, and component magazines were used for automatic feeds to provide individuals some freedom from the relentless pace of an assembly line (Aguren and Edgren 1980, Lindholm 1975). However, they soon also became recognized as repositories of inventory, and as JIT systems were increasingly adopted and "Japanese competition highlighted the negative impact of excessive buffers or slack within an operation" (Klein 1991 p. 25), pressures arose to remove the buffers and magazines and all other sources of excess work-in-process inventory.

Klein (1991) illustrated how the increased interdependence resulting from tighter process controls in manufacturing operations forced a reexamination of the autonomy dimension. Work teams could still have the autonomy to have a say in task design, but because of the tight coupling resulting from the removal of slack and buffers, they had to relinquish their autonomy over task *execution*. Coordination needs between tightly coupled units of the organization had to take precedence. "Decision making relative to daily activities can still be pushed downward, but those involved in the decision must understand the need for coordination within the system" (Klein 1991 p. 3).

She concluded that tighter process controls appeared to be most compatible with collective autonomy. She went on to ask whether collective autonomy is as effective as individual autonomy in creating a committed workforce. Adler (1989a) also asked for a reconsideration of the autonomy variable's role in the design of work in AMT situations. He suggested, ". . . when workers take a purely instrumental attitude to their work, autonomy as absence of constraint may be a good predictor of satisfaction; but if workers see instrinsic value in their work, autonomy is less salient than efficacy (which he defines as the power to accomplish significant objectives)" (p. 32).

The questions raised about the autonomy variable may apply to several other attributes of work design. It would appear that organization design in the context of computer-based technologies is forcing a significant rethinking of the relationship of work to technology.

HUMAN RESOURCE SUPPORT FOR AMT

Flattened organizational structures truncate the traditional upwards oriented career trajectories of many managers and professionals. Collapsing boundaries demand interdisciplinary skill attributes from professionals and challenge their long-held views about narrow specialist careers. Multiskilling and the truncation of job demarcations have similar consequences for operators and maintenance workers. All these changes demand a reexamination, and sometimes a first time examination, of the career system in manufacturing organizations. AMT brings accelerated technological change. Although traditional job security cannot be guaranteed, employment security can often be offered in exchange for flexibility in adapting to technological changes. The career psychological contract between a firm and its employees may require extensive revision.

AMT invites a wide range of changes in the human resource management (HRM) systems that support the design of organizations. Human resource management practices and assumptions should be informed by the STS design principles of compatibility and support congruence described in Chapter 7, inviting their greater integration with the design of new technologies in the manufacturing environment and an earlier involvement of human resource professionals in the social system design processes. This section takes a limited and selective look at several such HRM practices and assumptions—selection, training, reward systems, and union-management relationships. Others have devoted considerably more attention to the subject (Kochan, Katz, and McKersie 1988, Majchrzak 1988, Manufacturing Studies Board 1986, Rankin 1990, Wall et al. 1987, Walton and Lawrence 1985).

Selection

The teamwork and multiskilling characteristics that accompany AMT have prompted organizations to search for a new type of employee in their recruitment and selection procedures. Many organizations have developed extensive interviewing processes and assessment approaches to select for the interpersonal skills and problem solving abilities that they assume will be required for team based structures and work situations requiring flexibility. Typically, as little as 10 percent of the applicants interviewed have been selected on the basis of these requirements. Implicit, and sometimes explicit, in some of these perspectives is that older workers from unionized and/or traditional establishments are not the population likely to offer up desirable candidates for new work situations. Rothwell (1987 p. 67) illustrated this with an example from one organization where "A compromise between management preference for younger people and trade union preferences for seniority was achieved in an agreement to accept volunteers who met the (good eyesight, right attitude, and a willingness to accept training and complete it to a satisfactory standard) requirement."

Several recent experiences severely challenge these assumptions. General Motors closed its Fremont, California factory because of poor productivity and quality. The factory was reopened several years later by NUMMI (New United Motors Manufacturing Inc.), a joint venture of General Motors and Toyota, which hired back most of the same employees who had worked there previously, including most of the union executive. Under new management, their performance has been amongst the best of the American-based automobile assembly plants (Figure 7.1).

In a second example, Shell Canada made a commitment to the Energy and Chemical Workers Union (ECWU) to give first preference for employment at a new petrochemical refinery they were building in Scotsford, Alberta to ECWU members who would be losing their jobs as a result of the simultaneous closing of three other refineries in Canada. The Scotsford refinery had been designed to sociotechnical design principles as was an earlier refinery in Sarnia, Ontario that was designed in collaboration with the ECWU. The Sarnia refinery selected 180 employees from 2600 applicants per the process previously described (Halpern 1984 p. 54). The Scotsford refinery, using the supposedly "wrong" population base, has nevertheless been a consistently high performing organization.

Training and AMT

Training needs for automated manufacturing technologies are almost always underestimated (Majchrzak 1988). Although managers usually appreciate the need for technical training, for a variety of reasons, some of which we will explore shortly, they rarely provide enough technical training to operate the technology effectively. However, compared to the technical training, the social system training provided is almost nonexistent.

Majchrzak (1986) surveyed 133 manufacturing plants with CAD/CAM or flexible automation equipment. Only 45 percent of the plants in her survey offered a structured training program to prepare the work force for the new equipment. Most skills for learning about computer-based technologies were acquired through unstructured on-the-job training (OJT). However, OJT, although having the advantage of being inexpensive and hands-on, has several disadvantages. Unstructured OJT disrupts production work; it tends to focus on a specific set of tasks at the expense of workers acquiring a more holistic approach to technological change; and the trainee learns only those techniques taught by the particular instructor (Majchrzak 1988 p. 158).

Rothwell (1987) researched twenty case studies on the application of computerized technology in the United Kingdom. Most of the managers she spoke with, when asked what—with hindsight—they would have done differently or what they would do next time, usually said they would have provided more training and allowed more time for training (p. 67). This is an oft-repeated comment by managers and workers at various conferences and presentations where they have described their company's innovations in technology and work organization (Giles 1989). If there is universal agreement by managers that more training and more time for training is necessary to successfully innovate and implement new technologies, why is the need so consistently underestimated? In Majchrzak's survey, why did 55 percent of the plants who installed CAD/CAM equipment engage in no training other than OJT? Rothwell (1987 p. 67) provided a partial explanation:

> Most of the organizations we studied had ambitious intentions of training for information technology, although concepts of training ranged from propaganda, communications, and persuasion at the beginning in order to shape attitudes and encourage employees to come forward to change, to classroom lectures, and on-the-job practice at implementation. In practice, however, because of technical schedule slippages and management pressure to get systems up and running, training was often abandoned or crammed in at the last minute as part of the implementation process, adding to the stress of all involved, and to the delays in debugging.

The skills associated with new technologies, cited above, drive the requirement for enhanced training. Most of the 45 percent of the plants in Majchrzak's (1986) sample that did engage in CAD/CAM training taught skills in the areas of safety, specific machine operation, and a general knowledge of technological advances in manufacturing. Almost 70 percent of these plants also taught skills in maintenance, programming, problem solving, and knowledge of the manufacturing process.

In the past, skills, other than professional skills, have been seen in terms of the ability to *do things*, rather than in terms of understanding (Rothwell 1987). The skills sets for which training is pursued in AMT environments go beyond doing to include conceptualization and cognitive understanding and beyond the narrow domain of a specific operation to include the particular plant's integrated manufacturing system and its various interdependencies.

Operator-technology interactions in computer-based applications are not well understood, and much of the training that is devised has only limited empirical experience to inform the design of the training (Cavestro 1990). Training practices vary widely and they vary from company to company even for the same equipment (Wilkinson 1983). Many companies, see training as a completely ad hoc activity to be engaged in as necessary by whoever has the pressing need, so relatively little advance thinking is applied to the issue. Furthermore, new technology diffuses heterogeneously, so that many adopting organizations consider their applications unique and develop training practices as if their technological and organizational configurations are unique and quite unable to benefit from the experiences of others. "Not-invented-here" policies are not confined to R&D departments.

Training Sources. Majchrzak (1986) listed the most popular sources for delivering structured education and training programs cited by the respondents to her survey. Vendors or manufacturers of computer-automated equipment topped the list of sources of training programs for the 45 percent of her sample of plants that had CAD/CAM equipment training. Eighty-seven percent of them used vendor training. Three other sources received frequent mention: in-house instructors (80 percent), traditional educational institutions (54 percent), and training industry and management consultants (47 percent).

Vendors normally offer some training with the purchase of their equipment and are usually well qualified to teach about its operation and maintenance. However, AMT is often as new for the vendor as it is for the purchaser. In such instances, the vendor's experience base for teaching about their equipment can be quite limited. There are other disadvantages of vendor training: it tends to be general rather than specific to a customer's application; more often than not it is machine specific at a time when CAM technologies are becoming increasingly integrative; and because of its general nature, it usually requires a common minimum knowledge base from participants (Majchrzak 1988).

In-house training can be dedicated for a particular plant or provided by a corporate training unit to more than one specific plant. If the latter, it suffers from the same problems of generality as vendor training, although it may be available to many more people than the limited number of slots vendors offer to equipment purchasers. If in-house training is dedicated to a particular plant's needs, it can be quite expensive because of the cost of developing a dedicated program and keeping dedicated trainers on staff. Training and management consultants have the advantage of being able to be called back as often as they are needed and the plant is not burdened with permanent salaried staff. However, if the new installation is a substantial one, or if it is a greenfield site, most external management consultants will insist on the presence of a full time internal facilitator to maintain the level of activity, commitment, and training when the external person is not there. It is also a preferred approach to wean the plant away from a dependency relationship with an external consultant.

Local educational institutions often enter into training agreements with local manufacturing plants that work to the advantage of both institutions. The educational institutions have the facilities, the training technology, and the resources and by scheduling carefully with the local plant they can optimize the use of their resources so that they are fully utilized yet able to respond to the schedules and demands of other full-time students. The plant enjoys the good fortune of local resources who are almost always available at costs below those of dedicated in-house courses or management consultants, although the plant has to be prepared to invest in the development of training programs if they are to be tailored to the plant's needs. However, there are disadvantages here too. Training carried out remotely is often not innovative vis-à-vis the plant's requirements. It tends to address the machine's functions since each plant's manufacturing system is integrated differently; thus it must be augmented with OJT training if it is to deal with the real complexity of the situation. For example, NC training is often viewed as programming training instead of being oriented toward addressing the real problems that operators handling CNC machines will encounter (e.g., complex diagnoses and problem solving when difficulties are experienced, understanding the interactions between different cutting tools and the characteristics of the materials they operate on, and knowing how to judge if the correct pressure and force and cutting angle are being applied by the computer controlled program (Cavestro 1990)).

Setting up community colleges and training institutions has been used by some states as incentives to lure manufacturing plants to their locations. When the plant's peak training demands pass, the institution redirects its freed-up resources toward the community's needs. This usually leaves the community with a valuable legacy and the plant with a source of future employees trained to a set of standards it appreciates because it has had an active role in their development.

Some organizations have begun to understand that the training requirement is not a constant one and are prepared to staff heavily for peak training needs, which can be as long as two years after a greenfield startup or, in an ongoing plant, after the installation of new AMT equipment. They can then reduce the complement of trainers as skills acquisition improves. This is particularly the case for social systems training, where the acquisition of skills in meeting management, handling interpersonal differences and conflict resolution, and learning to function as a team can require intensive and long periods of process support from trainers working closely with operators and teams.

In the Cummins Engine Company's Jamestown plant, training is seen as one of the required second skills or "vertical tasks" (with finance, materials management, etc.) that must be staffed by one member of each engine assembly team. Some vertical tasks have regular activity. The training task is a more variable one, with peaks when a new employee joins the team or a member transfers to another role within the team for which training assistance is needed, or even when a temporary member joins to help the team through a tight period or a bottleneck (Kolodny and Dresner 1986a).

Who Should Be Trained. The content of training programs can be very broad given the range of occupations that are affected by AMT. Even subjects that may appear generic to the new technologies (e.g., programming or systems integration) will require different levels of curriculum development for different occupational groups. Managers, designers, operators, maintenance workers, and supervisors may all have a need to upgrade their computer knowledge—the CAD software interests of design engineers, for example, will differ significantly from the hardware and software diagnostic and trouble shooting interests of maintenance workers—but the needs will be difficult to fulfill from a single training course on the particular subject.

Majchrzak suggested that there is a core set of courses that are appropriate to all job positions:

- *Human relations training:* interpersonal skills, communications skills problem solving, decision making, team building, group processes
- *Knowledge of the manufacturing process:* will vary with each company's technology
- *Basic skills:* intended more as remedial for those who missed basic skills (e.g., in high school) or as refresher courses for subjects such as mathematics and reading

Percent of Time on Training. If training were to be given its due, how much should be invested in it? Although there is some evidence that training in general is becoming more appreciated by some North American manufacturing companies, the overall average remains very low. Hoerr (1990) cites the case of Corning Inc., which committed to spend 5 percent of all hours worked in 1991 to training, which he contrasts to the 1 to 2 percent range at most U.S. companies. C. Jackson Grayson, Director of the American Productivity and Quality Center, in Adler and Howard (1990), confirmed that on average U.S. companies invested less than 1 percent of their payroll in training, which he contrasted with Baldridge Award winners who averaged 3 percent and the Japanese who spend 7 percent.

For AMT in particular, training practices are numerous and disparate, varying from 30 to 300 hours/year (from 1.5 percent to 15 percent) (Cavestro 1990). Ettlie (1990) has suggested that training budgets in AMT should be at least 10 percent of the project cost. Bill Wiggenhorn, Corporate Vice-President of Training and Education at Motorola, Inc., in Adler and Howard (1990), estimated that the cost of training is a full 15 to 20 percent of the total cost of implementing new technology. Tektronix, Inc., in Forest Grove, Oregon in a new CAD/CAM plant producing printed circuit boards, spends 10 percent of its manpower budget for training and participation activities, whereas GE Canada's Bromont, Quebec plant, which uses AMT to produce aircraft engine blades and vanes, budgets 12 percent of employee hours for the same activities (Painter 1991). In a survey of CAM plants, Majchrzak (1986) found that the more integrated the equipment the larger than training needs.

Startup training in AMT, particularly in greenfield sites, has a high initial requirement. Jim Lewandowski, Vice President for People Systems at Saturn Corporation, stated that employees spend from 250 to 750 hours in training just to get ready for their jobs but noted that "Saturn has perhaps the most extensive training program in the world" (Adler and Howard 1990 p. 20). At General Motors' Vanguard plant, union employees received 1000 hours of startup training (Painter 1990).

Startup training can be even more substantial for employees hired to participate in the design phases of a greenfield site. At GE Canada's Bromont plant, some 30

"A" level employees were hired a year in advance of the plant's design to learn the technology from a sister plant in Rutland, Vermont and use that knowledge in concert with social systems training they received to participate in the subsequent STS design. General Motors followed a similar strategy for its STS/team-based greenfield engine and transmission plant in Aspern, Austria by sending a team of foremen to an Adam Opel AG sister plant in Russelshein, West Germany to study their technology and use their acquired knowledge to design a better work organization for the Aspern factory.

To sustain skills acquired, ongoing training can also be significant, particularly if multiskilling and pay for knowledge systems with their particular training requirements are part of the reward structure (see following). Jim Lewandowski noted that Saturn employees must spend a minimum of 5 percent of their time in training to sustain their knowledge base (Adler and Howard 1990). At L-S Electrogalvanizing Co. in Cleveland, Ohio the job progression system assures continuous training at a rate of 2.5 days/month (Painter 1990).

Design a Training Program. Majchrzak (1988 p. 167) proposed nine elements to consider for development of a comprehensive in-house training program for flexible automation.

1. Purpose—provision of the skills required
2. Format—on or off the job training and where
3. Individualization—tailored for individual needs or generalized
4. Extensiveness—initial training and continuous training needs
5. Trainers—professional instructor, technical expert, manager
6. Presentation—classroom, simulation, CAI
7. Timing—before installation, during installation
8. Curriculum—technical, human relations, manufacturing processes
9. Evaluation—assessment, follow-up

Two other elements, would probably enhance the list. *Participants*—refers to which groups of people would receive training and how much since AMT places its demands for comprehension on a wide range of occupations. *Sequence*—addresses the order in which participants would be trained. Supervisors, for example, are often left out of training or when included, are often grouped together with operators. The apparent team-building logic that informs this choice often results in more supervisory resistance than team cohesiveness. There is a strong argument in favor of their being trained first and then asked to train others.

Training can often serve as a vehicle for other objectives of the organization (e.g., team building) in which carefully selected teams may attend training programs together with the additional objective of building cohesiveness in a relatively nonthreatening environment.

The Manufacturing Studies Board (1986) cited some themes resulting from their study of implementation of AMT:

- Training programs should be customized.
- The effectiveness of training should be continually assessed.
- Training for computer-integrated systems should consider forms of cross-training.
- Good technical training is essential.

Reward Systems

Changes in the design of work should be reinforced with appropriate rewards. Lawler (1981) argued that reward systems could be used as a strong lever to initiate change and may be the best way to orient behavioral change in the direction desired. Practice, however, has tended to be quite different. In an in-depth study of change in six firms, Beer and Spector (1985) argued that rewards appeared to be a "lag policy" variable; reward systems must be revised to reflect other organizational changes, but not necessarily immediately.

Reward systems are difficult to design and usually require considerable adaption for the unique circumstances of the particular situation (e.g., the history of the unit with respect to AMT, relationships with the local union, and the implicit and explicit psychological contrast that has prevailed). Reward systems elicit strong feelings with respect to issues of fairness and equity. It is often preferable to experience the situation that has changed as a result of AMT before attempting to redesign the reward structure.

Skill-based Pay Systems (or Pay for Knowledge Systems). Two rewards systems that have received recent attention in AMT situations are pay for knowledge or skill based pay systems (Gupta, Jenkins, and Curington 1986, Lawler and Ledford 1985, Tosi and Tosi 1986) and gain sharing systems (Bullock and Lawler 1984, Moore and Ross 1978, Nightingale and Long 1984). Pay-for-knowledge (PFK) or skill-based pay systems, sometimes referred to as learn-earn systems, pay people for acquiring a broader range of skills than they may need in their positions at any particular time. They are paid for what they know, rather than what they do. The system rewards them for becoming multiskilled because this helps provide the organization with needed flexibility when product or process uncertainties arise.

PFK systems have a series of defined skill levels that people must learn, both on the job and through structured training, to advance along the pay system. Formal assessments of people's skills are usually carried out, often by peers, to ascertain proficiency at a level. "Time-in-grade" provisions are frequently used to ensure that people do not advance more quickly than they should because the formal assessment criteria are often difficult to establish. In many PFK systems, people "top out" (i.e., reach the top of the pay ladder) after five years or more. This creates a problem that has as yet no adequate solution. Some factories have added additional levels to extend the top and made them difficult to achieve, but this approach has its limitations. Some have installed "gates" that restrict people's upward movement unless they are needed for higher level roles in the plant.

Skills learned in a PFK system tend to be more "breadth" skills than "depth" skills. The latter are assessed as they would normally be in any performance evaluation process. The "breadth" skills are those that realize the multiskilling objective. Although horizontal skills dominate in PFK systems, most also have some vertical skill components such as administrative skills (e.g., work scheduling, vacation planning) and some also include coordination skills in the system (e.g., meeting management, group processes).

The expanded roles for operators in AMT environments cited earlier are forms of multiskilling that lend themselves well to being supported by a PFK system. Many AMT environments have adopted PFK systems or are in process of planning for one (Arnopoulos 1985, Beer et al. 1985).

Gain Sharing Systems. Gain sharing systems are designed to share productivity improvements made above some agreed-on standard with those who contributed to the improvements. The sharing can be in cash or in company shares or in some combination of both, and it can be disbursed in immediate payouts or in deferred ones or, again, in some combination. The benefits are generally shared between the company and the employees according to a formula that is built into the particular type of gain-sharing plan adopted. Three different plans have been the ones most often cited: Scanlon plans, Rucker plans, and Improshare (Nightingale and Long 1984). Each has a different approach to calculating the gains or productivity improvements produced by the unit in question and sharing it among the employees.

In addition to being a gain sharing plan, each is also a participative process that involves workers in the productivity improvement processes by actively soliciting their suggestions and making them members of the committees that evaluate and select suggestions and propose them for implementation. Here again, the three aforementioned plans differ in their approaches to the participative process.

Gain sharing systems have been and are being examined by many companies with PFK systems as a way to respond to the concerns expressed by employees about the topping out problem referred to above. It is felt that gain sharing is a way to add additional incentive when there is no more "earning" possible for additional "learning." However, there have not been many successful implementations in these situations. One reason might be because the basis of gain sharing systems is the ability to construct a standard against which improvements can be assessed. In new technology applications, particularly in greenfield sites, such standards are difficult if not impossible to establish and some new basis for gain sharing is needed. There are also questions about how to share improvements between the plant and the employees when a good portion of the improvement is attributable to the new technology. It is next to impossible to differentiate the portion of improvement that results from the technology from that which is a result of employee creativity and problem-solving. While at least one organization (Faux and Greiner 1973) credited all savings to the gain sharing account—even when they resulted from new technology investments—that was carried out in an era when the technological investments were relatively small compared to those involved with current advanced manufacturing technologies.

Union-Management Relationships and AMT

The role changes in computer-based technologies generally cut across traditional job demarcations and, in so doing, undermine many of the protections trade unions have established over the years. Although some North American unions have been willing to see job demarcations compressed and their traditional collective agreements changed (Kochan et al. 1988, Rankin 1990), others have staunchly resisted any incursions into traditional relationships. For those who have been willing, there have been several different incentives. The most significant has been the sheer threat of plant closure and job loss because of insufficient flexibility to adapt to market uncertainties. A second incentive is technology agreements between company and union that give unions advance information of intended changes to the technology and commit to work with the unions to minimize the dislocation resulting from the new technology; for example, by training dislocated people in new skills that have applicability in the plant and by committing to some form of employment security for those affected by the changing technology.

The implementation of technological change, discussed in the next section, is eased if commitments to employment security (Greenhalgh, McKersie, and Gilkey 1985, Peganoff 1984, Skeats 1987) can be made. Commitments have been made by many manufacturing organizations who have taken the time to understand the implications of making such commitments (Spector and Beer 1985), but too many companies have taken a traditional and adversarial view with respect to their employees and/or unions and have lost the opportunity to make a difficult change process easier.

Many unions have been prepared to enter into a parallel agreement with the employer with respect to changes in work organization as a result of new technology implementation, providing the basic collective agreement would not be affected in any way. It subsequently became clear that such was not possible. The changes resulting from new technology are profound and necessarily have impact on many aspects of the collective agreement (e.g., job classifications, promotion patterns, supervisory relationships). This situation has continued as a source of consternation and conflict for unions and employers alike.

SUMMARY AND CONCLUSIONS

Advanced manufacturing technologies are qualitatively different than traditional technologies. The computer basis of their operations that makes them programmable and adaptable also stimulates the development of levels of integration beyond anything previously experienced in manufacturing environments. AMT invites a reconceptualization of the design process to create a new manufacturing organization, particularly because it involves a reassessment of the roles that workers and professionals and managers will have vis-à-vis the technology. The nature of the skills needed requires a significant upgrading and professional engineers are increasingly taking employment as AMT operators. AMT is being designed not only to allow operators a significant amount of control over how technology functions but also to actually encourage operators to take that role because there is an increasing understanding that complex technology makes the organization more dependent than ever on its members. Every role is affected by AMT; none are spared.

In the turbulent external environment in which AMT is situated, choices about how to best design social and technical aspects must be made in light of business outcomes and factory effectiveness outcomes and not on technological parameters alone. Even a professional engineering degree may be insufficient training to operate in an AMT environment. A business understanding and some organizational training in addition to technical training may have to be part of the basic training for the operator of the future. Considering the investment in some CAM technologies (e.g., an FMS), it is not unreasonable to expect that backgrounds of this nature may be required.

AMT has changed the ground on which the deskilling/reskilling debate has raged to such an extent that one has to ask if the questions being asked are still valid. It has changed the concept of the role of the operator vis-à-vis the technology in such a way that it will likely encourage machine designers to spend more time on factory floors to understand how operators function and what philosophies inform them. AMT has forced human resource specialists to examine issues they never considered in the domain of the factory floor (e.g., careers). It would be somewhat ironic if advanced manufacturing technologies became a stimulus for humanizing the workplace.

PART III
INNOVATION

9
THE INNOVATION PROCESS: ADOPTION

The literature on organizational innovation has focused on new products and services as opposed to throughputs (Roberts 1988). By taking an organizational level approach, that is, considering innovativeness as an organizational level variable and relying on correlation analysis, it has deemphasized internal organizational processes (Tornatzky, Eveland, Boylan, Hetzner, Johnson, Roitman, and Schneider 1983). It has unavoidably concentrated on incremental as opposed to radical innovations due to the latter's rarity (Roberts 1988). Adoption, the decision on whether or not to have an innovation, has received a great deal more attention than implementation, the first use of an innovation (Tornatzky et al. 1983). Consequently, we know very little about the manner in which new methods for converting inputs into outputs are adopted and, in particular, implemented, especially when they are significant departures from current practice.

Correcting this deficiency is not important solely to researchers. Faced with an urgent need for revitalization, more and more American firms are turning to computerized manufacturing technologies in hopes of stimulating performance. Often they are not ready for the difficulties in deciding whether to purchase, in preparing the organization's support functions, and in operating, maintaining, and controlling the technology. Because it is conducted at a detailed level of analysis, a process-oriented approach can better identify specific problems, isolate their causes, and suggest ways of solving them than can an organizational-level approach. Once more, examination of the decision process is well suited to uncovering the factors that determine the innovation of a particular kind of technology such as AMT. Organizational level analysis is meant to identify characteristics which influence a firm's general level of innovative activity. Meyer and Goes (1988) demonstrated empirically that the factors influential at the detailed level do not coincide with those operating at the aggregate level.

230 THE INNOVATION PROCESS: ADOPTION

Although the needed perspective does not currently exist, relevant bits and pieces are found in three different settings. The product innovation literature contains pertinent concepts, but they must be selected bearing in mind that conclusions applying to one type of situation may not apply to another. There is also the burgeoning AMT literature much of which is focused on the management of CAM technology. A third source is interviews conducted in American and European firms that had installed or were in the process of installing some type of integrated CAM system.

This last database was drawn from seven companies in four different countries and included:

- A diversified American manufacturer with a tractor division that was making a product line for which the major housings are machined on a flexible manufacturing system
- A British producer of medium-size and large electrical motors and generators that was in the process of installing a direct numerical control system (an FMS without automated material handling) for the machining of prime components
- A British manufacturer of plain bearings for industrial engines with a rudimentary direct numerical control system that was machining thin wall bearings
- A German aircraft manufacturer that was installing a flexible manufacturing system designed to machine over 200 different parts
- A German producer of transmission systems for industrial and agricultural vehicles that was developing its own flexible manufacturing system to machine gears and other rotary parts
- A French manufacturer of industrial vehicles that began installing a flexible manufacturing system for gear boxes
- A large British machine tool manufacturer that was in the process of designing an FMS for its own use

Semistructured interviews were held with over thirty-five managers and engineers representing various levels in manufacturing management, accounting, quality control, manufacturing engineering, maintenance, production preparation, and data processing. Access problems made it impossible to interview people in the same function in all companies. Respondents were asked questions in three general areas: descriptive contextual information, the nature of the adoption process, and implementation problems. Emphasis was placed on how CAM affected the respondents' roles. These probes stimulated discussion, which led to more specific *ad hoc* questions and more discussions.

The theory is oriented towards CAM innovation, given the sources on which it is based. The next part of this chapter discusses the theory's conceptual underpinnings. Starting from an idea mentioned by Tornatzky et al. (1983), a single theme—uncertainty—is proposed as a way of integrating a good deal of the disparate material in the literature and the interviews. The chapter's concluding part traces the process by which a company becomes aware of CAM and then decides whether to purchase it. Chapter 10 is concerned with the manner in which a firm prepares to support the new technology. Chapter 11 deals with the problems arising in the initial use of the technology. Implications for researchers and practicing managers are discussed in Chapter 12.

CONCEPTUAL FRAMEWORK

Uncertainty, defined as lack of information on goals, alternatives, and consequences, is the starting point. Thompson (1967) viewed organizations as open systems faced by environmental uncertainty but requiring certainty in order to function. Every organization has a technical core devoted to efficient performance of some processing function. Management's role is to handle uncertainty so that the core can operate as efficiently as possible. Various tactics that regulate inputs to and outputs from the core make this possible.

Innovation in manufacturing processes is traditionally considered to have objectives such as improved productivity, better quality, or faster delivery time. It is also a tactic which, in the long run, handles uncertainties affecting the technical core. Conventional automation reduces uncertainty. Abernathy (1978) found that American auto companies' technologies evolved from fluid, uncertain characteristics to rigid, known specifications. Woodward (1965) argued that firms that change from batch production to mass production or mass production to process production experience an increase in predictability. CAM technology can adapt to uncertainties facing manufacturing managers by virtue of its flexibility aspect and to some extent reduce them by virtue of its automation aspect. Gerwin and Tarondeau (1982) discovered that about half the reasons four firms gave for adopting CAM reflected desires to adjust to or reduce production related uncertainties.

Process innovation, however, threatens to pierce the core with new uncertainties in the short run during adoption and implementation.

- *Technical uncertainty* refers to difficulty in determining the precision, reliability, and capacity of new processes, and whether still newer technology may soon appear to make the equipment obsolete.
- *Financial uncertainty* includes whether return on investment should be the major criterion and whether net future returns can be accurately forecasted.
- *Social uncertainty* is exemplified by questions concerning the nature of the required human support system, and by the possibility of conflict.

The tension created between the core's need for certainty and new equipment's generation of uncertainty is likely to account for many of the problems that arise during the innovation process. A good deal of human behavior can be analyzed in terms of efforts to deal with these problems by developing coping strategies which either avoid, adjust to, reduce, or take advantage of the uncertainties. These strategies may be beneficial or detrimental depending on the values of the organizational faction evaluating them.

Basic Assumptions

It is assumed that the innovation being considered is radical with respect to at least the focal organization and perhaps the industry or the society as well. In other words, it is a significant departure from existing manufacturing technology (Zaltman, Duncan, and Holbek 1973). Consequently, uncertainty and risk are relatively high. Computerized manufacturing technologies are likely to furnish good examples, particularly integrated manufacturing and assembly systems. The technology for each of the seven firms interviewed fits the definition of radical as a significant departure from existing conditions in the industry or society. The American firm purchased its nation's second FMS for its tractor division. The British machine tool builder was designing the country's first locally produced FMS. The British motor producer was one of the first there to be installing a full-fledged direct numerical control system. The British manufacturer of industrial bearings was one of two firms in the country with a rudimentary direct numerical control system. One German company, a producer of transmission systems, was designing the country's first FMS for rotary parts. The other, an aircraft manufacturer, was one of the first German firms to install an FMS. The French transportation vehicle manufacturer was designing its nation's first FMS for a plant producing gear boxes.

Based on fairly solid evidence from large sample studies it is also assumed that the organization in question has a large size. Collins, Hull, and Hage (1989) found a statistically significant positive linear relationship between number of full-time employees and whether a plant was a nonadopter, low adopter, or high adopter of programmable automation. Kelley and Brooks (1988) using factor analysis defined a variable representing the degree to which an enterprise has large plants that are part of a multiplant system. Results from a logistic regression indicated that the fraction of plants adopting programmable automation was significantly related to this variable. Munro and Noori (1988) found that the extent to which a firm's manufacturing process contained flexible automation was significantly related to the number of its employees. The Survey of Manufacturing Technology conducted by the Bureau of the Census (U.S. Department of Commerce 1989) found that 93.7 percent of establishments with 500 or more employees had at least one of the advanced technologies surveyed, whereas only 60.9 percent of those with less than 100 employees had at least one.

A number of reasons have been offered to explain the relationship. Small companies lack the necessary financial resources (Collins et al. 1989; Kelley and Brooks 1988). Their scales of operation are insufficient (Collins et al. 1989). Small companies do not have the technical and human resources to support the technology (Kelley and Brooks 1988). For example, to justify and utilize CAM properly, training is needed for managerial, technical, and operating personnel. Finally, the technology is not designed with the needs of the small firm in mind, although this may be changing (Hicks 1986). Hicks believed that, unless these problems are overcome, CAM will polarize a given industry into competitive large and uncompetitive small operations.

Defining stages is an essential first step when developing a process-oriented approach to innovation. It provides the framework within which theorizing takes place. Unfortunately, there is no agreement on what are the stages of the innovation process or if they differ depending upon the innovation or organization being considered. An influential identification of stages advocated by Hage (1980), which he justified on theoretical grounds, is followed here. The innovation process is assumed to consist of adoption, preparation, implementation, and routinization.

As we will demonstrate, these steps do not necessarily exist in a strict sequential order. Overlaps in time or movements back to a previous stage may occur. During adoption the organization becomes aware of an innovation that meets some specific rational or nonrational need, and a decision is made on whether to have it. The participants in this stage cope with the most uncertainty; decisions must be made on the basis of scanty information. Participants include vendors who attempt to sell technology. They are a critical element of the innovation process for manufacturing equipment that has not been sufficiently treated in the past. The organization's innovating group or technical task force under the leadership of the new technology's champion makes recommendations on hardware, software, and vendors. The organization's strategic management makes the final decisions on the recommendations.

During preparation the infrastructure needed to support the new manufacturing process is developed. Computerized technology requires a highly sophisticated support system in order to function effectively. New skills, attitudes, systems, procedures, and social structures are needed by the people responsible for operating, maintaining, and controlling the innovation. However, the prevailing uncertainty precludes a sound basis for determining what is needed; preparation may never be successfully completed.

Implementation, as defined here, is the initial tryout and evaluation until full scale operation is attained. The participants—operating management and technical people—deal with less uncertainty because some performance data are available that can be compared to expectations. Discrepancies are attacked through adjustments in the technology, the organization, and in expectations. At the same time, intraorganizational conflicts caused by the changes and the remaining ambiguities must be handled.

Eventually the innovation becomes part of daily practice during routinization. There has not been enough research on computerized manufacturing processes to consider these longer term aspects.

Participants in the Process: The Champion

No participant in the theory is more critical than the champion. The need for such a person in order to give a manufacturing innovation a reasonable chance of being considered is well established (Dean 1987, Gerwin 1981a, Meredith 1986). This individual so believes in the innovation that he or she is willing to assume responsibility for the success or failure of the project. Such a person usually assembles and leads the project team, sells the project within the organization, gathers needed resources, deals with organizational and technical obstacles, and is chief liaison with the vendor. In so doing, the champion absorbs a great deal of the prevailing uncertainty that others do not want to handle. Although the theory considers the champion to be head of the task force, this is not always true. Sometimes this person's boss or a higher official assumes the role.

What personal characteristics facilitate the champion's mission? Above all, commitment to the innovation as a good for the organization is necessary. Pfeffer (1981) defined commitment as a state of being in which beliefs and actions reinforce each other so that an individual's involvement is maintained. It is enhanced by at least three conditions. First, the individual must have had a choice initially; he or she voluntarily accepted the responsibility rather than being forced into it. Second, the decision must be made public so that the person cannot deny or forget it. Third, it helps if the decision is irrevocable.

Although not often mentioned in the literature, the champion's commitment is also filled with dangers for an organization. He or she chooses the wrong innovation much more frequently than the right one. Adherence to erroneous beliefs and actions may persist even in the face of evidence that an innovation is inappropriate. This inability to see or admit a mistake is reinforced by various individual and organizational phenomena (Carpenter 1988, Pfeffer 1981).

1. Commitment may exist for the wrong reasons. The champion is seduced by the innovation's technical novelty and is less concerned with whether or not it meets organizational objectives.
2. Counterarguments are discounted because it is always easier to come up with criticisms of rather than support for something new.
3. An irrational belief exists that sunk financial and psychological costs force continuance of the project.
4. Questioning the innovation is seen as part of the problem so that conformity and loyalty are strictly enforced.
5. The innovation's problems are assumed to be due to insufficient means ("We haven't devoted enough effort to the project.") rather than inappropriate ends ("We did the wrong thing.").
6. As more resources and control are acquired by the innovation's advocates to correct problems, they amass more power, which is used to sustain their point of view.

In the final analysis the champion's commitment may steer a firm into making a disastrous choice unless a counterweight to excessive zeal exists. One can now begin to appreciate the depth of uncertainty faced by decision makers.

Other personal characteristics are important for the champion. He or she needs to have a balance between administrative and technical skills (Boer and During 1987, Graham and Rosenthal 1986b). Project management experience under highly uncertain conditions is essential to avoid a narrow, parochial point of view. A technical background in manufacturing and computer-aided approaches is needed so the individual is not put at the mercy of technical experts. The champion should also have a network of personal contacts in the firm and a willingness to take risks (Meredith 1986).

Various organizational conditions facilitate the champion's efforts. This person needs to be relieved of other operating responsibilities and given an opportunity to survey manufacturing facilities elsewhere (Gerwin and Tarondeau 1982). Reporting to a high level executive emphasizes to others the significance of the champion's role (Graham and Rosenthal 1986b). Equipment vendors find it easier to work through a champion who has a good deal of organizational influence (Hayes, Wheelwright, and Clark 1988). A reward system that compensates for success and does not punish for failure is a well-established condition.

The champion acts as a liaison between equipment vendors on one side and strategic management on the other. Based on his experiences with an FMS in a British company Carpenter (1988) identified other factions that must also be managed. "Disciples" consist of the project team supporters of the innovation. The champion must keep their faith from wavering. "Opponents" are individuals who believe the project is not in the organization's interests or who stand to lose if it is accepted. They need to be convinced or neutralized. "Indifferents" are doing too many other things to get involved or are waiting for signs of top management's commitment to the project. Their inertia must be overcome.

Limited evidence suggests that in large corporations, at least, there is more than one kind of champion (Gerwin 1981a, Maidique 1980, Meredith 1986, Peters and Waterman 1982). The individual who has been discussed up to this point can be referred to as the manufacturing process champion. If this person is low in the hierarchy and technically oriented he or she may need a middle manager to act as a translator in wooing corporate headquarters. A division manager, for example, can explain the project in language that senior executives understand. A particular need exists for this "executive" champion when senior management is unfamiliar with the technology, a situation that is quite common when CAM is on the agenda. The process champion may also need a sponsor, a senior-level person who helps in obtaining resources, acts as a consultant when problems arise, and protects the project if necessary.

THE ADOPTION PROCESS

One way an organization becomes aware of a new manufacturing technology is by searching for ways to improve performance, but other less rational methods also exist. Awareness is refined by determining which aspects of the technology could be developed in-house. A firm must then decide whether or not to select the innovation. Given that the firm knows so little about the technology the decision process is filled with political and social elements.

Awareness of the Innovation

The concept of a performance gap is often stressed as the motivation for innovation (Hage 1980, Zaltman, Duncan, and Holbek 1973). An organization assesses its survival capability by identifying positive differences between the values of objectives and performances on relevant dimensions. In order to eliminate these differences it develops new action plans including manufacturing process innovation, for example.

Ideally, performance gaps should serve as useful control devices for helping an organization to survive. In practice, uncertainty and other factors hamper effective control (Hage 1980). There is often not enough information to verify the existence of a gap. Frequent measuring is costly particularly if there are many dimensions to be considered. As a result, gaps sometimes go undetected until very large discrepancies in several dimensions make it obvious that something drastic needs to be done. The stage is set for adoption of a radical innovation.

Even then there is plenty of room for interpreting the nature of the problem, that is, deciding which gaps to focus upon. Suppose a problem is defined at the operational level in terms of finding a way to manufacture a certain set of parts. The associated gaps are relatively easy to close and fairly well defined. If CAM is chosen as the answer it is likely to be a success from the company's viewpoint. Yet the technology's strategic advantages are not utilized. Suppose the problem is conceptualized at the strategic level such as finding a way to facilitate product innovation. The associated gaps are harder to eliminate and less well defined. CAM is less likely to be a success but the company will utilize the innovation to its fullest extent. Gerwin and Tarondeau (1982) gave examples of companies fitting into both categories.

A performance gap not only results from a crisis, that is, a reduction in performance, but also from an opportunity, that is, an increase in aspirations due to a change in environmental stimuli. Gerwin (1984) found that both conditions stimulate the adoption of CAM technology. The role of crisis is exemplified by the experience of the tractor division of an American company. Basic changes in the nature of farming were creating a demand for large sophisticated customized tractors. However, the division was still producing small- and medium-size vehicles with antiquated manufacturing processes. Declining sales and profits necessitated that the corporation either sell the tractor business or regenerate the division. The division's proposal for a new product line and new manufacturing equipment were developed in an atmosphere in which survival was at stake.

Most of the other companies studied by Gerwin (1984) were reacting to environmental opportunities at least as much as crises. The British, German, and French firms that were designing their own FMSs took advantage of their respective governments' desires to stimulate the development of advanced manufacturing systems through partial funding of the projects. Once more, all three companies had ambitions beyond merely installing the new technology in their own factories. They saw an opportunity to become leading vendors in what was believed to be a profitable new market. The German aircraft enterprise purchased an FMS in conjunction with being awarded a large military contract that almost doubled the capacity of its main plant.

Once performance gaps are identified, a search is conducted for alternative ways of eliminating them. This raises the question of how firms learn about CAM. Kelley and Brooks (1988) asked plant managers to rate how important was each of ten different sources of information in learning about new manufacturing technology.

They classified the responses into four categories including written media such as trade journal articles and advertisements, participation in professional or trade association conferences, direct contact with vendors through plant visits by their sales representatives, and informal contacts with users of the technology. The last category was most frequently cited as being an important source of information.

Including CAM as an alternative necessitates a decision on what Pelz and Munson (1982) called originality level. The organization must decide whether to develop its own solution (invention), modify features of an outside source's solution (adaption), or essentially take another source's ready-made solution (borrowing). For a large complex innovation such as CAM it is possible that some components may be invented, others adapted, and the rest borrowed.

What factors determine originality level? Invention is selected by firms planning to become suppliers of new technology. Three of the four companies studied by Gerwin (1984), which developed substantial portions of their manufacturing systems, fall into this category. Companies that want specialized equipment select invention because a vendor's profits depend upon furnishing off-the-shelf solutions. Their manufacturing processes may demand customized technology as was the situation for Gerwin's fourth inventing company, or they may be seeking a competitive advantage over firms using standard equipment (Hayes and Abernathy 1980). Invention is also employed when the substantial costs and risks can be reduced. The three prospective vendors had partial financial support from their governments. They also could install the system in one of their plants to demonstrate viability to prospective customers. It was not necessary to risk trying to sell it in order to show it was successful. An enterprise's accounting system influences the extent of invention (Foster and Horngren 1988). Where internal development costs are charged to the year the expenses occur, the tendency is to use external vendors because their charges are normally capitalized.

In some U.S. industries invention is widely used for major manufacturing innovations, that is users develop their own equipment (von Hippel 1988). It has been a minor source of FMSs however. A U.S. Department of Commerce study (1985) identified forty-seven FMSs in operation as of 1985. Ten systems from five different companies were developed in-house although in some instances the machine tools were purchased. Some of these companies were large multinationals considering whether or not to become FMS suppliers and one was a machine tool vendor. The thirty-seven remaining systems were all purchased from machine tool vendors, which represents borrowing with varying degrees of adaption in each instance.

The foregoing process, sketched for the way in which a firm begins to consider CAM, is perhaps too rational for a situation characterized by a great deal of uncertainty. Problems are identified in terms of performance gaps and alternative solutions are found using likely sources of information. CAM is evaluated because it is applicable to many kinds of problems. Mohr (1987) labeled this process "search" but went on to speculate about other means more rooted in organizational reality. In other words, CAM technology may be found through methods that are essentially political or social in nature. Since these routines are continually used by an organization for many purposes, Mohr believed that firms become acquainted with radical technology in rather ordinary ways. He also believed that CAM is likely to be an output from more than one routine, which explains why it is currently such a popular alternative.

We can divide Mohr's processes into those occurring mainly within the organization and those arising from interorganizational linkages. The first category includes "craft modernization" in which individuals look for more challenges or new ways to improve their skills. "Status enhancement" involves people looking for ways to increase their social ranking. One avenue is to find new technology whose operation, maintenance, or control will provide them with esoteric knowledge important to others in the firm. "Play" is a routine that involves technical people scanning current opportunities for new toys. "Slack distribution" occurs when excess resources, over and above the amounts necessary for business reasons, are allocated to powerful departments. The resources may be in the form of CAM technology.

The second category includes "imitation" in which a company slavishly follows what comparable organizations are doing. "Recruitment" is a process in which newly hired people who identify with CAM push its consideration. "Socialization" refers to taking the advice of external agents such as consultants, particularly when an organization is unable to come up with its own course of action. "Constituency satisfaction" occurs when a key customer suggests consideration of new technology. Kelley and Brooks (1988) reported that this practice is fairly common in Japan. It does not appear to be as widely used in the United States. Only 9 percent of the roughly 750 firms in their sample that sold machined parts directly to customers indicated that those customers required or requested use of CAM. It is, however, becoming increasingly salient for firms seeking orders from the Department of Defense and automobile manufacturers (Bennett et al. 1987). Kelley and Brooks found that industry sales to the Department of Defense as a percentage of total industry sales was significantly related to the chances that a plant would adopt CAM.

Selection of the Innovation

Many American firms have a short-term orientation that stresses immediate profits over long-run returns (Ayres 1984, Choate and Linger 1986, Dertouzos et al. 1989, Hayes and Abernathy 1980, Johnson and Kaplan 1987, Reich 1983). Utilizing existing assets more efficiently is emphasized. Little consideration is given to investing in new ones if returns are not expected rapidly, do not have high probabilities of occurring, and are not easy to measure. Long-term investments in computerized manufacturing technologies are consequently avoided. Reliance on traditional financial justification techniques that consider only measurable revenues and costs is one manifestation of this syndrome.

There is a great deal of agreement among American managers that a short-term orientation exists. For example, Johnson and Kaplan (1987) mentioned a 1982 survey of 230 chief executives by *Dun's Business Month*, in which a "thumping" majority agreed that U.S. managers are excessively concerned with the short run at the expense of longer-range issues. In Howell, Brown, Soucy, and Seed's (1987) study of preparers and users of managerial accounting information, 47 percent of the respondents believed that an orientation toward short-term returns was an obstacle to improving capital justification procedures. This consensus has been achieved mainly on the basis of personal experience. Little research evidence exists on the degree to which the capital decision making process is affected by short-term thinking (Baily and Chakrabarti 1988).

Why do firms adopt these self-defeating policies? A great deal of speculation exists but not much hard evidence. Blame has been placed on:

- The cost of capital
- America's financial institutions
- Organizational reward systems
- Managers' backgrounds and experiences
- The increasing technical complexity of capital projects

Firms with a high cost of capital, as compared to companies facing a low cost of capital, are at a moderate disadvantage in making short-term investments but at an insuparable disadvantage for long-term investments. The disparity in required cash flows between two such firms is much greater in the long run than in the short run. The cost of capital is relatively high to American firms versus say Japanese companies (Dertouzos et al. 1989). A low private savings rate and high government deficits are, in part, to blame. Between 1981 and 1986, net U.S. savings averaged 3.6 percent of gross domestic product compared with 17.3 percent in Japan. Tax policies are also held to be a contributing factor.

America's financial institutions now collectively control many large domestic enterprises (Choate and Linger 1986). Their fund managers, judged on the basis of short-term earnings, sell stocks that do not produce quick results. Companies that choose not to pursue short-run profits will suffer from falling stock prices. They will find it difficult to raise equity capital and be susceptible to takeovers. If they attempt to repurchase their own stock using retained earnings, not enough will be available to finance long-term projects. Borrowing to repurchase stock will further reduce the ability to obtain funds for long-term investment.

Factors operating within U.S. corporations contribute to a short-run orientation. Ayres (1984) reviewed the arguments that point to an organization's reward system as the culprit. A manager's compensation and promotion are based on short-term financial performance. Most firms try to identify potential top managers early in their careers and then move them quickly from job to job to give them wide experience. A typical candidate assigned to a new job may have just two to three years to show results. Under these conditions long-term investments in process technology make little sense. The expenses start immediately, whereas the major benefits are not available until after the manager hopes to be promoted.

Consider a manager's background and experience (Hayes and Abernathy 1980). From the mid 1950s to the mid 1970s, the backgrounds of the presidents of the 100 largest U.S. corporations shifted dramatically. Financial and legal people increased by 50 percent to command one-third of the total. Technical people decreased by about 12 percent to also rest at one-third of the total. Meanwhile, training of MBAs has stressed a financial versus a manufacturing orientation and application of analytical models over judgment based on experience. A company's influential decision makers are therefore prone to rely on discounted cash flow techniques.

A manager's background and experience has effects that are reinforced by the technical complexity of CAM. An FMS, after all, involves interactions between several machine tools, one or more computers, software, an automated material-handling system, fixtures, the parts being processed, and workers. The greater this complexity, the less likely is strategic management to inquire into the technical aspects of recommendations from the task force. Issues such as the type of equipment, capabilities, limitations, functioning, and requirements will not be considered in sufficient detail. Consequently, it is difficult to judge whether the proposed technology has features that are compatible with the company's needs, and whether the technology's needs can be met by the firm's resources. Strategic management is faced with a great deal of technical uncertainty. Not surprisingly, there is an absence of computer-aided manufacturing expertise among senior officials of American companies (Baily and Chakrabarti 1988, Gold 1980).

A financially oriented strategic management trained in the sanctity of analytical methods may opt for a coping strategy that avoids technical uncertainty. It defers to the task force's conclusions on technical issues and relies solely on financial evaluation, its main area of expertise. Proposals from the task force become potential investments in a financial portfolio rather than means of production with implications for a company's strategy.

Do Japanese firms exhibit long-term orientations? Sakurai (1990) conducted mail surveys of management accounting practices in Japanese companies with advanced manufacturing technology such as FMS and CAD/CAM. In one survey of about 277 AMT users he discovered surprisingly that 66 percent used the payback method to justify investments in the technology. Of those companies using payback 41 percent used a threshold of three years or less and 84 percent used five years or less. However, in another survey of about 196 firms he found that about 72 percent supplement financial measures with other quantitative and qualitative criteria when evaluating AMT. Japanese enterprises do exhibit characteristics of a long-term orientation but perhaps not as much as is currently believed.

Direct Impacts of a Short-term Orientation. One would expect a short-term orientation to have a negative direct effect on the chances of adopting CAM due to the technology's significant intangible returns. Flexibility, in terms of a broader range of capabilities and faster response times, is one intangible benefit. Others include better control over shop operations, reductions in inventory, improvements in quality, and savings on floor space. Even direct labor savings have important intangible components (Bennett et al. 1987). Of course, a firm must pay attention to increases in intangible costs as well.

A few studies demonstrate that a short-term orientation bars CAM technology from getting a fair hearing in the capital appropriation process.

1. Rosenthal (1984) surveyed users and vendors of computer-aided manufacturing equipment. Seventy-six percent of vendor personnel said inability to quantify returns was either significant or very significant in the decisions of potential users not to purchase their technology.
2. Bennett et al. (1987) indicated that some companies attempt to postpone acquiring new NC equipment because the depreciation expense in the near term will be much greater than depreciation from their existing machines. This postponement enhances a business unit's annual profits.
3. Farley et al. (1987) interviewed twenty-nine senior manufacturing managers from large U.S. companies concerning anticipated expansions of flexible automation. They ranked on ten-point scales the perceived importance of four general problem areas: gaining initial approval, employee resistance, starting up the system, and keeping the system running. Gaining initial approval was considered the most important due in part to insufficient short-term benefits.

Large organizational size, through its impact on bureaucratic specialization, may be an important factor in the negative effect of a short-term orientation on the chances of adopting CAM. In the larger firms studied by Bessant and Grunt (1985) production and financial people existed in separate compartments with little opportunity to iron out their differences over CNC technology. The former group, believing that CNC was the wave of the future, wanted strategic as well as financial factors considered in appropriation decisions. The latter group, noting that CNC required larger investments and had higher operating costs than traditional equipment, paid attention only to financial considerations. The only way for production people to get their message across was to use the chain of command to attract the attention of upper management.

In the smaller firms it was easier for members of the two functions to meet and work together. They learned to appreciate the position of the other side and were often able to settle their differences. The tendency was for each capital proposal to be considered on its own merits using strategic and financial criteria.

Corroboration of the findings in small firms comes from Howell et al.'s (1987) mail survey in which 70 percent of the respondents came from the group, division, or plant level, and 51 percent worked in business units with annual sales less than $100 million. Many respondents believed there was significant pressure for short-term results. Seventy-one percent reported using a payback period to justify advanced manufacturing technology and of these about 70 percent used three years or less. Sixty-nine percent used discounted cash flow techniques. Yet they also indicated that qualitative factors, associated with a long-run orientation, were important in the justification process. Over 80 percent considered nonmeasurable performance improvements in quality, delivery, and technological capability. Seventy-eight percent considered whether the investment would increase the ability to compete and 68 percent inquired into its relationship to manufacturing strategy. The ensuing conflict was handled on an *ad hoc* basis for each investment. These results are remarkably similar to those in Sakurai's (1990) investigation of Japanese companies.

Indirect Impacts of a Short-Term Orientation. Strategic managements with a short-term orientation and consequent reliance on financial evaluation techniques are influenced by hard quantitative information known with a high degree of certainty. The prevailing technical and financial uncertainty means that most future benefits and costs are difficult to predict and hard to control. Acquisition cost therefore assumes a key role in discounted cash flow or payback calculations. It is the one significant financial quantity known with little risk and negotiable in the short run. Another reason that a strategic management with a short-term orientation focuses on acquisition cost is that the higher this quantity the greater is yearly depreciation expense and hence the less is annual profit.

In an uncertain capital budgeting framework the innovators will try to discover which criteria are important to strategic management and then develop recommendations that will fit them (Bower 1970, Carter 1971). Subordinates tend to request projects that are likely to be accepted, withhold projects that have little chance of approval, and where possible adopt proposals to fit existing goals (Carter 1971). It follows that where a short-term orientation exists the task force is likely to adopt a coping strategy of using acquisition cost as one of its main criteria. As will be explained shortly, new computerized manufacturing innovations may have relatively low acquisition costs creating one favorable condition for their adoption.

In Gerwin (1981a), acquisition cost was one of the three major criteria used by the task force in making its recommendations in part because it was of concern to strategic management. The FMSs relatively small acquisition cost compared to other alternatives represented a useful selling point.

Just as the technical task force is searching to find strategic management's decision premises, potential vendors hope to uncover those of the task force. Then they can tailor the flow of information to meet the premises. If vendors find the task force reacting to a short-term orientation their coping strategy will be to emphasize concrete advantages and downplay intangible benefits. Ultimately, strategic management will receive only tangible data, the kind of information that it wants to receive. The chances of adopting computerized technology will be diminished further.

As evidence, Boddy and Buchanan (1986) noted that vendors of information technology recognize the criteria a firm uses to evaluate projects, often easily measured benefits such as productivity improvement and cost reduction, and direct their promotional activity toward stressing them. Primrose and Leonard's (1988) research into the economics of CAM found that companies invested in systems that could be justified in terms of direct labor savings, and vendors responded by developing systems to meet these specifications, thus limiting the new technology's scope and potential.

The newer the process technology the more anxious will be a vendor to place a unit in some company. Vendors are aware of the high degree of technical uncertainty surrounding new technology and the consequent resistance of potential customers. By placing a unit they hope to demonstrate its viability and obtain useful feedback information to lower the uncertainty for subsequent customers. Frequently, potential purchasers of a manufacturing system will want to visit a company where it is being used.

If the vendor's analysis of decision premises reveals a short-term orientation, then offering a low acquisition cost will be a useful coping strategy. A vendor of new computerized technology who is competing against conventional processes the reliability of which has already been established is likely to use this strategy. A relatively low acquisition cost will help offset the higher risk for the customer. It will also be favorably received because of its concrete nature.

The Confidence Game. Since the adoption decision cannot be made purely on rational grounds, one of strategic management's coping strategies is to assess its confidence in the task force (Bower 1970, Carter 1971). Degree of confidence is determined using objective data where available as well as social and political information. Where confidence is lacking, intensive analysis of proposals, requests for rechecking, and possibly outright rejection occur. The task force therefore has an interest in defining a social and political reality that makes it appear trustworthy.

How does strategic management gauge its degree of confidence in the task force? It checks into the reliability of the group's leader who here is considered the technology's champion. Have previous recommendations been successful? Have past projections turned out to be highly exaggerated? A reliability check is effective because there is little the champion can do to alter senior managers' perceptions of his or her record. It also tempers the champion's desire to manipulate strategic management in the current situation out of fear that his or her reliability will ultimately suffer. This phenomenon requires further explanation.

In an organizational setting individuals and groups which transmit information under uncertain circumstances bias their communications in ways which produce favorable (or avoid unfavorable) results (Cyert and March 1963). Biasing assumes at least three different forms (Dean 1987). Measurable benefits such as direct labor savings are overestimated and measurable costs are underestimated. Intangible benefits are quantified, whereas intangible costs are not. Arguments relating to more effective competition such as "Other companies in the industry are certain to adopt CAM." or "Customers' perceptions of the firm will be enhanced." are advocated.

The task force will bias its recommendations to favor the manufacturing process of its choice, assumed here to be CAM technology. If initial objective calculations of expected net returns hover just around the minimal acceptable level, biasing is more likely than if expected net returns are clearly acceptable or unacceptable (Dean 1987). Due to its large proportion of intangible benefits a proposal for CAM may very well be on the borderline. A project's technical complexity and technical uncertainty influence the amount of adjustment that is made since strategic management's ability to comprehend is affected by these two factors (Carter 1971). CAM is associated with high values of both variables so financial projections for it may contain a relatively large amount of bias.

There is a limit, however, to the extent of biasing. Adjustments increase the chances of the project being selected, but approval eventually allows determination of whether or not the calculations were realistic. The champion's reliability, a significant personal asset, will be at risk (Dean 1987). In practice, of course, there are ways of avoiding the reliability test. In some firms the same individuals who developed the original CAM proposal are allowed to do the postaudit. Some firms do not conduct postaudits for computerized manufacturing technology (Howell et al. 1987, Winch 1989). Winch conducted in-depth studies of fifteen British metalworking firms that had implemented CAD/CAM. Only one attempted to measure the performance of its system against the criteria laid down in the initial justification.

Due to CAMs widespread organizational impacts, strategic management's confidence in the task force is also increased when a coalition of different functional groups exists that supports the proposal (Dean 1987). The more widespread is the coalition and the more powerful its members, the more top managers will be impressed. An especially influential example includes engineering design, responsible for a new CAM system's products and manufacturing engineering, responsible for the new system (Gerwin 1981a, Dean 1987). Once more, there is an intangible benefit in having these two often conflicting units working together.

Members of the task force employ the following tactics to build coalitions (Dean 1987). They share information, form joint steering committees on specific issues, and request senior managers to exert pressure. They also agree to support an issue of relevance to a functional unit in return for cooperation. The danger in this last coping strategy is that tangential issues get attached to the proposal which may lengthen the approval process. In one company, advocates of manufacturing resource planning (MRP II) gained the help of a key group by agreeing to support its objective of a mainframe computer. Unfortunately, hardware decentralization was so politically sensitive that the MRP II advocates some time later decided to separate the two issues.

Strategic management's confidence is augmented in another way, when it perceives that the task force is confident about the proposal. In fact, there is likely to be various kinds of testing for signs of doubt, confusion, and indecision. For example, senior managers ask extremely detailed questions and assess the thoroughness of the answers and the confidence with which they are delivered (Dean 1987).

A useful coping strategy for the task force is therefore to appear to be confident in its recommendations. There should, for example, not be too many changes in them. In one company the merchandising department, under pressure from a hard-to-convince general manager, more than once changed its sales forecasts for new products to be manufactured on a contemplated FMS. He eventually lost confidence in the forecasts. In contrast, the person generally acknowledged as the FMS champion, the head of manufacturing engineering, took the lead in selling the FMS to divisional management. His solid commitment was characterized in the interviews as having a positive impact on the decision to accept the proposal (Gerwin 1981a).

What impacts does strategic management's degree of confidence in the task force have? It affects the use by the task force of certain coping strategies, which advance its interests by seizing on the prevailing financial and technical uncertainty. If strategic management's confidence is very low, proposals will be exhaustively reviewed and the task force's strategies will be of little use. If confidence is very high, there is little need for them. A curvilinear relationship exists in which, as confidence builds from a very low point, the strategies are used more frequently until a peak is hit and their use begins to subside.

One coping strategy for the task force is to reduce the visibility of its preferred technological alternative to keep it from being intensively reviewed. A proposal for a radical computerized system is buried in a large capital improvement program as was done in at least two of the firms studied by Gerwin and Tarondeau (1982). The sheer number of projects combined with their technical complexity will hinder an exhaustive analysis of any one, especially if none represents a major fraction of the program's total cost. Alternatively, a proposal for a CAM system is divided into smaller components such as individual work stations that are treated as separate projects. A request for integration appears in a subsequent year. This shields the technology from the more intensive higher management review usually associated with expensive investments.

The chances of a firm adopting CAM have now been discussed in terms of several causal factors. Normally adoption involves the two simple alternatives of yes or no. However, when a high degree of uncertainty exists, there is a third alternative of hesitation. Delay of the decision when it is unclear what to do opens up the possibility that more information will become available in the future or that the need for a choice will evaporate. In one of the firms studied by Gerwin and Tarondeau (1982), strategic management sent back the task force's proposals over thirty times for revision before a decision was made.

A number of studies indicate that hesitation is a typical coping strategy in risky situations (MacCrimmon and Wehrung 1986). Collecting more information, bargaining, or requesting further study are often employed in hopes that the situation will eventually become more attractive. MacCrimmon and Wehrung studied hundreds of executives making hypothetical personal and business choices including capital budgeting. In one set of experiments—a "risk in-basket"—managers delayed decisions in 28 percent of the situations handled. Although delay is a valuable means of reducing risks, excessive amounts are a symptom of risk avoidance. Once more, delay is not costless; it may reduce one's options if key deadlines are missed.

SUMMARY AND CONCLUSIONS

In order to bring together what has been said about the adoption process some major observations will be stated in the form of propositions. A flowchart that displays the propositions for innovation selection appears in Figure 9.1.

During the innovation awareness subphase a decision is made on originality level:

Proposition A1. The more specialized is the technology, the more risks and costs are reduced, and the greater the firm's intention to sell the technology; then the greater the chances of invention.

During the selection of an innovation strategic management's short-term orientation is a critical factor:

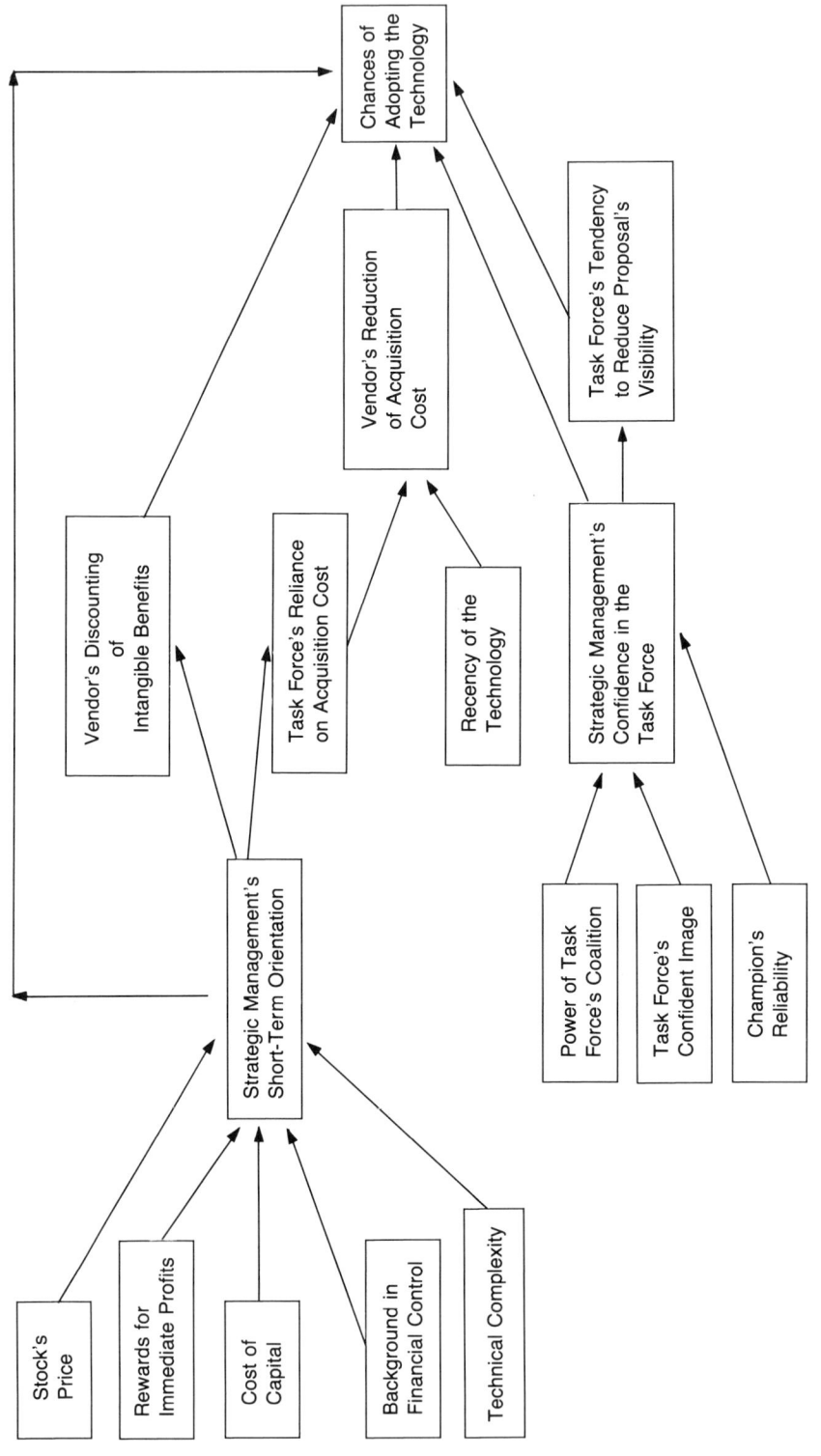

FIGURE 9.1. The adoption process.

SUMMARY AND CONCLUSIONS   247

Proposition A2. The lower the price of the firm's stock, the more rewards are based on immediate profits, the higher the cost of capital, the more strategic management's background and experience reflect financial control, and the greater the proposed equipment's technical complexity; then the greater strategic management's short-term orientation.

The direct impact of a short-term orientation at least in large corporations is considered in the last proposition. Its indirect effects are handled below:

Proposition A3. The shorter the time orientation of strategic management, the more likely the task force will adopt acquisition cost as a financial decision premise.

Proposition A4. The shorter the time orientation of strategic management, the less the vendor will emphasize the manufacturing technology's intangible benefits to the task force.

Proposition A5. The more the task force relies on acquisition cost and the greater the recency of the manufacturing technology, then the more likely the vendor will offer a low acquisition cost.

Another influential variable during innovation selection is strategic management's confidence in the task force:

Proposition A6. The greater the champion's reliability, the more powerful the task force's political coalition, and the more confident the task force appears to be in its recommendations; then the greater strategic management's confidence in the task force.

Proposition A7. As strategic management's confidence in the task force increases up to some point, the latter will try more and more to reduce the visibility of its preferred alternative. Beyond this point the task force will be less and less inclined to reduce visibility.

The sequence of cause and effect relations indicated by Propositions A2 through A7 results in a set of variables which directly influence the chances of adopting CAM:

Proposition A8. The less strategic management has a short-term orientation, the less the vendor discounts intangible benefits, the more the vendor reduces acquisition cost, the greater strategic management's confidence in the task force, and the more the task force reduces the proposal's visibility; then the greater the chances of adopting computerized technology.

10
THE PREPARATION PROCESS

Preparation means developing an infrastructure or support system sophisticated enough to operate, maintain, and control a technological innovation. According to Peters and Waterman (1982) and Skinner (1985) the infrastructure represents a factory's managerial aspects including people's skills (e.g., electronic maintenance), attitudes (e.g., willingness to work on second and third shifts), systems and procedures (e.g., cost accounting controls), structures (e.g., work organization), leadership styles (e.g., participative), strategies (e.g., product innovation), and shared values (e.g., improving product quality). Increasing the infrastructure's sophistication often means instituting administrative changes and innovations to support the technological ones represented by CAM.

Preparation is less well understood than adoption or even implementation. For example, Hage's (1980) detailed theory of the innovation process contains only one relevant hypothesis. It merely states that the more radical the technological innovation the greater the need for new personnel, funds, and associated technologies. A general theme, however, does seem to have emerged from the literature. Although a sophisticated infrastructure is crucial for the success of CAM, many firms for one reason or another fail to put together the necessary ingredients. In Rosenthal's (1984) survey, for example, 68 percent of vendor personnel agreed that customers do not have a realistic sense of the resources needed to get CAM operating smoothly.

Two general requirements determine whether a company develops an infrastructure capable of supporting CAM operations. First, it must have knowledge of the necessary administrative changes and innovations. Are electronic maintenance skills required? Will a semiautonomous work group operate the system? Is a product costing system based on machine hours instead of direct labor hours necessary? Unfortunately, social and technical uncertainties exist that render the pertinent information biased and incomplete.

Second, a company needs to have facilitating conditions that allow the revised infrastructure to be designed. These conditions often represent certain characteristics of the innovation process such as allowing affected functional units to participate in selecting the new technology. Then it is more feasible to develop an electronic maintenance training program, design a semiautonomous work group, or develop a new product costing system. Altering the innovation process to facilitate infrastructure development represents a second type of administrative change. Unfortunately, factors connected with uncertainty limit the extent to which such changes occur.

KNOWLEDGE OF THE REQUIRED INFRASTRUCTURE

New knowledge is required by almost every functional unit and hierarchical level in the factory and perhaps in the business unit as well. All units and levels must have an understanding in varying degrees of the new technology's capabilities and limitations if they are to have any hope of revising pertinent infrastructure elements. Technically oriented departments such as manufacturing engineering and product design need the most detailed understanding, but even senior management must grasp something about what the technology can and cannot do if new strategies are to be developed.

A leading vendor of FMSs and FMCs has summarized the knowledge requirements of various functional units (Kearney and Trecker Corp., no date).

1. Manufacturing engineers have to write sophisticated parts programs and be able to provide vendors with specifications for new tools and fixtures.
2. The quality control department must determine the degree to which inspection will be integrated into the new technology versus performed off-line. It must establish tolerance specifications for the parts.
3. The maintenance unit has to add electronic skills in order to help solve complex breakdown problems.
4. Engineering designers must learn how to modify the parts' designs to make them compatible with automation.
5. The production scheduling department has to grapple with the problems created for downstream activities by random processing.
6. Accounting has to develop new product costing and performance appraisal systems not based on direct labor hours.
7. Purchasing needs to develop a list of vendors who can provide tooling, fixtures, and raw materials to rigid specifications.
8. Industrial relations people must create a work organization and accompanying payment plan which reflects the activities workers will perform.
9. The information systems unit must determine how to interface the company's computers with the new technology's and what information is to be transmitted back and forth.

10. To take advantage of market oriented flexibilities the marketing department must quickly identify changing customers' needs and communicate the information to Production Scheduling.

Hierarchical levels in the chain of command are also affected. Skilled operators are needed to monitor machine operations so that the expensive equipment is not damaged. Production supervisors must have extensive technical knowledge to react properly in the face of breakdowns. They also need interpersonal skills to motivate workers and to solicit cooperation from other units (Blumberg and Gerwin 1984). Senior managers must convert potential uses of the new technology into strategic opportunities for the firm (Boddy and Buchanan 1986).

Due to the radical nature of the technological innovation the required knowledge often does not exist. When it does exist it must then be embedded in specific new skills, systems and procedures, structures, and other infrastructure components in order to be potentially useful. Ettlie (1988b) provided some examples that are being developed to support CAM technology. They include teams of programmers and operators, decentralization of corporate manufacturing engineering to plants, a common supervisor for the product design and manufacturing engineering departments, new policies for dealing with suppliers of components, and semiautonomous work groups.

Where does a company go in order to obtain the necessary new knowledge? It is unlikely to reside within the firm or comparable organizations because of the technological innovation's radical nature. Five of the seven companies studied by Bennett et al. (1987) introduced new CAD/CAM systems. The system managers had not known beforehand what performance measures they needed. Some accountants admitted that their lack of familiarity with CAD/CAM hindered them from contributing to the development of measures.

The vendor suffers from lack of knowledge for at least three reasons. First, it may not have introduced the new technology into its own facilities. According to a U.S. Department of Commerce report (1985), there is a general belief that most Japanese vendors developed an FMS for internal use before marketing them, whereas American suppliers did not. Second, the technology's recency hinders the vendor from having accumulated experience via placements. Third, the technology's complexity may prevent temporary installation of a finished system in the vendor's test facilities prior to shipment. The first investigation of the system's capabilities, limitations and impacts may be at the customer's factory. Respondents in all six firms studied by Boer and During (1987) indicated there was insufficient knowledge of fixturing methods, software, and system control within their firms and at their vendors.

Transferring Knowledge

The very process of transferring information from the vendor to the task force impedes acquiring knowledge of the required infrastructure. Each time a transfer occurs the possibility of distortion and leakage in information exists. Some knowledge

KNOWLEDGE OF THE REQUIRED INFRASTRUCTURE   251

is tacit and therefore difficult to communicate even in an experiential fashion. Often differences in status, skills, and objectives hinder effective communications.

Transferring knowledge from the vendor to the task force is only the first step in a process that continues throughout the life of the new technology. Some typical knowledge transfer points are indicated in Figure 10.1. The task force may prepare for a special transition team consisting of technical specialists to debug the system during the start of implementation (Graham and Rosenthal 1986b). This group in turn eventually transfers responsibility to an operating team which handles normal functioning. Differences in objectives are a factor here. The engineers are more concerned with learning about the system's capabilities and limitations. The operating people want to know how to meet production deadlines.

The operating team contends with at least two kinds of knowledge transfers. First, knowledge must be supplied within the team to replacements as individuals leave or are transferred. Second, an innovation is often introduced into a single organizational subunit on a pilot basis and then attempts are made to diffuse it internally. Operating people, in conjunction with task force and transition team members must communicate their knowledge to other parts of the company if internal diffusion is to succeed. Geographical and organizational separation, the need to assign experienced people to other tasks, and differences in the nature of the current and new projects impede the flow of information (Hayes, Wheelwright, and Clark 1988).

Finally, the vendor, in an effort to augment its own expertise for future clients, solicits feedback from the task force, transition team, and operating group. This information, however, is often colored by the degree to which the customer's personnel perceive the technology to be a success.

Various alternatives exist for improving the effectiveness of knowledge transfer. Eliminating a transfer point is always a possibility but requires a careful weighing

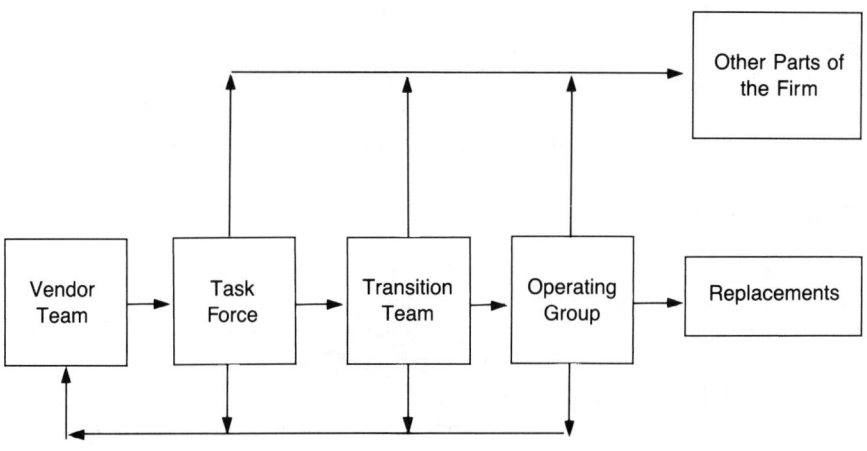

FIGURE 10.1. Knowledge transfer points.

of the costs and benefits. Consider the transition team. Having one gives technical people a chance to learn about the new system, and eases the problems of the operating group. On the other hand it is one more opportunity for information distortion and leakage, it delays the moment when operating people must come to grips with the system, and it is sometimes used as a justification for hiring low paid, low skilled workers on the operating team.

Overlapping group membership is commonly used to improve communication. Certain members of one team may participate in the activities of a subsequent team. The vendor may send a representative to work with the transition team. Individuals from the transition team are often asked to play a role in the operating group. A person from the operating group, familiar with and committed to the innovation, may be transferred to aid in its internal diffusion. Designated members of a subsequent group may be stationed in an existing group as when future operating people are put on the task force. Two groups functioning simultaneously in time may be integrated as when liaison people are stationed between the vendor's team and the task force.

Roberts (1988) called these "human bridges" the most effective transfer mechanism. The upstream movement of people brings a group information on intended project use, establishes direct person-to-person contacts that will be helpful in posttransfer troubleshooting, and reinforces the belief that the receiving unit is prepared to accept ownership. Movement of people downstream brings expertise for subsequent problem solving, and creates the risk-reducing impression that the receiving group will not be stuck with solving problems by itself.

OPPORTUNITY TO DESIGN THE REQUIRED INFRASTRUCTURE

Knowing about the required infrastructure is not enough. Conditions must exist in the innovation process that facilitate designing the elements. Some conditions represent innovation process characteristics over which management has control in the short run, such as the amount of dependence upon the vendor, the level of strategic management's pressures for immediate full-scale production, and the composition of the task force. Others, such as strategic management's beliefs about the need for managerial change, the infrastructure's initial degree of sophistication, and external institutional constraints, are only manipulated in the long run.

Short-Run Issues

Given the uncertainty facing an innovating company it may have a coping strategy of depending heavily on the vendor for making important decisions and performing major activities. This tendency is increased when confidence in the vendor is high and when the infrastructure's initial level of sophistication is low. It is decreased when the vendor is located far from the plant. The vendor is tempted to accept responsibility out of its desire to ensure that the innovation is successful. The company unfortunately loses a significant opportunity to gain knowledge needed for infrastructure development. Once more, the vendor's personnel are beyond its direct control which augments uncertainty.

Confidence in the vendor is based primarily on satisfactory prior relationships. A supplier with whom the company has had no prior experience represents more of a risk for delivering what was promised. Research studies indicate a general tendency for machine tool users to favor previous vendors for current purchases. Cunningham and White (1974) found that 60 percent of the machine tool purchases they studied involved suppliers with whom the buying firm had previous transactions. Cardozo and Cagley (1971) asked industrial buyers to participate in a controlled experiment using a buying game. Over 60 percent of bid selections and purchase decisions were characterized by suppliers having a prior relationship with the purchaser. The tendency is so strong that Graham and Rosenthal (1986a) discovered FMS users who relied on previous vendors even when those vendors had little experience with integrated systems.

When a customer's infrastructure has a low level of sophistication the firm is more or less pushed into depending upon the vendor to solve problems. Moreover, the members of the task force will not have sufficient expertise to accurately gauge the vendor's strengths and weaknesses. Instead of attempting to supplement the vendor's skills they will be readily convinced that it is omniscient. Inevitably, when things turn out differently they will be late in detecting it and unable to react quickly (Graham and Rosenthal 1986b).

Strategic management's short-term orientation results in pressures for immediate financial returns particularly if the acquisition cost was large. Preparation and implementation are curtailed in favor of achieving full-scale production as soon as possible. Employees are prevented from learning about the technology's capabilities, limitations, and impacts. Two of the firms in the Gerwin and Tarondeau (1982) study noted this problem. In Japan, however, the project team remains with a new FMS long after installation. Members continually improve its functioning as they gain valuable experience (Jaikumar 1986).

Determination of the task force's composition involves decisions on the degree of participation by various groups affected by the new technology. The term participation here refers more to a sharing of influence among various functional units than between bosses and their subordinates. Even if workers are included in the task force most of the other members are not likely to have authority over them. One implication is that the literature on cross-functional integration is more relevant than research on superior-subordinate relations.

What are the benefits of participation on the task force?

- It counteracts the effects of immediate production by ensuring an earlier starting point for infrastructure development.
- It allows people to gather information on the technology's consequences for their specialties.
- Process design is more systemic because it involves more relevant people. For example manufacturing engineers can select or design a system that is in

agreement with current shop practices rather than their idealized view of what is occurring. Often specialists have made undocumented changes in procedures of which engineers are unaware.
- It leads to design changes in the technology that facilitate development of a compatible infrastructure.
- It increases the commitment of individuals and unions to bring about all the necessary technical and administrative changes.

Most studies of CAM innovation support participation in the task force by affected parties. Graham and Rosenthal (1986a) rated the effectiveness of eight FMS project teams based on members' evaluations of problems and achievements. The most effective one was a small group with technical (manufacturing engineering, software), organizational (industrial relations, personnel), business function (purchasing, accounting), and line (operating, supervisory) experience. Less effective teams had either too few people to cover all aspects of the relevant experience, or too many people to manage. Boer and During (1987) found that project teams chosen on the basis of just technical knowledge were a bottleneck to effective FMS innovation. In Rosenthal's (1984) survey, 81 percent of vendor personnel considered it important to involve factory floor people in the decision to purchase.

Levi, Slem, and Young (1989) surveyed five electronics companies that successfully introduced AMT. In three that allowed employees to participate in the introduction 398 people responded. In two that did not have participation in the introduction 378 individuals were queried. Employees in the former companies had a significantly greater belief in their firms' ability to manage technological change. They were significantly more likely to believe that technological change would help their careers and improve the quality of their jobs. They were significantly less likely to believe that technological change would increase job stress and job insecurity. In most categories, however, the results for production workers were not as strong as those for professionals, managers, and supervisors taken as a group.

Additional evidence comes from a 1983 survey by the British National Economic Development Office which examined approaches used by electronic companies to introduce new technology (Boddy and Buchanan 1986). Of forty-six plants studied, twenty-nine believed that using participation had helped particularly when it was introduced early in the innovation process. Early involvement is necessary to prevent a major decision from becoming a *fait accompli*. Some specific benefits cited included greater personal commitment, reduction in the amount of time to introduce the technology, and greater personal understanding of decisions.

Given CAM technology's widespread impacts, how is it possible to ensure participation without making the project team too large? Graham and Rosenthal (1986a) suggested that members have some cross-functional experience, that they be phased in and out of the team as they are needed, and that some people serve as consultants to the group. Majchrzak (1988) suggested a two-tiered system in which the upper tier—a steering committee—reviews plans and makes strategic recommendations and the lower tier—a technical committee—makes design plans.

Another approach involving a steering committee, a design committee, and technical study teams was described in Chapter 7.

Long-Run Issues

Strategic managers in some companies do not realize the necessity for developing infrastructures. They assume that the introduction of new manufacturing technology will automatically lead to productivity increases; the more sophisticated the technology the greater the increase. Support systems need not be considered or can be dealt with in an *ad hoc* manner during implementation. They do not understand that the latest technology in an unprepared setting will not function as effectively as unsophisticated equipment in a compatible environment (Schumacher 1973, Trist 1981). In one firm studied by Gerwin and Tarondeau (1982) the executive responsible for the project saw only engineering problems. He believed that installing the FMS in a plant with a good record of union-management relations was sufficient to handle infrastructure issues. Most of the companies researched by Boer and During (1987) perceived introduction of an FMS as essentially a technical problem with no implications for people, systems, and structures.

The infrastructure's degree of sophistication at the start of the preparation stage influences whether the required support system eventually develops. If the initial sophistication level is high the chances of attaining the required level in a reasonable amount of time improve. It is not only a question of having less distance to go. Generalized problem-solving methods are available to aid in handling new difficulties. Coping mechanisms have been developed to augment tolerance of further changes. Confidence has been accumulated in being able to handle uncertainties.

A number of studies have indicated the importance of a sophisticated infrastructure. Munro and Noori (1988) found that the extent of computerized automation in a manufacturing process and the amount of investment in computerized automation relative to other capital expenditures were both significantly and positively related to the availability of in-house expertise. Gerwin and Tarondeau (1982) found that the more experience a company has had with NC and CNC machine tools the better equipped it is to cope with an integrated system. The British motor producer had virtually no such exposure at the site chosen for its DNC system. It had to forge from scratch a comprehensive development plan to accompany implementation. The German aircraft manufacturer had considerable NC experience so fewer adjustments were needed. However, the research also suggests that there is a qualitative leap in complexity from individual machines to an integrated system. No amount of NC and CNC experience can completely handle the problems.

The Japanese companies studied by Jaikumar (1986) had more sophisticated infrastructures than their American counterparts. More than 40 percent of the work force in the Japanese firms were college educated engineers who had been trained in the use of CNC. In U.S. firms only 8 percent of the workers were engineers, of whom less than 25 percent had CNC knowledge. On average the Japanese companies had two and one-half times more CNC machines. It is little wonder that the average Japanese FMS took between one and one-quarter and one and three-fourths years to develop with 6000 man-hours of input, whereas the average U.S. system took between two and one-half and three years to develop with 25,000 man-hours of input.

Collins, Hull, and Hage (1989), who studied fifty-four manufacturing plants in New Jersey, divided them into three categories—high, low, and nonadopters—based on the percentage of production capacity devoted to computer controlled equipment. They found a number of statistically significant differences between the infrastructures of high adopters versus low and nonadopters. High adopters had greater proportions of professional, technical, and skilled production employees. They had greater proportions of graduate degree and Bachelor of Engineering degree holders. They also had a higher ratio of engineering to nonengineering degree people.

Various indications usually exist that an infrastructure's initial level of sophistication is too low. A company may not have enough capable people to design the necessary elements or these individuals may leave during the innovation process. At least three firms studied by Gerwin and Tarondeau (1982) lost key managerial, operating, or staff people, in some instances after they were trained. A company may not be able to afford the expense of changing the support system or be willing to sacrifice the time of the few available individuals with specialized skills. The prospect of large-scale simultaneous technical and administrative changes may be too much for the organization to handle. Finally, the state of the art in a specialized field such as accounting may not be advanced enough to meet the demands of the new technology.

External institutional constraints are usually under control or subject to negotiation only in the long run. The right of certain occupational groups to do particular tasks is provided wholly or in part by sources outside a company or one of its factories (Child, Ganter, and Kieser 1987). Work rules including job classifications and internal hiring procedures are determined through collective bargaining. This may prevent FMS operators from handling routine maintenance activities or compel FMS operators to be chosen on the basis of seniority rather than CNC experience. Often these constraints form part of the basis for an occupation's role, power, and status in an organization. They serve to buttress resistance to change during implementation.

A second type of external institutional constraint is regulations by various governmental agencies on the cost accounting practices of American defense contractors. These directives prevent defense contractors from making significant changes in their cost accounting systems even though they have some of the most automated manufacturing facilities in the country (Foster and Horngren 1988, Howell et al. 1987). The regulations also create a powerful incentive for contractors to maintain the status quo. Prices for most of their products are based on recoupable costs so that a unit of cost yields at least a unit of revenues. Many defense industry managers believe that changing the way costs are reported can only lower revenues since the government will not accept upward cost revisions.

As an example, due to the relatively large size of a defense plant and the variety of products manufactured there, use of multiple overhead pools, bases, and rates would significantly contribute to more accurate product costing. Yet 52 percent of the defense contractors in Howell et al.'s (1987) sample used single plantwide overhead rates. One accounting manager explained why more sophisticated approaches have not been widely accepted. "Gray area" costs, presumably those subject to intense negotiation with the government concerning inclusion in recoupment, would become more visible and less defensible.

SUMMARY AND CONCLUSIONS

Organizational lag refers to the tendency for the rate of administrative innovation in an organization to be less than the rate of technical innovation. Damanpour and Evan (1984), who defined organizational lag, conducted a mail survey of eighty-five public libraries in six northeastern states. Data were collected within three different time periods between 1970 and 1982 for small-, medium-, and large-size libraries. In eight of the nine resulting categories the average of the number of technical innovations per library divided by the number available for adoption exceeded the corresponding figure for administrative innovations. Differences in six of the cells reached statistical significance. Child et al. (1987) conducted case studies of microelectronics technology adoption in the British service sector including automated blood analyzers in hospitals, electronic point-of-sales terminals in retail stores, and automated tellers in banks. They concluded that the new technology was generally introduced with a minimum of change in organizational patterns. Karmarkar, Lederer, and Zimmerman (1990) studied how the nature of cost accounting and production control systems is affected by the characteristics of a firm's production technology such as complexity. They collected data in plants with small and medium batch sizes using a structured questionnaire and found few statistically significant associations. These results, they concluded, were compatible with production and accounting systems adjusting with a lag to changes in production technology.

Our discussion of preparation highlights some possible reasons for the organizational lag phenomenon at least in firms adopting CAM. Administrative innovations must go through their own adoption processes during which they can be blocked from further consideration. Unfortunately, little is known about adoption for administrative innovations associated with new manufacturing technology. It is apparent that dangers lurk in the knowledge awareness subphase. Companies have a great deal of trouble learning exactly what is needed and then developing applications for their context. During the selection subphase it is undoubtedly hard to conduct a formal justification due to the preponderance of intangible benefits and costs. One powerful argument is that the very expensive manufacturing technology will fail unless the administrative innovation in question is adopted. Due to the unavailability of evidence this is essentially an appeal to strategic management's fears.

Once more, short- and long-term characteristics of the innovation process for the new technology conspire against introducing administrative innovations which augment the infrastructure's sophistication. The former include dependence on the vendor, pressures for immediate production, and lack of participation by affected parties. The latter are strategic management's belief that change is unnecessary, a low initial level of sophistication for the infrastructure, and external institutional constraints against change.

Administrative changes in the innovation process are necessary to correct these deficiencies. The low magnitude and frequency of occurrence of these revisions in American companies unfortunately retard their effectiveness (Hayes, Wheelwright, and Clark 1988). Many firms adhere to a given set of innovation procedures over several projects and then use the accumulated experience to make major improvements in the process. During the long periods between major alterations cumbersome new rules are developed on an ad hoc basis to keep new problems from recurring. Their effect is to hamper the process from functioning adequately. When the major changes occur they can only improve the process's functioning back to the level of the previous major alteration. Smaller and more frequent planned changes, say after each project is completed, make for continual, gradual improvement based on immediate experience.

Some hypotheses for the preparation process are depicted in the flowchart of Figure 10.2. The chances of having by the start of implementation an infrastructure that can support the new technology depend on an accumulation of knowledge concerning what is needed and whether opportunities exist to turn available knowledge into operational infrastructure elements:

Proposition P1. The greater the knowledge of the required infrastructure and the greater the opportunity to design the required infrastructure, then the greater the chances of developing the required infrastructure.

Knowledge accumulation is governed by the following propositions:

Proposition P2. The greater the vendor's knowledge of the required infrastructure, the greater is the company's knowledge.

Proposition P3. The less the vendor's own use of the technology, and the greater the technology's recency and complexity; then the less the vendor's knowledge of the required infrastructure.

Whether opportunities exist to convert available knowledge into an operational infrastructure depends upon the following:

Proposition P4. The less the dependence on the vendor, the lower the pressure by strategic management for immediate production, the more widespread the participation in the task force, the more strategic management believes infrastructure changes are needed, the greater the infrastructure's initial sophistication level, and the less the impact of external institutional constraints; then the greater the opportunity to develop the required infrastructure.

Proposition P5. The shorter the time orientation of strategic management and the larger the acquisition cost of the innovation, then the greater the pressures for immediate production.

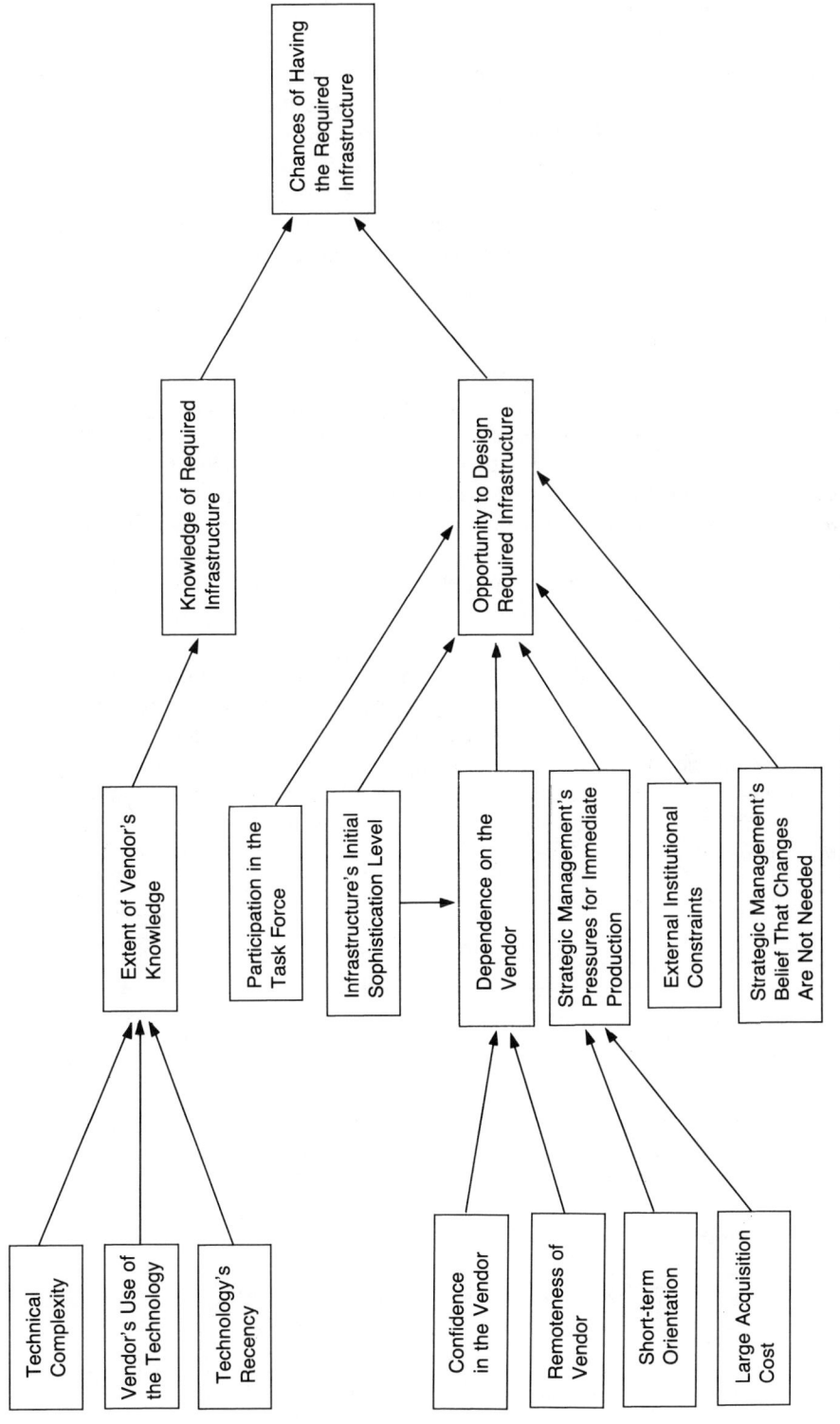

FIGURE 10.2. The preparation process.

Proposition P6. The greater the confidence in the vendor, the less the infrastructure's initial sophistication level, and the less remote the vendor from the factory site; then the more the dependence on the vendor.

For all the indicated reasons the infrastructure's sophistication level is usually not high enough at the end of preparation. Further development must occur during implementation, a situation that creates new problems and opportunities. Difficulties occurring during the third stage can be handled only after a considerable time lag. It takes time to train people, develop a new procedure, or reorganize. However, the existence of operational difficulties serves to allocate resources to where they are most needed. It is also possible to obtain some crude feedback information on how solutions are working. In other words, during implementation infrastructure development proceeds according to a rudimentary trial and error learning process.

The significance of implementation for infrastructure development was indicated by Boer and During (1987). Most organizational changes that they observed, occurred after FMS installation when the companies discovered the need for changes. Similarly in Gerwin (1981a) a new American FMS processed high quality castings with narrow tolerance limits. Only when it became clear that external suppliers were not meeting this need could a decision be made to have the business unit's foundry make the castings.

One theoretical implication is that the innovation process for radical manufacturing technology does not conform to a rigid sequence of stages. Contrary to Hage's (1980) views there is no sharp boundary between preparation and implementation. Infrastructure development continues during implementation subject to virtually the same hindrances. Once more, some infrastructure development may even occur during the adoption phase when, for example, there is early participation in the task force.

11
THE IMPLEMENTATION PROCESS

Implementation is viewed here as a control process since it represents the first opportunity to regulate an innovation. The control systems collect performance data and make comparisons to expectations. Attempts are made to eliminate any discrepancies by altering the technology and/or the infrastructure. However, the ability to exercise control over a radical manufacturing technology is problematic.

Ordinarily the information on revenues, costs, capabilities, and limitations collected by the control systems should contribute to reducing financial and technical uncertainty. The systems, however, which are part of the infrastructure, are imperfectly developed at the beginning of implementation. Once more, changes in environmental uncertainties, that is, market conditions, necessitate reevaluating the systems. The nature and levels of standards and performance measures are open to question. As an organization tries to reduce apparent discrepancies between expectations and performance it must also improve the validity and reliability of the control system. Decisions on these improvements are influenced by wishes to make the innovation look as good as possible.

The ability to exercise control is also hampered by the growth of social uncertainty. Based in part on unreliable information from the control systems, the innovation's consequences for roles, power, and status are estimated. Individuals and groups that believe they are adversely affected will stubbornly resist.

The chances of having a successful innovation therefore depend on at least three factors; expectations for the innovation, the innovation's performance, and the extent of organizational conflict. Before going into more detail about them it is necessary to discuss just what is meant by success. One meaning refers to whether the new technology is working. Control systems designed to make this evaluation tend to be formalized. Accounting, quality, and maintenance systems employ expectations in the form of operational standards.

The second definition of success is whether the innovation is solving the problems that originally propelled it into existence. In other words, investigation begins of whether the new technology will contribute to closing the previously identified performance gaps. Control systems for this purpose are relatively informal reflecting difficulties in measuring expectations and determining cause-effect relationships.

Existence of a dual set of criteria raises some unusual possibilities for evaluating the innovation (Tornatzky et al. 1983). It may be a successful failure, that is, it works but does not help remove existing gaps or it opens new ones. Presumably a failed success may also occur in which the innovation's not being able to work produces a better organization than if it had worked.

Most attention during implementation probably centers on the first set of criteria. After all, if the innovation does not work, its ability to meet the second set cannot be established. Once more, due to more formalized control systems, evaluations of whether the innovation works are more readily conducted. Voss (1988), who studied advanced manufacturing innovations in the United States, England, and Australia, found that getting the innovation to work was more frequently achieved than using it to improve the firm's competitiveness. If these results are indicative of the general situation in these countries many firms are not realizing the technology's most important benefits.

EXPECTATIONS FOR THE INNOVATION

In constructing expectations measures for evaluation of whether the innovation works, a firm must deal with at least two sources of uncertainty. It is unclear whether the correct type of standards are being used and whether the right initial levels have been chosen for the standards.

One complication in selecting appropriate standards is that different business functions and hierarchical levels have their own criteria based on their roles in the organization. Consider, for example, strategic, middle, and lower management (Boddy and Buchanan 1986). Strategic managers evaluate a manufacturing innovation on whether it facilitates movement into new product markets, improves competitive position, or enhances the firm's image. Middle managers, concerned with the control of operations, value consistency of performance, ease in determining accountability, and impact on control of the workflow. Supervisors want minimal disruptions, that is, smoother day-to-day operations, fewer labor problems, and no additional tasks to perform.

At the same time, different intraorganizational political factions with various stakes in the outcome of the innovation process are motivated to develop their own standards. Proponents of the innovation are likely to advocate standards on which the technology will do well such as inventory reduction, lead time improvement, and flexibility enhancement. Opponents will push for criteria on which the innovation will do less well such as cost efficiency and machine utilization perhaps.

Gerwin's (1981a) case study illustrates some technical dilemmas involved in selecting appropriate standards for CAM. The business unit, which bought an FMS, evaluates operations using a standard cost accounting system. Standard costs per unit for a given part had previously been expressed in terms of direct labor hours. Due to the FMSs high degree of automation almost all labor hours consumed were indirect. It was finally decided to express cost standards for the new equipment in terms of machining hours. Even then uncertainty persisted because machining hours can be defined in alternative ways, each of which has different implications for computing and analyzing accounting variances.

Foster and Horngren (1988), who studied twenty-five implementors of FMS and FAS in America and England, found direct labor hours or dollars were still the most frequently used base for allocating indirect manufacturing costs. Companies justified their reluctance to choose a more appropriate base in terms of tradition or the desire to have a single product costing system. In fabrication the most common alternative was machining hours, but problems were experienced in devising operational definitions. In assembly some printed circuit board manufacturers were experimenting with alternatives such as number of inserts per board or number of boards.

The level of expectations in an organization is a function of past performance, the performance of comparable organizations, and past expectations (Cyert and March 1963). Virtually no information is available from any of these sources to determine the initial levels of standards for a radical manufacturing technology. A company must rely upon estimates supplied by the vendor, which are undoubtedly biased, and on its own intuitive judgments.

Once the company studied by Gerwin (1981a) decided on the nature of cost standards it had to select the initial levels of certain cost parameters. They reflect what per unit costs should be under normal conditions. No relevant information existed in the factory because the FMS was so different from other equipment and because the new product line could only be manufactured on the new technology. Data could not be obtained from comparable facilities due to the customized nature of the manufacturing system and because it was the first in the industry. Standards were based on problematical intuitive estimates, which seriously limited the utility of variance analysis.

THE INNOVATION'S PERFORMANCE

Measures of the innovation's performance suffer from uncertainty. Gerwin's (1981a) case study again provides an example. Machine utilization is one of the key variables that manufacturing management tries to control in its efforts to lower costs. This quantity represents the number of hours equipment is actually used divided by the number of hours it could have been used in a given time period. Typically, the denominator reflects an adjustment for time lost due to normal machine breakdowns. Determining the FMSs utilization is complicated because if one of its machines breaks down a less efficient alternative exists. Parts can be rerouted to other machines in the system that take longer to perform the same functions. It is far too complex to calculate the adjustment for normal breakdowns under these conditions. Since no determination is made machine utilization for the FMS tends to be understated by some unknown amount. The system appears to be less efficient that it really is.

A critical aspect of CAMs performance is its unanticipated indirect impact on factory activities. Attempts by the control system to determine these impacts are usually doomed to failure. Obtaining resources to conduct a study, measuring variables, and controlling variables represent insurmountable problems. It is little wonder that most companies rarely attempt this type of investigation.

Consider the following list of indirect impacts from the companies interviewed by Gerwin (1984):

- The American company's foundry was modernized to provide the FMS with high quality castings, which in turn upgraded the quality of castings provided to the rest of the division.
- The British motor manufacturer found production scheduling had become more difficult. Products designed for machining on the new equipment could not easily be run on the shop's conventional equipment.
- In the British bearing producer, machining activities had been organized according to type of operation performed. The new equipment had to be organized in a single unit because each machine performs a number of different operations.
- The German aircraft manufacturer found that in order to keep pace with the FMSs need for material inputs, premachining activities had to quicken the pace of their production. This created a quality control problem.

These impacts are not easily subjected to formal study.

Introduction of a radical manufacturing technology often occurs simultaneously with other changes meant to bolster a firm's competitiveness. Some of these changes represent modifications in the infrastructure designed to properly support the new equipment. Others, which might have occurred anyway, such as efforts to improve quality, lower inventories, or reduce lead times, still affect the technology's performance. It is not possible to determine the extent to which observed improvements using the technology are due to these other factors. Due to this inability to run controlled experiments performance measures are relatively unsophisticated as compared to what is needed.

A special case occurs when new parts are manufactured on the technological innovation. Current performance cannot be compared meaningfully to past performance when both the parts and the manufacturing process have changed. There is no way to tell if reductions in inventory, downtime, or lead time have taken place. Even if a firm makes comparisons against a hypothetical collection of conventional equipment that would have been needed to manufacture the new product, the precise configuration and its performance consequences cannot be made very explicit. Uncertainty is so high that the question of how much improvement has taken place may be unanswerable.

Given measurement difficulties, performance levels depend on the compatibility between the innovation and the infrastructure. As incompatibility widens the technology becomes less effective. This argument derives from economic development theory

(Schumacher 1973) and sociotechnical systems theory (Trist 1981). It has been employed in the context of computerized technology by Bessant and Dickson (1982), Blumberg and Gerwin (1984), and Skinner (1985).

Schumacher reviewed attempts to economically develop Third World countries. Approaches based on the assumption that rapid progress resulted from sophisticated plant and equipment inevitably failed when a nation's infrastructure components such as people's skills and attitudes were lacking. He concluded that unsophisticated technology in an appropriate environment would function more effectively than sophisticated equipment in an unprepared situation.

At the factory level, sociotechnical systems theory stresses that optimal effectiveness occurs from a good fit between technology and social organization. When the technology and its work group are jointly designed, compatibility is achieved in terms of a semiautonomous group that operates in a fairly self-contained workshop atmosphere. In practice, designs often loosely approximate this ideal but still can result in improved productivity and satisfaction (Rice 1958, Trist, Higgin, Murray, and Pollock 1963, Thorsrud 1970).

Moreover, managerial and staff activities must be compatible with the work floor design if it is to be successful. This seems to be one of the lessons from the Norwegian Industrial-Democracy Project (Thorsrud 1970). Experiments with semiautonomous groups indicated that roles should change in a number of supporting activities. The need for shop supervisors is greatly reduced, middle management acquires responsibility for controlling boundary relations between the groups, industrial engineers are no longer concerned with the efficiency of manual work, and training units' responsibilities are augmented. The experiments also demonstrated that traditional incentive schemes have to be revised. In one plant, some employees in the new groups began to earn more than the most highly skilled workers on the premises. The resulting resentment led to the experiment's suspension. In the Shell U.K. project, new performance-appraisal procedures for judging managers based on both human resource and technical considerations had to be developed (Hill 1972).

Some hard evidence exists that performance is directly associated with compatibility between an innovation and the infrastructure. Ettlie (1988b) measured compatibility in terms of the similarity between the degree of radicalism of technical and administrative innovations. Similarity exists when both types of innovations have either little or a lot of novelty. A particular innovation's novelty was determined by an expert panel rather than through a separate determination in each plant in which it was employed. It was found that significantly higher performance from CAM technology was attained on four out of nine measures for plants that initiated associated administrative innovations of similar novelty. The measures were system uptime, system utilization rate, time to install the system, and return on investment.

Damanpour and Evan (1984) measured compatibility in terms of the organizational lag phenomenon, the difference between the rates of technical and administrative innovation where each is expressed as a percentage of the total innovations of that type available for adoption. They uncovered a significant inverse correlation between the efficiency of public libraries and the magnitude of organizational lag. In other words the higher was efficiency the smaller was lag. A further test found that in a group of high efficiency libraries the mean organizational lag was significantly lower than in a group of low efficiency libraries.

The compatibility between innovation and infrastructure is defined here in terms of the former's technical complexity and the latter's sophistication level. To the extent there is a match, that is, the infrastructure is sophisticated enough to handle the innovation's complexity, compatibility exists. Now consider the determinants of the two variables.

Technical complexity is usually specified in the adoption phase when an innovation is designed and/or selected. Answers to the following kinds of questions are needed.

1. What will be the system's magnitude, that is, will we have an FMC or an FMS?
2. Will the system's components be on the frontier of knowledge or well-established technologies?
3. What will be the system's degree of automation, for example, how much will inspection be taken out of human hands?
4. What will be the extent of integration of the system in terms of centralized control?
5. Will the system handle new or existing parts? This decision is especially critical in determining the amount of uncertainty created in the design. General Motor's Hamtramck, Michigan's plant, built in the mid 1980s with a state-of-the-art manufacturing process to assemble new car models, ran into costly debugging problems due to the resulting high uncertainty (Keller 1989).
6. What will be the system's degree of customization? Can the vendor pull standardized designs off the shelf or will it have to start essentially from scratch?

As a system's magnitude, state-of-the-art technology, automation, integration, and new parts increase, technical complexity goes up much faster because of interaction effects.

Ordinarily, during implementation, technical complexity is a constraint to which the participants must adjust. It can be finessed, however, by taking advantage of an innovation's divisibility. Considered one of the most significant attributes of an innovation (Hage 1980), divisibility is the extent to which it can be broken down into components that are implemented in steps. As a result, the technical complexity faced at any one point in time is reduced considerably. The problems handled at each step are more manageable.

CAM technology, an integrated system at least, possesses this characteristic. It is possible to install a system one module or work station at a time, operate each on an individual basis, and ultimately integrate them. All seven of the companies interviewed by Gerwin (1984) had installed or planned to install their systems in steps. Rosenthal (1984) found that in a list of thirteen reasons why companies selected their CAM system, future add-on capability was the second most popular, and modularity was the third. Bessant and Dickson (1982) saw divisibility as a key implementation strategy for microelectronic manufacturing technology.

Attempts to continue raising the infrastructure's sophistication level are thwarted by some of the same factors that operated in the preparation stage. It is true that lack of participation and absence of interest in developing the infrastructure may diminish as feedback reveals that managerial adjustments are required. Pressure for immediate full-scale production however may increase as the time since installation grows. Dependence on the vendor and external institutional constraints may continue as before. Although the level of the infrastructure's sophistication is higher at the start of implementation than at the start of preparation it is still not high enough.

CONFLICT AND RESISTANCE TO CHANGE

The value-laden nature of innovation research surfaces with the topic of conflict and resistance to change. Many studies assume from the start that new technology will improve organizational performance and help make an organization more responsive to a changing environment (Kimberly 1981). Management acting out of the organization's interest is seen as the advocate of innovation. Resistance stems from the rank and file's attempt to protect its own special interests. Therefore, ways must be found to control resistance.

This scenario represents a one-sided view of organizational reality. An organization may adopt inappropriate innovations and have difficulty discarding those which don't meet expectations (Kimberly 1981). Perhaps management did not evaluate the need critically enough or acted out of its own special interests. Then resistance by the rank and file may be in the organization's interest and should be encouraged.

Once more, the innovation may come from the bottom of the firm and management may resist it. Frequently, lowly technical people are the inspiration for CAM and management does not understand or feels threatened by the technology. Majchrzak (1988) identified examples of resistance to CAM technology from managers and technical people as well as production workers.

Uncertainty as to the innovation's capabilities may lead the vendor and its allies in the customer's organization to overestimate what it can do. Or perhaps the champion, blinded by his or her commitment, is oblivious to arguments that the innovation isn't working (Pfeffer 1981). Other groups responsible for achieving what they consider to be unrealistic targets may resist. Difficulties in collecting performance information may drag out the conflict over a long period of time. The innovation's advocates may regard the resisters as trying to block needed change, but their resistance is actually in the organization's interests.

How does one interpret the various forms that resistance may take? There is defensive resistance which attempts to block the innovation from seeing the light of day or to prevent it from receiving an objective evaluation. There is proactive resistance which seeks to change the innovation. The new technology's advocates see attempts to water down what is needed. The resisters see useful ideas for achieving a better fit between the organization and the technology.

Why is conflict over the innovation such a problem? When organizational tasks are uncertain, technical and interdependent, coordination is achieved through mutual adjustment and team efforts (Mintzberg 1983). Accordingly, the implementation problems of radical manufacturing innovations require considerable interpersonal cooperation to solve. Handling defects in the parts produced, for example, involves quality control people, shop management, operators, maintenance workers, and suppliers of raw material. Intraorganizational conflict hinders successful performance by preventing the necessary cooperation from occurring.

The extent of conflict depends on the amount of change that takes place in the infrastructure. New people, occupational specialties, and procedures create a need to change the roles, statuses, and powers attached to existing organizational positions. Individuals and groups acting out of the organization's interests or perceiving themselves as adversely affected often resist, as is well documented in the innovation literature (Hage 1980, Zaltman, Duncan, and Holbek 1973).

In the context of advanced manufacturing innovations Gerwin (1984) found several examples of role, status, and power conflict. Consider role conflict. During the design of the American FMS, the innovators consisting of the vendor's and customer's engineers believed that it would be possible to have the finished parts adhere to rather close tolerances. Their high expectations imposed tough behavioral requirements that were resisted by operating personnel who had to attain the tolerances and quality control people who had to measure conformance.

The British bearing manufacturer furnished an instance of status conflict. The process planning department, which had the task of choosing the machines, operations, and tools for each order, was also assigned the task of preparing parts programs. In order to encourage people to take on the new challenge, a new higher top level was added to the pay scale. It could be attained by excelling in at least one of the two tasks. The older men in the department, however, were unwilling to learn programming whereas two younger men became proficient in both tasks. Once more, the older men who had been on top of the original pay scale found themselves below the younger men on the new scale. The resulting status incongruence has been the cause of a great deal of friction in the department.

Power conflict occurred in the American firm as lower-level manufacturing managers on the shop floor became more dependent on staff and service departments such as accounting and maintenance. These managers had always used informal procedures based on direct labor hours for controlling operations as they made their rounds of the shop. The procedures were of little use in controlling FMS operations because of the diminished relationship between how much people worked and the amount of production. In addition, manufacturing managers could not readily interpret the meaning and significance of new cost accounting concepts based on machining hours. Due to the FMSs technical complexity supervisors could no longer contribute to the repair of machines. They had to rely on electronic and electrical maintenance people without having much control over the timing and quality of their work.

Analyzing the Reasons for Conflict

Based on twelve in-depth case studies of new production or production-related technologies Leonard-Barton (1988) suggested a condition under which role changes produce resistance. Are the impacted activities a significant part of one's job, and does the impact increase the difficulty of meeting job performance criteria? In one case Leonard-Barton studied an internally generated structured methodology for developing software (SSA) that was provided to systems analysts throughout a large corporation. SSA changed the way they performed one of their significant activities, "scoping out" the business a system was to serve. Many analysts believed, however, that it added a great deal of time to the beginning of the development process. Supervisors judged the analysts' performance chiefly on whether they completed their projects on time. About three years after its release, and despite corporate headquarters' requirement to use it for all software development, SSA was still used by only 60 percent of those analysts for whom it was highly relevant.

Leonard-Barton's analysis suggests one reason why a new FMS does not necessarily cause a shift from high volume production of a few parts to flexibility. The technology can significantly change the activities of plant personnel but in a way that makes performance goals harder to attain. Shifting to low volume production of many parts makes little sense if performance is judged on the basis of output and cost. Conflict and resistance will persist until the criteria are changed.

Strategic contingencies theory (Hickson, Hinings, Lee, Schneck, and Pennings 1971, Hinings, Hickson, Pennings, and Schneck 1974, Salancik and Pfeffer 1977) helps us understand how changes in the infrastructure produce shifts in the organizational balance of power and ultimately lead to conflict. The theory focuses on horizontal relations between business functions rather than vertical relations among individual superiors and subordinates. It attributes power to structural factors, mainly the division of labor, versus psychological or interpersonal variables such as personality or influential allies.

According to the theory, every organization is faced with uncertainties with which it must deal. Some of the specific ones identified—order mix, design requirements, volume, equipment operation, and material quality—correspond directly to those considered here. One can imagine handling of the uncertainties becoming the responsibility of certain subunits as a consequence of the division of labor. These subunits have the opportunity to acquire power to the extent the uncertainties affect the activities of other departments.

Taking advantage of the opportunity depends on three factors, all of which must be present if a subunit is to be powerful. Most important is the subunit's ability to effectively adapt to its assigned uncertainties, thereby shielding other departments from uncontrollable events. In so doing it controls contingencies or requirements for their activities. A subunit is therefore reluctant to eliminate uncertainties because this alternative reduces its ability to exercise power. Second, as the centrality of a department's activities increases its power goes up. Centrality refers to the degree the activities are linked with those of other subunits, and the extent to which curtailing the activities would quickly and significantly affect the organization. Third, as substitutability decreases power goes up. Substitutability occurs when alternative units exist within or outside an organization which can perform the same activities.

How can strategic contingencies theory help us understand the shifts in power which occur when computerized technology is introduced? Consider, as an example, its use here to analyze the results of Bjorn-Andersen and Pedersen's (1980) study of the power implications of a new computerized production scheduling system. In investigating three assembly plants of a Danish radio and television manufacturer, they collected data from eighteen production planners, factory managers, and works managers. Using a five-point scale the researchers assessed the power of each group over daily production decisions. Respondents were asked for the perceived change in influence of their occupation and the other two occupations as a result of the introduction of the new system. Planners viewed themselves and were viewed by others as having the largest increase. Next, respondents indicated the influence level of their occupation and the others after the introduction of the system. Planners perceived themselves and were perceived by the others as having the most influence.

Strategic contingencies theory explains the reasons for these findings. The new system allowed planners to gain control over a significant contingency for managers, the overall supply of components to the line. The system was too complex and inaccessible for the managers to use directly. Data collected on perceived changes in workflow centrality for one's own group indicated that on average the planners' centrality increased the most. Data were not collected on substitutability, but, to the extent that a department acquires special skills and knowledge, this factor decreases. Planners reported the largest gains in both expert and up-to-date knowledge. Other possible explanatory variables not considered in strategic contingencies theory hardly changed at all.

We are now in a better position to understand something about the power implications of CAM technology. Building flexibility into a manufacturing process through CAM represents an adaptation to significant long-run market and process uncertainties faced by an enterprise. Those who handle the technology are in a position to control important contingencies affecting the activities of many departments. They stand to gain power whereas those who had handled uncertainties through obsolete means, inventories, for example, stand to lose power.

Whether these potential gains and losses actually occur depends on certain changes made in the infrastructure. These changes impact on the three variables accounting for power. New skills required to support CAM such as electronic maintenance are frequently in short supply and hence not readily substitutable. New structures, involving redirection of information flows, changes in the workflow, and realignment of departments affect centrality. Subunits that employ new systems and procedures have augmented coping ability and can add to their control of other units' activities.

New and revised systems and procedures, however, reflect an increase in routinization, the standardization of recurrent activities through rules and programs. Routinization in turn increases substitutability allowing lesser-trained people or computers to handle the standardized functions. Parts programmers and process planners will have less influence as CIM computerizes their tasks. We can therefore expect these specialists to resist changes which routinize their work.

Strategic contingencies theory does not consider the short-run financial, technical, and social uncertainties associated with the introduction of a new technology. They may create an opportunity for unusually adept factions to augment their influence at least temporarily. Yet, due to their intractable nature, it is more likely that the uncertainties will be a hindrance to units seeking increased power through the innovation. As an example, effective coping, the most critical source of power in the theory, is hampered by ambiguous control systems. An inability to handle short-run uncertainties therefore has political as well as effectiveness implications in an organization.

Underlying strategic contingencies theory is the assumption that organizational power arises from fortuitous circumstances. It is bestowed on subunits by virtue of conditions over which they have little control. Changes in technology and the infrastructure shift the division of labor, coping ability, centrality, and substitutability. Conflict arises as the result of or to prevent the ensuing changes in power.

One must consider, however, that power also stems from opportunism; it is acquired as the result of a subunit's intentional efforts. Attempts to change the division of labor, coping ability, centrality, and substitutability may lead to adjustments in technology and the infrastructure. This amounts to a reversal of the causal sequence. A department may for example pursue a certain technology in order to augment its own coping ability. Alternatively, it may act out of a desire to weaken another subunit in order to decrease its own substitutability. Thus competition among subunits to increase their power by changing technologies and the infrastructure is also a source of conflict. Subunits attempting to augment their power in this manner are taking large risks since the ensuing short-run uncertainties threaten to impair their ability to handle other departments' long-run requirements.

Methods for Reducing Conflict

Hage (1980) identified three general approaches for handling conflict and resistance to change—revolution, evolution, and isolation. He investigated radical innovations in health and welfare organizations. Nutt (1986) fleshed out Hage's categories although it was not his explicit intention. Over a seven-year period Nutt conducted ninety-one case studies of important changes in service organizations. Some represented innovations and presumably some of these were radical. While only 16% of the changes involved equipment versus construction or programs, there was no significant correlation between type of change and method of implementation.

Before discussing their work the following caveats are in order. Both investigators eschewed radical manufacturing technology in business firms so the applicability of their results is open to question. Both considered management as the motivator of change and affected parties as potential resistors. Hage in particular suffers from an innovation bias, the implicit assumption that innovation is good and resistance is bad. These viewpoints are not necessarily wrong but they offer only one portion of the total picture.

Revolution, the first general approach, involves strategic management using its institutional and personal power to push through a radical innovation. There is no participation by affected parties; their interests are not considered. The entire innovation is quickly adopted, but implementation proceeds very slowly. Conflict is handled by steamrolling over it. In general, the amount of change actually implemented is small as resistance blocks the innovation from effective utilization.

Revolution is effective only under special circumstances (Hage 1980). First, abundant financial resources are needed to implement a full-scale radical manufacturing technology all at once. Second, suppose a crisis exists for which new process technology is a possible solution. Employees are willing to set aside their own immediate interests to cooperate in the organization's survival. They are willing to submit to management's authority because time is of the essence. If the organization also has high centralization then power is effectively wielded by management, and if there is a low proportion of technical specialists, challenges to the innovation on technical grounds are not forthcoming. Nutt, however, found that time pressure and perceived importance of a change to managers, both signs of a crisis, were not associated with the use of a particular method such as revolution.

Nutt identified two variations of revolution that he labeled edict and persuasion. In the former management plays an omnipotent role and the burden of change rests on it. Managers define the problem, select the innovation, and announce the change. Expected behavior is prescribed through memos and presentations. Little flexibility exists to make revisions in technology and infrastructure (Bessant and Grunt 1985). Management cannot afford to be seen as making mistakes and affected parties have no inputs.

Persuasion involves experts preparing rational justifications to convince management versus honing arguments to gain the acceptance of affected parties. Managers may initiate the need for change but due to a lack of knowledge make little effort to direct the process. Internal or external technical specialists come up with alternatives during adoption and revisions during implementation which they try to sell based on their costs and benefits. Managers usually demand elaborate documentation and some reevaluation as they fight to understand and to gain time. Flexibility to make revisions is high initially but diminishes as resistance mounts (Bessant and Grunt 1985).

The essence of evolution, the second general approach, is participation by affected parties in decision making. Adoption is lengthy as it takes time for all factions to reach compromises. Implementation is speedy due to the lack of conflict, but there are good reasons why the amount of change actually put into effect is low. The innovation is implemented incrementally over a long period of time. Its radicalism is whittled away as the affected parties make it more palatable for themselves. Although flexibility for revisions exists throughout, it is technically easier to modify carly when the design is fluid rather than late when it is firm (Bessant and Grunt 1985).

Under what conditions is evolution appropriate? If the organization has high decentralization and a high concentration of technical specialists then structural conditions are ripe for power sharing. The innovation needs to be divisible in order to be implemented incrementally. Its advocates must remain with the organization for a long time and be able to recruit individuals who favor the change.

Nutt found two variations of evolution called intervention and participation. The former is based on consultation. Management generally defines the problem. It or a task force identifies a solution and subsequent revisions. The task force explores the consequences of the alternatives. Its recommendations are subject to a management veto leaving open some possibility of resistance. Quality circles work in this manner.

Participation involves true power sharing. Managers define the problem, whereas a task force comes up with a solution and revisions. The task force works under managerial constraints such as budgetary or policy limitations, but its recommendations are not subject to a veto. Sometimes the task force only frames recommendations and the details are worked out by specialists. As long as the team consists of representatives of various subunits they still must persuade their colleagues in the subunits to accept the recommendations. The possibility of resistance is therefore not completely eliminated.

In utilizing isolation, the third general approach, a new organizational unit with its own financial resources and newly recruited personnel is created. It may be on a greenfield site or in a separate part of an existing factory. Resistance is lessened because during innovation there is not much need to make concessions to existing factions. Once more, the new hires are selected for their commitment to the technology. In the pilot plant version the isolated unit functions until implementation is over. Then the innovation is incorporated into the organization's mainstream for routine functioning. Resistance may occur at this point especially if those responsible for routine operations had little contact with the pilot plant. Another alternative is a permanent separate unit responsible for innovation and routine functioning.

Figure 11.1 helps us understand the issues surrounding a decision to use isolation. It indicates that there are varying degrees of organizational and geographical isolation among which a choice is made. An expectation of resistance from existing personnel increases the degree of isolation. Bessant, Braun, and Moseley (1981) found that creating a new unit was used when labor relations are poor. It is less threatening to the existing workforce. To implement its computerized system the British bearing manufacturer studied by Gerwin (1984) desired an autonomous unit in a separate location of its factory with new operators and infrastructure components. It wanted to avoid conflict with existing workers who were dedicated to hand skills. An

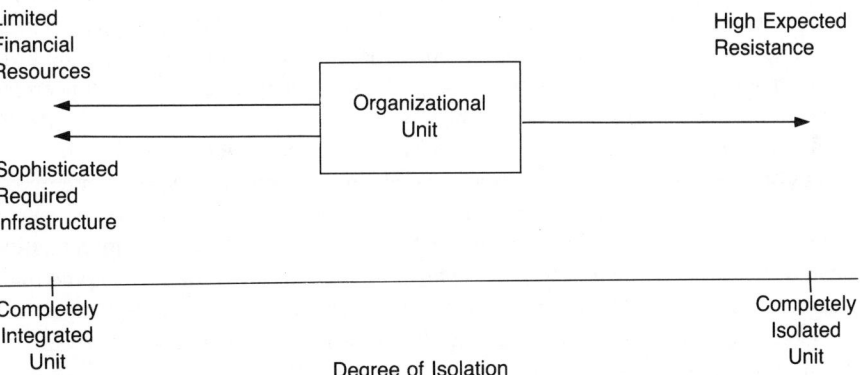

FIGURE 11.1. Factors influencing the degree of isolation.

automative parts firm studied by Beatty and Gordon (1988) installed a new robotic welding line with six operators next to a manual line employing over thirty welders. The project was abandoned because of resistance by workers who feared losing their jobs. Ultimately, the line became part of a new factory with new workers at a distant location. It was up and running smoothly within a few months.

Isolation is particularly sought when invidious comparisons are a factor in resistance. Individuals not associated with the innovation compare their organizational arrangements, amount of attention received, and degree of success to those of the people involved with the innovation. Unfavorable balances lead to efforts to stop the innovation from spreading and to stamp it out. Part of the reason General Motors did not locate its Saturn plant in Michigan was to prevent the bulk of its production employees from complaining about the new pay system and work group structure.

Financial resource limitations tend to reduce the contemplated degree of isolation. It is expensive to provide the human and material elements that comprise a separate entity especially when existing elements are not being utilized efficiently. The British bearing manufacturer's plan for an autonomous unit never materialized because funding was withdrawn during an economic downturn.

The extraordinary problems in creating a sophisticated infrastructure from scratch also tend to reduce the degree of isolation. Managers, technical specialists, and workers with experience in handling CAM are extremely difficult to find. New people have no experiential knowledge of working in the firm's context and must learn how to cooperate with each other. The advantage is with the existing infrastructure people, who have already worked together in a similar factory situation, especially if they have acquired relevant experience. Where implementation of a radical innovation requires a large, sophisticated infrastructure with many different occupational specialties, and where the required infrastructure already exists, the tendency will be to use an existing unit.

The German aircraft manufacturer studied by Gerwin (1984) considered building a new plant because it was almost doubling its capacity, but ultimately ruled it out precisely because the existing factory already had a well developed support system. This was a sound choice because the support system was instrumental in reducing the trauma of implementation. For example, the process planning department in selecting machines, operations, and tooling for specific parts to be run on the FMS took advantage of similarities with the decisions that had been made concerning parts run on the shop's NC equipment.

Thus, the chosen degree of isolation is often a compromise between the three forces of Figure 11.1. For CAM technology the result tends more toward integration with the existing organization because of the need for a large amount of financial resources and a highly sophisticated infrastructure. There may be a common facility, shared infrastructure elements, geographical proximity, and some overlapping management. These are all points at which resistance egged on by invidious comparisons may break out. Isolation is emphasized mainly when significant resistance is likely from existing personnel.

Overall, one must be pessimistic about the chances of implementing a radical manufacturing technology without resistance and conflict. Revolution demands abundant financial resources and leads to sabotage unless a crisis exists, the organization is centralized, and there are few technical specialists. Evolution produces meaningful results only over a long time period, and then only if the innovation has not been watered down. It requires that the innovation is divisible, the organization is decentralized, and the organization has many technical specialists. Isolation is often infeasible due to financial constraints and the need for a sophisticated infrastructure. Current evidence as reflected in the process-oriented theory developed in this book suggests many firms lean toward evolution. Participation and divisibility are important components of the propositions.

More can be said about a method's frequency of use and degree of success at least in service organizations by considering Nutt's (1986) data. Table 11.1 indicates that revolutionary methods were the most frequently used amounting to about two-thirds of the cases. Persuasion was used 42 percent of the time and edict 23 percent of the time. Persuasion was also employed in a majority of the firms in Bessant and Grunt's (1985) research on incremental manufacturing innovation in Great Britain and West Germany. Evolution characterized the remaining one-third of Nutt's cases with intervention and participation about evenly divided. No data appear on isolation.

Even though revolution was used most frequently, evolutionary methods were more successful. To measure success Nutt asked respondents to indicate whether the changes were actually used. Intervention and participation had higher success rates than persuasion and edict. The differences between intervention versus participation and persuasion, and between participation and persuasion versus edict were statistically significant.

No matter which method a firm uses, strategic management's support for a new technology is generally believed to influence the rank and file not to resist. We can analyze the effectiveness of this coping strategy for reducing social uncertainty by considering the process by which influence is exerted. Given the particular situation, issues arise concerning the initiators of influence (strategic management), the methods

TABLE 11.1 Use and Success of Implementation Methods

Method	Frequency of Use (% of cases)	Degree of Success (% of cases)
Revolution		
Persuasion	42	73
Edict	23	43
Evolution		
Intervention	19	100
Participation	17	84

Source: Nutt (1986).

employed, and the recipients (the rank and file). Initiators must be committed to the innovation in their own minds and choose appropriate influence methods. Recipients must perceive the influence attempt and be open to changing their behavior.

What determines the extent to which top management commits itself to the innovation's implementation? Factors that plagued the adoption decision continue to operate here. Commitment is dampened by technical complexity, which inhibits understanding, and by uncertainty, which creates fear that the risks are too great. On the other hand, top managers who supported adoption may become more committed as the result of their efforts to relieve postdecision dissonance.

Verbal communication through standard organizational channels is a typical means of demonstrating support (Fidler and Johnson 1984). Messages may employ persuasion, expertise, authority, or sanctions. When both complexity and uncertainty are high, as is the situation for CAM, persuasion is the most influential but also the costliest in terms of communicating the necessary information. With high complexity and uncertainty interpersonal channels are more effective than written ones, but there may not be enough available channel capacity for the influence attempt to succeed.

Support may also be demonstrated in other ways. Top management may provide resources for implementation such as information, budgets, or legitimacy. It may participate in establishing overall objectives for the implementation program. These activities also send "messages" to the rest of the organization albeit of an indirect nature. Indeed, the use of verbal communication without meaningful provision of resources or participation in overall goal setting may be viewed as hypocritical. For those being influenced it is the *perceived* amount of managerial support which counts.

Does perceived managerial support always increase the chances of an individual's taking part in implementation? Recent research by Leonard-Barton and Deschamps (1988) casts doubt on this assertion. The computer manufacturer they investigated had recently adopted a new expert system to aid sales people in configuring complex computer designs. Telephone interviews were held with ninety-three of the sales people concerning the innovation. Overall, no significant relationship was found between the amount of perceived managerial support and the extent of the system's use. For those individuals who had little need for the system or who displayed a low job competence a significant relationship did exist. For those with high need or high competence there was no significant relationship and the extent of their use depended upon the innovation's degree of accessibility. In sum, the ability of perceived managerial support to foster implementation is mediated by the characteristics of those individuals being influenced such as their need and competence.

REVISING THE TECHNOLOGY AND INFRASTRUCTURE

Comparisons between initial expectations and performance are likely to reveal large discrepancies. Boer, Hill, and Krabbendam (1988), who studied seven European firms that were operating FMSs, found that the percentage of goals a company considered to be achieved ranged from a high of 75 percent to a low of 33 percent. One reason is that expectations spurred on by the innovation's advocates are high, whereas performance affected by the mismatch between technical complexity and the infrastructure is low. Once more, during implementation environmental uncertainties may enlarge existing or create new discrepancies. As an example, Boer, Hill, and Krabbendam found unanticipated market changes to be an important factor hindering goal achievement for new FMSs. In two of the seven companies studied the markets for the new products collapsed during implementation. Discrepancies may also materialize during implementation as the result of improved functioning of the control system. Three kinds of coping strategies in the form of feedback processes attempt to reduce the gaps. The technology is altered to better fit the organization, the organization is changed to better match the technology, and expectations are adjusted.

This part of the implementation control loop—taking remedial action to remove discrepancies—is an exercise in organizational learning. By observing the results of previous acts an understanding is acquired of which new acts to try (Hedberg 1981). Given a situation that does not change too frequently, a valid and reliable control system, and knowledgeable individuals, learning should proceed quickly and efficiently. The prevailing uncertainty and other factors, however, force learning to occur slowly and wastefully if at all. Boer, Hill, and Krabbendam (1988) confirmed this observation in their study of new FMSs.

To understand why this occurs let us refer to March and Olsen's (1976) account of the organizational learning process. Individuals' actions lead to organizational actions, which create environmental responses. These responses in turn change individuals' beliefs and hence influence their future actions completing the cycle. In our situation individuals' actions include new behaviors (attempts to close gaps), efforts to preliminarily test new behaviors and attempts to have new behaviors accepted by the organization. New behaviors become organizational actions when they are given official sanction and backing. Environmental responses are the discrepancies between expectations and performances, and individual beliefs include accumulated knowledge on how to eliminate the discrepancies.

With March and Olson's aid it is possible to identify several ways in which the learning process breaks down. Individuals are blocked by organizational constraints from taking needed action as when workers are not allowed to meaningfully participate in the production process. Power conflicts and resistance to change hinder recommended actions from being converted into organizational actions. The consequences of organizational actions in terms of changes in discrepancies are hard to decipher in part because so many different actions occur simultaneously. Once more, it is not clear whether appropriate standards or performance measures exist or whether deviations are accurately indicated. Too rapid fluctuations in environmental conditions alter what is expected from the CAM system. Individuals' beliefs therefore become obsolete rapidly. Finally, due to insufficient knowledge, employees do not know what actions to take. No proven adjustments exist elsewhere to imitate, existing employees lack prior experience, and new people with the necessary skills are hard to find.

Competency traps (Levitt and March 1988) represent another reason for breakdowns in the learning process. An existing inferior procedure is maintained because people through experience have become competent in its use. Even if a superior procedure exists employees will not have accumulated experience with it, making its use unrewarding at least initially. Although this tendency to specialize in an inferior procedure decreases as the superiority of the new routine increases, it is often difficult to measure the difference in performance potentials.

The first learning strategy tries to increase performance by making adjustments in the technology or the production process within which it is embedded. This engineering viewpoint as to the causes and solutions of problems may involve a minor adaptation such as a change in some software code or a major reversal such as a rethinking of the entire system's design. Due to the financial and human cost the chances that fundamental changes occur are low.

As a result of technical uncertainty (the causal links between any adjustments and subsequent performance may not be completely understood) the desired results may not occur. Once more, even if there is a positive direct impact, technical complexity has been increased resulting in a negative indirect impact on performance. The imbalance between the technical and support systems has grown. Consequently, using terms coined by Tornatzky et al. (1983), either a benign adjustment (the discrepancy between expectations and performance decreases) or a malign adjustment (the discrepancy increases) may result.

Leonard-Barton (1988) provided examples of benign and malign technological adjustments. One company experimented with substituting copper for silver in hybrid circuit board production. Several technical changes, which proved to be benign, were necessary once the innovation was transferred into a production facility. Since a copper alloy paste was found to have poor adhesion, manufacturing engineers altered the process by testing and then purchasing a new vendor's alloy. They also slowed some of the steps in the process. In part, due to these changes the innovation was eventually implemented successfully. Another company purchased an MRP II system. In adjusting the program to local conditions software technicians made it very difficult to install later versions of the vendor's system. These malign changes had to be removed and stored in separate programs capable of being accessed by the MRP II system.

The second learning strategy tries to further adapt the organization to the new technology. Having a discrepancy leads to increasing the infrastructure's sophistication in order to improve performance. This is a continuation of the steps begun during preparation to better maintain the innovation or to improve the control system. The adjustments might vary from a small redesign of a particular role to a major shift in strategy for a plant, but once again the probability of a fundamental reorientation is low.

Compared to the technologically oriented feedback process the time to make adjustments and produce results is longer and the connection between adjustments and performance is even less well understood. Once more, increasing the infrastructure's sophistication also leads to a growth in conflict, which indirectly reduces performance. Consequently, benign or malign adjustments may result here too.

Leonard-Barton (1988) also illustrated benign and malign organizational changes. A computer program for detecting the source of irregularities during the rolling of aluminum sheeting was originated in the research labs of one corporation. Process engineers did not have the skills or time for the additional monitoring and analyses. An engineer was hired to do the work. He also trained process technicians whose job performance criteria were changed to encourage their gaining proficiency with the new technology. These adjustments proved to be benign; they helped make the innovation highly successful. Another firm developed a CAE system to aid in the design of prototype circuit boards. Designers, however, regarded it as forcing more attention to detail in aspects of their activities not particularly significant to them. "Shame Sheets" were instituted to encourage use of the system. They indicated which designers were sending the drafting department manually prepared schematics. By causing a great deal of resentment this negative inducement turned out to be malign.

There is not much evidence to suggest the way in which specific technical and organizational changes occur. In Gerwin (1981a) the initial discrepancy between expectations and performance for quality defects enabled the quality control department to introduce procedures that had previously been blocked from spreading to automated processes. This example of a solution that found a problem suggests that garbage can processes (March and Olsen 1976) may have explanatory value.

A third learning strategy reduces expectations or removes existing gaps from further consideration. One motivation is the need to improve the control system due to its initial uncertainty. A reduction in over-optimistic aspirations in the face of low performance is an example. A second motivation stems from environmental uncertainty. Changes in market conditions may obviate the need to worry about a particular gap. A third reflects attempts to seize upon opportunities presented by the control system's imperfect development or environmental uncertainty to make the innovation look as good as possible. With this last political motivation, decision makers may remove gaps that are hard to close and add gaps that are easy to eliminate. Four examples of the learning strategy are given as follows.

Improving the control system by reducing aspirations is illustrated in Gerwin's (1981a) case study. The vendor's and customer's engineers believed the new FMS could attain tight tolerance specifications. Operating and quality control people did not concur. Exhaustive testing finally revealed that greater machine capability than available was needed to meet the engineers' targets. Expectations were lowered causing a reduction in the radicalism of the innovation. This example also illustrates that resistance is sometimes in the organization's interests.

An intertwining of the control system's development and political considerations occurred in one of Leonard-Barton's (1988) cases. Unattained original specifications were subsequently judged as irrelevant because of the realization of unanticipated benefits. In implementing process controls in a steel mill the original specifications were not met. The software failed to adequately set the steel preshaping function. Yet, when managers realized that the output's quality increased far more than anticipated, they changed the specifications to exclude the automatic preshape function (Leonard-Barton 1988).

Winch (1989) gave an example of seizing upon market changes to eliminate a standard not being met. A company stopped measuring the productivity of a new CAD/CAM system when it became apparent the expected gains would not materialize. By then management had decided that the real purpose of the system was to reduce lead times. It was unclear how much the shift was due to new environmental pressures and how much to the failure to meet the original standard.

Making the innovation a success by changing the standards against which it is judged extends to using it for other purposes than originally intended. Ettlie (1988a) cited a multimillion dollar flexible assembly system which, due to technical difficulties, never performed up to the levels predicted by the vendor or customer. Its yield and labor requirements were unsatisfactory. Eventually the FAS became part of a JIT delivery system with four-hour lead times to a final assembly plant. Its success is now measured by the amount of flexibility it provides as part of the delivery system.

In the final analysis conditions may be too uncertain to objectively determine whether the implementation of a radical manufacturing innovation is a success. The mismatch between technical complexity and the infrastructure, the ambiguity of expectations and performance, and incomplete learning all conspire against the making of definitive judgments. A firm may continue to utilize the technology without being clear if it is successful. If so, it will probably not learn much from the entire experience; in the final analysis learning depends upon an ability to evaluate the outcomes of actions.

Case study results illustrate these points. Boer and During (1987) found that in all companies they studied implementation lasted longer than expected and it was unclear whether operational goals would be achieved in the long run. The firm studied by Gerwin (1981a) had not found a completely reliable basis for computing accounting standards after seven years of FMS operations. Overall planned costs were a fairly reliable benchmark. The planned values of important cost components were still uncertain making it difficult to learn whether their actual costs were within acceptable limits of control. The accounting department attended to those items such as rework and maintenance, which represented large fractions of the budget and over which some control could be exerted.

Why is stopping the innovation an unlikely alternative? The heavy financial and psychological investment already incurred by the implementation stage forces continuation even though these are sunk costs. Cancellation does seem to occur when large discrepancies persist, only a fundamental change in technology and/or organization will remove them, and the firm is unwilling to commit the necessary resources (Leonard-Barton 1988).

Routinization, in which the innovation becomes part of daily practice, follows the implementation phase. Although there is little research it is probably more dynamic than one would think with problems that are similar to those of earlier stages. Internal diffusion, expansion, and renovation may occur.

The problems of innovation, introducing the first example of CAM, may be less severe than the problems of internal diffusion, getting CAM positioned throughout a plant, business unit, or company (Hicks 1986). Hicks conducted a mail survey of over 1200 small- and medium-size U.S. metal working establishments with at least one example of NC or CNC. He found that only 6 percent of all the plants doing turning operations had greater than 50 percent of their lathes controlled by NC. Only 18 percent had more than 50 percent of their lathes controlled by CNC. Similar patterns existed for milling, drilling, and boring operations. Kelley and Brooks (1988) also found that programmable automation, defined in terms of NC, CNC, and FMS, is not likely to be the dominant technology in a plant. Of the over 1000 metal working and engineering firms they studied, only 44 percent had at least one such machine tool.

In large multidivisional corporations a special organization is needed to disseminate information about promising innovations. Individual units do not ordinarily seek ideas from headquarters or share information with each other. In one firm a corporate manufacturing services unit provides this service by conducting training seminars. It also arranges meetings of functional specialists to identify the best internal practices. Group vice-presidents for technology are expected to move ideas around the corporation through their networking activities. Task forces at the group and divisional levels help realize specific objectives such as setup time reduction. They set goals and share ideas on means such as CAM.

During or after routinization the original pilot technology may undergo expansion or renovation. Meredith (1989) gave an example of the latter situation for an FMS that had been in service for many years. As the system's throughput steadily increased over time some components became bottlenecks. Accuracy problems multiplied with age, which conflicted with an increased stress on quality. There were added maintenance problems at the same time that it became harder to find replacement parts. Once more, a new or slightly different replacement part created a heavier load on existing parts leading even more of them to fail. It was eventually imperative to renovate the machines and software to bring them up to date.

The capital justification process is one stumbling block to expansion or renovation (Bennett et al. 1987). Even if top management approves an original investment using qualitative criteria, additional investment will not be considered unless claimed benefits are quantified. Middle managers are forced to create new financial measures to help justify the added investment, but since these measures are unorthodox and unfamiliar, top management is often reluctant to see them used. To avoid arguments it is necessary to have top management's assent to the measures at the outset, that is, before the original technology is justified.

SUMMARY AND CONCLUSIONS

Implementation of a radical manufacturing technology is a very imperfect process of control. Financial and technical uncertainty still operate due to questions concerning the validity and reliability of standards and performance measures. Social uncertainty grows into a formidable concern as resistance to the innovation spreads. The chances for successful implementation, equated with reducing discrepancies revealed by the control system, are not high. As indicated in Figure 11.2:

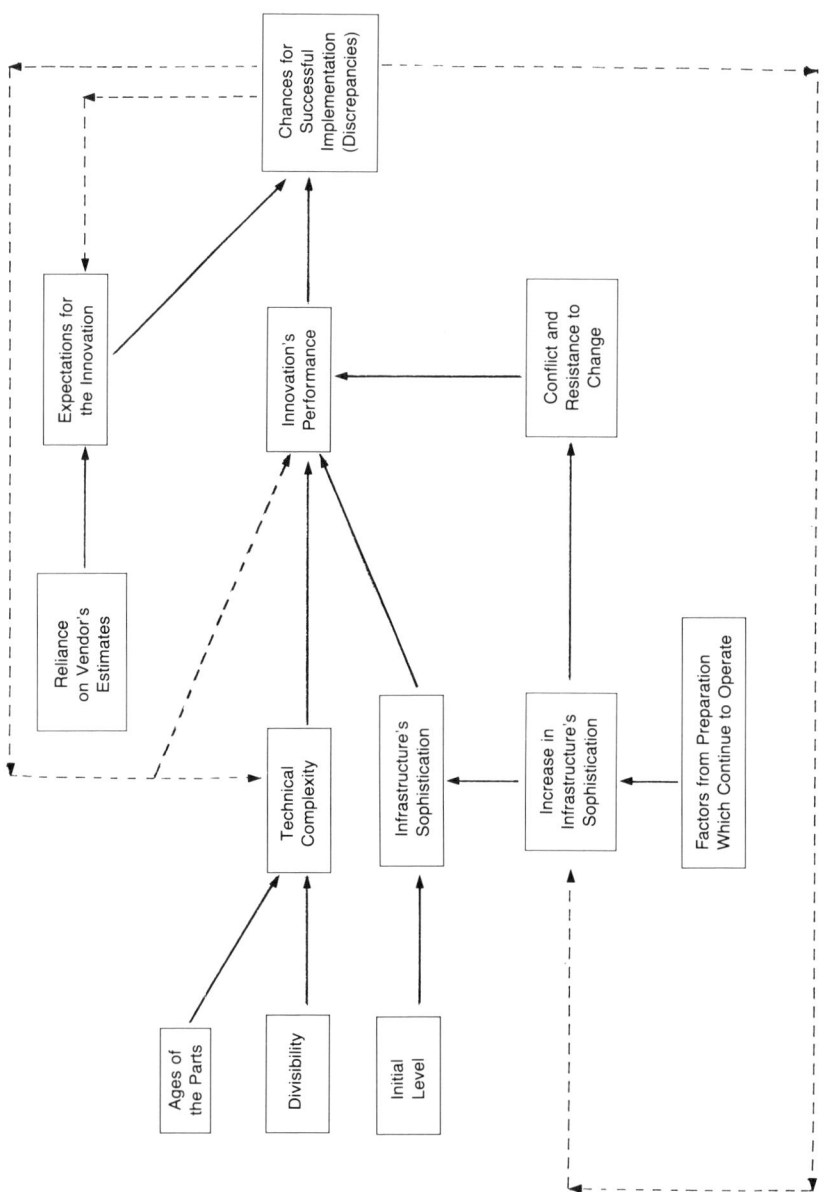

FIGURE 11.2. The implementation process.

Definition D1. Discrepancies are the positive differences between expectations and performance.

Initial levels of expectations are difficult to estimate. The firm must rely upon the vendor who has a vested interest in making the innovation look good:

Proposition I1. The greater the reliance on information from the vendor, the higher the expectations levels for the innovation.

The innovation's performance depends upon the match between technology and organization. It is also influenced by the degree of conflict and resistance to change:

Proposition I2. The greater the innovation's technical complexity, the lower the infrastructure's sophistication, and the higher the intraorganizational conflict; then the lower the innovation's performance.

A CAM system's technical complexity is influenced by several factors not the least of which is whether new parts will be processed. If, however, the innovation can be broken up into separate modules which are implemented sequentially technical complexity is reduced:

Proposition I3. The greater the innovation's divisibility and the older the parts which are manufactured, then the less the innovation's technical complexity.

The infrastructure's level of sophistication is given by the following definition:

Definition D2. The infrastructure's sophistication level at any point in time is the sum of its initial level at the start of implementation and the change which has taken place up until that time.

Development of the infrastructure hinges upon the same variables which were operating during preparation. Some of these factors have become more influential, others less.

Changes in the infrastructure lead to shifts in the roles, powers, and statuses of existing employees. Those who are negatively affected or who see a threat to the organization resist. Resistance is handled by revolution (crush it), evolution (reduce it), or isolation (avoid it). Limited evidence suggests that participation and divisibility, important elements of evolution, are used during CAM implementation. We have:

Proposition I4. The greater the increase in the infrastructure's sophistication, the greater the intraorganizational conflict.

A company must learn how to reduce gaps between expectations and performance that arise from initial conditions, changing market uncertainties, and efforts to improve the control system. It may try to increase performance by adjusting the technology to fit the organization or by adjusting the organization to fit the technology. Reductions in expectations or elimination of gaps from consideration may also occur as the result of improvements in the control system or changes in market conditions. The desire of the innovation's advocates to make it appear a success, acts as a reinforcement. All of these forces cause the very structure within which implementation takes place to continually change. Actions taken at one time are irrelevant at some later time. In sum:

Proposition I5. The greater the discrepancies between expectations and performance, then

the greater the direct increases in both the innovation's technical complexity and the innovation's performance,

the greater the direct increases in both the infrastructure's sophistication and the innovation's performance,

the greater the decrease in expectations.

Due to the prevailing uncertainty the learning implied by the above proposition proceeds very slowly. A firm may never know if a radical manufacturing technology is a success. The heavy costs already incurred by the implementation stage, although sunk, make stopping the project unlikely.

12
IMPLICATIONS OF THE INNOVATION THEORY

The theory's implications will be explored in three different ways. First, implications exist for understanding the innovation process, specifically, why it is so difficult to introduce radical manufacturing technology. Second, implications exist for designing the innovation process with coping strategies that will help insure the technology gets a fair hearing. Third, implications exist for the content and conduct of innovation research. Finally, one particular research implication is considered at length: a shift in focus from the user organization to the interrelationships between vendor and user. It is a logical next step in a process-oriented approach to innovation.

UNDERSTANDING THE INNOVATION PROCESS

Using stage models of the innovation process may create the erroneous implication that a strict temporal sequence exists. Considerable evidence demonstrates that innovation is much more complex than that (Tornatzky and Fleischer 1990). There are often overlapping stages, skipped stages, and backward movement to previous stages. The three phases of adoption, preparation, and implementation, which are employed here, do exhibit some of these complicating properties. During adoption individuals are beginning to gather information and experience, in other words, preparation is also commencing. A good deal of preparation overlaps into implementation. Feedback during implementation may indicate a need for movement back to preparation or even adoption.

If, in fact, a more detailed set of stages and substages had been defined, even more evidence of this behavior would have appeared. At the same time there does exist some order in the process; the stages cannot occur in a random sequence. It is hard to believe that a firm will invest a great deal of time and resources in preparation prior to a commitment to adopt. Implementation, the innovation's first use, logically follows adoption.

Why is it so difficult to innovate radical manufacturing technologies and computerized processes in particular? Uncertainty is the culprit. Organizations do not have the information to make decisions on whether to adopt radical technology. They cannot adequately prepare themselves to support it. They may not be able to determine if it is being successfully implemented.

During adoption the long run consequences (usually net benefits) of computerized technology cannot be precisely determined whereas the short run effects (usually net costs) are readily calculated. If strategic management has a short-term orientation, in part, brought on by the technology's complexity, the quantifiable considerations will play an inordinate role in decision making. The task force and vendor reinforce the orientation by supplying information that fits strategic management's preconceptions. Then if top managers lose confidence in the task force even worthwhile proposals will be rejected.

The forces acting during preparation are indicated in Figure 12.1. First, a new CAM system creates a need to change the infrastructure to a higher level of sophistication. Second, due to the technology's complexity and recency there is a lack of knowledge on what kinds of changes to make. Neither the company, the vendor, nor any comparable organization possesses the necessary information. Third, conditions are not appropriate for developing the infrastructure even with the in-

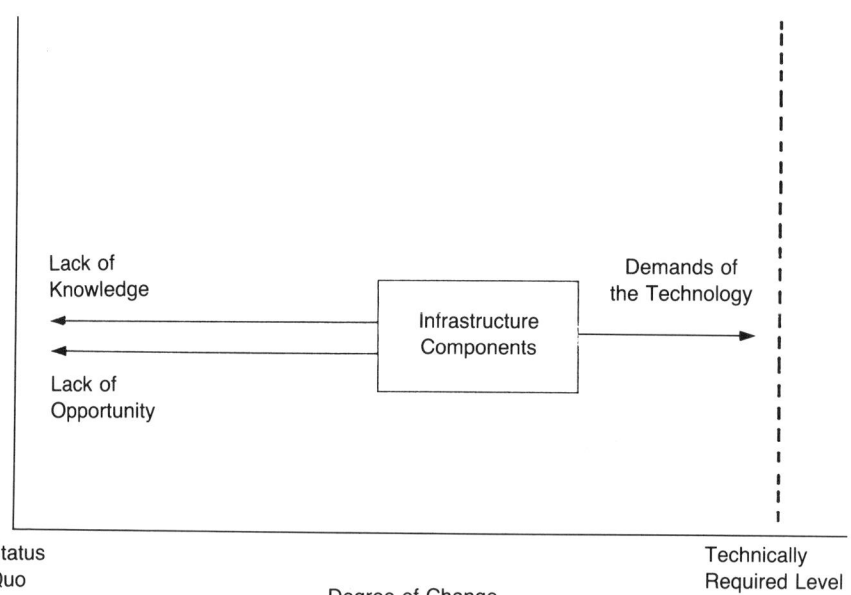

FIGURE 12.1. Factors affecting the infrastructure's change.

formation available. Its existing level of sophistication may be too low to attain the required level in a reasonable amount of time. Strategic management may pressure for immediate full-scale production in part due to the technology's costs. The company may become overdependent on the vendor. Subunits affected by the new technology may not have representatives on the task force. As a result of the conflict among the three forces in Figure 12.1 a compromise often ensues in which there are many incremental adjustments but few if any true administrative innovations. A good deal of preparation, instead of being ready when the technology arrives, will have to be done during implementation.

Implementation is viewed here as a control process gone awry. Expectations are initially high due to the arguments of the innovation's advocates, whereas performance is low due to the inconsistency between technology and infrastructure and the existence of intraorganizational conflict. Control activities are hampered by ambiguity in standards and performance measures. It is not clear how to adjust the technology and organization to reduce the performance gaps. Benign or malign adaptations may occur. Eventually, an organization may try to lower aspirations or change the standards against which the technology is judged. Ultimately, however, there may be too much uncertainty to objectively determine whether success has occurred.

DESIGNING THE INNOVATION PROCESS

In the face of these significant obstacles, what coping strategies are available to see that radical manufacturing innovations get a fair hearing? In other words, how can an organization decide whether or not to purchase and then whether or not to permanently integrate these systems? The theory offers a few suggestions, but the lack of a significant number emphasizes the dilemma faced by most firms. We can, however, have some confidence in recommendations which are grounded in theory.

Overall, the compelling requirement to deal with uncertainty indicates that a special role of process manager might be created based on the product manager concept. This position would cope with financial, technical, and social contingencies from the time the need for a radical technology is recognized until it is phased out. The role might fit in the hierarchy just under the senior executive in charge of manufacturing technology which itself is a new role being adopted in many companies (Voss 1984).

The activities of the process manager could subsume those of the champion. They could also include those of the "reorganizer" (Boer and During 1987) who initiates, realizes, and consolidates accompanying organizational changes. Due to the range of assigned duties, different individuals might have to fill the role at different stages of the technology's life. Another overall recommendation is for the process manager to incorporate significant time allowances for unforeseen events into his or her plans.

With respect to adoption, a key short-term strategy for the task force is to enhance strategic management's confidence in the recommendations. Keeping biased communications within reasonable limits and not making too many changes in the proposal will help. Prior to submitting a formal proposal requestors sometimes use a "softening" process in which through formal and informal contacts with senior executives they demonstrate their confidence in the project, learn of their superiors' concerns, and give superiors a chance to think about the project (Dean 1987).

For their part, strategic managers should determine the task force's composition utilizing track records as an important criterion. Due to the widespread organizational impacts of CAM they should also look for the existence of a coalition of key departments which supports the proposal.

In the intermediate term redesign of the justification process needs to be considered. First, the strategy section of this book recommends "experimental AMT justification" as one way of handling intangible benefits and costs. This gradual incremental approach owes as much to feedback at each step as to a preconceived overall plan. Second, large corporations have established special funds outside of the normal capital appropriation process to finance pilot projects that do not have to meet the usual financial justification criteria. The Ford Motor Company set up a five-year $75 million fund to encourage the development of new manufacturing technologies. It allows operating units to try new ideas on a pilot scale in order to contain risk. If the ideas work well they are diffused throughout the company.

A compelling long-run need is to lengthen strategic management's time orientation. Then the vendor and the task force are more likely to transmit meaningful information. The task force is less inclined to lower a proposal's visibility and strategic management is less likely to delay a decision. Lengthening the time orientation is accomplished by selecting new members of strategic management with manufacturing backgrounds provided they are not subsequently coopted by incumbents. It is also accomplished by building up the organization's support system or infrastructure to handle technical complexity.

Attempts have been made by some corporations to redesign their reward and reporting systems to create an improved balance between short- and long-term considerations. One expedient is to link executive bonuses to yearly rather than quarterly performance (Dertouzos, Lester, and Solow 1989). Another option is to reward managers for short- and long-term accomplishments as exemplified by Texas Instruments' Objectives, Strategies, and Tactics (OST) System (Jelinek 1979, Jelinek and Schoonhoven 1990). To meet short-term operational targets each hierarchical level is measured in terms of yearly profits and losses. In order to meet long-term strategic needs simultaneously, each level also has responsibility for developing product, process, and administrative innovations. The operating and strategic modes in a manager's budget are separated to eliminate the temptation to use strategic funds for solving operational crises. As originally conceived the OST System employed a matrix structure at the basic unit level. Managers had to implement their plans by pulling together resources from across the company.

As is well known, after years of substantial growth, Texas Instruments suffered heavy losses in the early and mid 1980s. The extent to which the OST System was a contributing factor is not known, but changes were made in it. Managers of the basic operating units complained that their authority to implement plans was much less than their responsibility. The matrix structure was scrapped and these individuals were given control over the resources needed to get their jobs done. Managers also complained of being smothered in bureaucracy. The number of variables that had to be projected for each of ten years was considerably reduced. Most important, managers argued that what had originally been intended as a facilitator of innovation had turned into a means of control. To avoid punishment for failure they stopped taking the kinds of risks that made innovation possible. By 1988 the original purpose had been restored. Overall, the company believed that the OST System was based on sound principles, but that over time its implementation had gone awry.

The key requirement during preparation and implementation is to have the innovation's technical complexity matched by the infrastructure's sophistication. One approach is to weaken the two forces in Figure 12.1 that make for the status quo. Due to the innovation's radicalism little can be done about increasing the amount of information on how to change the infrastructure. Conditions can be brought about, however, that facilitate employees learning through their own experiences. It is necessary to encourage participation by affected subunits in the task force. To avoid undue pressures for immediate full-scale production there needs to be a plan containing slow, deliberate "ramp-ups," which is worked out in advance (Meredith 1989). The dangers of turnkey projects are mitigated when the customer organization assigns its own employees to work side by side with the vendor's personnel.

A second approach is to shift the infrastructure's required level of sophistication to the left in Figure 12.1 by reducing the new technology's demands. Firms can acquire less complex equipment and install in stages, approaches which call for FMCs instead of FMSs. In fact, by the mid 1980s the market for FMCs began growing much faster than that for FMSs (Palframan 1987). Companies can also limit the number of new parts that are manufactured using the new technology. When these prescriptions are followed the resulting decrease in the amount of change required from the infrastructure also mitigates conflict and resistance which has a further salutary impact on performance.

As an example of what can be done the British motor and generator producer studied by Gerwin (1984) forged a comprehensive development plan that was put into action before installation of its new CAM technology and that called for:

1. Purchase of the equipment sequentially so the factory could better absorb it
2. Termination of dependence on corporate headquarter's mainframe computer and installation of a minicomputer at the local level
3. Assignment, whenever possible, of the unit's technical people to handle the new machines' operating problems
4. Creation of in-house training courses in electronics and other areas
5. Encouragement of workers to learn a variety of tasks so they could more readily adapt to the coming changes

6. Negotiation with the union to change its negative attitude toward added shift work

There is not enough research on how to stimulate learning during the feedback processes of the implementation stage. Jaikumar (1986) suggested that allowing the task force to remain with a new FMS long after installation and ensuring that a high percentage of operating personnel are college educated engineers with multifunctional responsibilities will encourage changes being made in the system. Jonsson (1987) believed that cost improvements while debugging new production technology depended on the proper design of managerial reporting systems. Based on studies of a European auto manufacturer, he suggested that central management needs a report that uses hard data for comparisons among operating units and control of performance. When something is out of control, that is, a discrepancy exists, the information should be communicated to local managers. They must change operating procedures to bring performance back in line, which requires a reporting system based on soft data gathered from experience. An information system designed for central coordination purposes will not satisfy local needs for learning, whereas emphasizing numerous local systems can not fulfill coordination needs.

Learning is also facilitated by not experimenting with new computer-aided inspection (CAI) equipment as part of a control system used to implement CAM. An auto assembly plant, for example, may introduce a vision system to measure dimensional quality just after a new computerized body-framing system is installed. First, this action will undoubtedly augment the uncertainty already inherent in the control system. Goodman, Griffith, and Fenner (1990) studied three new auto assembly plants with computerized body framing and vision systems. After a year operators were still trying to verify whether a signaled problem indicated defective auto bodies or a deficiency in the vision systems. Downtime, troubles with the cameras, and unreliable measurements hurt the vision systems' credibility. Second, because CAI is an example of CAM, one must struggle to develop a meaningful control system for its own implementation. In the same study it was noted that no mechanism existed within the vision systems, operators, or plants for comparing the systems' performance to expectations. Some changes occurred in the inspection technology, but they emerged over a long period and in a chaotic fashion. Although this study provides an excellent example of how implementing a new CAM conversion process goes on simultaneously with efforts to improve the control system, a key recommendation is to try to limit this simultaneity.

IMPLICATIONS FOR INNOVATION RESEARCH

One of the innovation literature's dominant trends compares characteristics of innovations to their rates of adoption using multivariate statistical analysis at the organizational level. Downs and Mohr (1976) argued for a more process-oriented approach, claiming that so-called innovation characteristics are also dependent on organizational context. A given attribute will exhibit different relationships to adoption in a sample of diverse organizations. Tornatzky and Klein's (1982) meta-analysis, however, found consistent relationships for a few characteristics.

Although it appears from this continuing debate that the comparative and process-oriented approaches are incompatible, they are really complementary. The theory developed here indicates the manner in which innovation characteristics produce outcomes such as adoption. Consider the attributes that Tornatzky and Klein (1982) reported as being frequently studied. Most play a role in the theory including compatibility (technical complexity versus the infrastructure's sophistication), complexity, cost, and divisibility. According to the theory, computerized manufacturing innovations exhibit low compatibility and high complexity—but moderate cost and high divisibility.

As an example of the manner in which these attributes affect the innovation process consider complexity; Tornatzky and Klein identified it as having a consistently negative relationship with adoption across different innovations and organizations. According to the theory this is because *ceteris paribus* complexity pressures strategic management into a short-term orientation that first reduces the chances of adoption directly, and second, by encouraging vendors to discount intangible benefits, reduces the chances of adoption indirectly. The ability of vendors to reduce acquisition costs mitigates these effects.

The theory is also a source of new hypotheses for innovation characteristics research, because it does not stop short of studying preparation and implementation. We have seen that *ceteris paribus* characteristics such as complexity, recency, and acquisition cost are negatively related to the chances of successful preparation. *Ceteris paribus* divisibility and compatibility are positively related, and complexity negatively related, to the chances of successful implementation. Confidence in these hypotheses is strengthened by the process-oriented explanations of the linkages between characteristics and outcomes.

One implication of the theory for research on innovation characteristics is that due to uncertainty it is difficult to know the values of certain attributes of a radical manufacturing technology until well into the implementation stage. In part the innovation process represents an attempt to discover the technology's basic attributes. Compatibility, the relation of an innovation's technical complexity to the infrastructure's sophistication, is one example. Preliminary measures of these characteristics are therefore of dubious value in trying to predict adoption. They are also of little use in expressing the degree of implementation through the concept of fidelity. Fidelity, which indicates the extent to which the core attributes of an innovation at the start of implementation are present at the end of implementation, could not be determined if there is no reliable set of measures at the start.

The theory has relevance for studies of technical versus administrative innovations. Daft's (1978) dual core theory emphasizes the distinctions between the two, whereas Van de Ven (1986) observed that significant linkages exist between them. Here, adoption of a technical innovation creates a need for administrative innovations during preparation which in turn, by raising the infrastructure's sophistication level, facilitate more technical innovations at some future time. In addition, implementation of the technical innovation may produce more technical and administrative innovations in an effort to reduce gaps between expectations and performance.

Synergistic diffusion, a term coined by Kimberly (1981) helps to describe the nature of these linkages. The adoption of one innovation is greatly facilitated by prior adoption of another. Fennell (1984) provided empirical support for the existence of synergistic diffusion but observed that research is needed on the processes by which an organization moves through sequences of linked innovations. The theory presented here begins to supply the necessary explanations.

How can we account for the literature's support for both the dual core and synergistic diffusion concepts? The former has been supported in organizations with dual authority structures such as school systems (Daft 1978) and hospitals (Kimberly and Evanisko 1981). It may be less appropriate for pyramidal organizations such as the business firms studied here. Once more, the innovations studied here are more radical than those that supported the dual core theory. Radical innovations undoubtedly have more synergistic potential.

The eclectic nature of the theory represents a useful guideline to researchers. It contains the following kinds of characteristics: innovation (e.g., complexity, divisibility), individual (e.g., strategic managers' backgrounds, the champion's reliability), organizational (e.g., decentralization in terms of the infrastructure's participation, formalization reflected in the infrastructure's sophistication), interorganizational (e.g., dependence on the vendor), and contextual (e.g., the cost of capital, large organizational size). It also contains less frequently considered types of factors such as strategic management's policy orientation. Consequently, exclusive concentration on just innovation characteristics or just organizational variables in empirical research is unlikely to produce valid results. With few exceptions (Baldridge and Burnham 1975, Kimberly and Evanisko 1981, Meyer and Goes 1988) empirical work has not included or controlled for different classes of variables.

Innovation researchers need to put more emphasis on the study of time dependent relationships among variables. The traditional cross-sectional study which portrays situations at a given slice of time obscures the following types of relationships. First, variables within an innovation stage interact over time. A discrepancy between expectations and performance during implementation creates feedback processes that modify compatibility thus reducing the gap. From a dynamic viewpoint, compatibility is not a given independent variable. Second, variables from prior stages affect variables in subsequent stages. The infrastructure's sophistication level augmented during preparation influences the innovation's performance during implementation. Third, although not indicated in the theory, variables from the current innovation process impact on variables in subsequent processes. The implementation gaps and organizational conflict resulting from the current process will determine confidence in the vendor and the champion's reliability later on. The infrastructure's final sophistication level will affect the opportunity to develop the required infrastructure next time.

At least one other time dependent process is illuminated by the theory—the manner in which a manufacturing innovation shifts over time from radical to routine status. This little studied process is important for understanding what is occurring to examples of CAM technology such as FMS, which were first introduced in the 1970s. We can examine the issue by asking how the model's main variables change in value from when there are no other FMSs in the same or similar industries to when there are several and the company in question has considerable NC, CNC, and perhaps even FMC experience.

The critical difference according to the theory is a decrease in technical, financial, and social uncertainty. During adoption there is no impact on the key variable, strategic management's short-term orientation. Its causal factors are not changed implying no lengthening of the time horizon for decision making. Due to the firm's similar previous investments and the experiences of comparable organizations it is possible to more accurately estimate longer term benefits and costs. Kelley and Brooks (1988) found informal contacts with experienced users were the most frequently cited source of very important information in deciding to purchase CAM. Beatty and Gordon (1988), who studied the adoption and implementation of CAD/CAM, found that the purchase decisions of senior managers were influenced by information on the experiences of other companies obtained through informal networks. Perhaps the company is also in a better position to justify the investment on the basis of short-term financial considerations. In sum, there is somewhat more emphasis on rational versus social and political factors, but adoption is probably not affected as much as subsequent stages.

During preparation more information on how to change the infrastructure is available from other institutions as compared to learning from experience. At least three general processes exist for diffusing information among organizations (Levitt and March 1988). First, a single source may broadcast to a population. Rules devised by governmental agencies, trade associations, and professional associations fall into this category. Second, an organization that has some information is in contact with an organization desiring it. Occasionally, this process is mediated by third parties such as consultants or employees who are changing jobs. Third, information is spread initially within a small group of educational institutions, trade publications, or experts, and then broadcast to a larger population.

Information brokers, that is, sellers or publicizers of an administrative innovation (Kimberly 1981), illustrate the third category. When a new technology is first introduced only the vendor usually advocates an associated administrative innovation even if it usually cannot do a very satisfactory job of it. Later, as the technology begins to diffuse, more information is available as sufficient demand develops for needed administrative changes. Other firms, consultants, professional associations, business schools, and consortia become brokers. For example, Computer-aided Manufacturing-International (CAM-I) is developing new cost accounting procedures for computerized manufacturing environments (Berliner and Brimson 1988).

Although not in the context of CAM, there is some evidence that learning transfers across manufacturing firms. Argote, Beckman, and Epple (1990) examined data from thirteen shipyards that constructed Liberty ships during World War II. Since the product was built according to a standardized design and most workers had the same nonexistent level of prior experience and the yards were all new, the data are free of important confounding factors. They found that a yard beginning production at a given date was more productive than those with earlier start dates *ceteris paribus*, but that after production commenced the yard no longer benefited from learning in the other ones. The authors attributed the transfer of learning to a central organization that was responsible for purchasing, approving plant layout and technology, and supervising construction, and that had engineers, auditors, and inspectors stationed at each yard. Admittedly, this means of knowledge transfer is rare.

What happens to the opportunity to create the needed infrastructure? The support system's initial sophistication level is at a reasonably high level due to the company's experience with similar technologies, and this also means there is less need to depend on the vendor. Other conditions including lack of participation in the task force, pressures for full-scale production, strategic management's beliefs about the need for developing the infrastructure, and external institutional constraints may continue to hinder putting the available information into action.

During implementation the control system is a more valid and reliable guide to action. Initial expectations are fairly realistic since they are based more on the firm's experience than on the vendor's estimates. Due to the infrastructure's relatively high level of sophistication initial performance estimates are more accurate. Firmer estimates exist of the impacts on performance of changes in the innovation and the infrastructure during the feedback cycles.

The control system deals with problems that are smaller in magnitude. Initially, the sizes of the gaps between expectations and performance are less since the infrastructure's sophistication at the start of implementation is higher. Since the support system's sophistication has to be changed by a smaller amount during implementation there is less occasion for conflict and resistance to change, which further augments performance. Implementation in sum moves in the direction of turnkey status.

RADICAL MANUFACTURING INNOVATION AS AN INTERORGANIZATIONAL EXCHANGE

Virtually all the research on manufacturing innovation examines it from the viewpoint of the user organization. Vendors are a given part of the environment or are not considered at all. As an example, the process-oriented theory just discussed focuses on managers' coping strategies in the user organization. They are, in part, determined by a vendor's uncontrollable behavior. Why not consider vendor and customer as caught up in a set of interactions in which the behavior of one is influenced by the other? Perhaps a symbiotic relationship exists manifested in an exchange transaction in which each acquires critical resources. This section conducts an exploration of an interorganizational approach based on interviews conducted among vendors and their customers. Some of the process-oriented theory's key variables play a role again but in new ways.

Studying CAM innovation from just the user's viewpoint may give a partial, biased account of what occurs since both parties are involved in a relationship which has significant political aspects. Consider the following examples:

- Customers say vendors are not responsive to their needs and vendors believe customers do not know what they need.
- A customer indicates that a certain objective was not achieved because the vendor did not have enough expertise and the vendor claims the objective was unrealistic.
- Customers indicate that vendors overestimate the performance of a new manufacturing technology in order to make a sale. Vendors argue that customers do not provide the data required to make accurate estimates in their contexts.
- Customers expect turnkey systems, whereas vendors say participation is essential.
- Customers say they are not prepared by vendors for the problems that will surface. Vendors find that users are unwilling to make organizational adjustments needed to handle the problems.

Studying CAM innovation through the customer-vendor relationship has practical benefits for the two parties. It can lead to more widespread deployment and more effective use of the technology. According to a U.S. Department of Commerce (1985) report, an ineffective customer-vendor relationship is one of the important factors impeding the diffusion of FMSs. A study of suppliers and users of computerized automation found that the most frequently mentioned item accounting for successful implementation was the quality of the relationship between the two (Ettlie 1986). Based on case studies, More (1988) concluded that when firms developing a new technology interact early and closely with customers the payoffs to both parties are greater and the risks are less.

Research on interorganizational relations (Benson 1975, Pfeffer and Salancik 1978) is used here to help analyze the customer-vendor relationship. In this view an organization survives through its ability to acquire critical resources from other institutions. One method of acquisition is exchange transactions, which have economic and political characteristics. An exchange creates mutual dependence since each party controls something that the other needs. They must work together in order to accomplish the transfer but also oppose each other due to having incompatible interests. It follows that each party seeks to control aspects of the relationship in order to obtain favorable terms for itself. In sum, mutual dependence produces both cooperation and conflict among organizations.

At least two caveats are in order. First, our knowledge of interorganizational relations amounts to more of a conceptual framework than a well-developed theory (Zeitz 1980). Second, the framework is meant to apply to organizations as they continuously compete for different kinds of resources over a long time frame. Purchase of a radical manufacturing technology is a single transaction of one major resource that occurs over a relatively brief period. Consequently, there is no set of propositions ready to be applied, but there do exist some useful insights around which to build propositions.

Nature of the Relationship

Although many actual interorganizational relationships, distribution channels for example, consist of formally structured long-term arrangements among several institutions, an even larger number occur as short-term ad hoc efforts among pairs of organizations such as client referrals (Van de Ven and Walker 1984). Long-term and short-term relationships represent two extremes with respect to stability, structure, and mutual control. The innovation process for a radical manufacturing technology falls somewhere in between. There is an ad hoc relationship between a pair of organizations that may last for two or three years. It may lead to a stable, long-lasting association between vendor and customer and is influenced by prior transactions between the two parties. It has some structure in terms of a joint project organization and a contract. There are mutual attempts to control the activities that must be performed and the vendor at least tries to influence the customer's organizational arrangements.

The resources being exchanged are at the heart of any interorganizational relationship. A vendor has the innovation that the customer needs to eliminate performance gaps or for some less overtly rational reason. It has knowledge about the manufacturing system that is imparted through training, maintenance, troubleshooting, software revision, and updating of the hardware. It may provide employees to the customer if they can be enticed away. A customer provides financial resources to the vendor. It has knowledge that the vendor requires in designing and implementing the system. It may provide useful feedback about the innovation's performance, which leads to an improved product. Eventually, it may agree to future transactions and/or to act as a demonstration site for potential customers.

Uncertainty is one defining characteristic of the relationship. Vendor and customer share in the technical uncertainty concerning the innovation's performance. Financial uncertainty exists for the vendor, who is unclear if the contracted payments will cover costs. It also exists for the user, who is unsure if discounted net returns will cover investments. Social uncertainty—the difficulty in predicting organizational impacts—exists mainly for the customer.

Another significant factor, derived from the interorganizational relations literature, is the desire of each party to maintain control over the project and refrain from becoming dependent on the other. The customer wants to avoid purchasing a system that does not perform effectively, whereas the vendor wants to ensure that the system's objectives are in a reasonable balance with what can be achieved. Control is exercised in several ways of which performing key activities is the most direct. If, however, critical activities are executed by the other party, it is still possible to have indirect control through the contract for example.

The basic dilemma for each party is that in order to reduce its share of the uncertainty, it must give up some of its direct control by relinquishing its performance of key activities. Uncertainty's main import is the risk of failure with consequent implications for the company and the individuals assigned to the project. If risk can be shifted to the other organization by making it more responsible for the project's activities, then the other organization will absorb more of the blame if something goes wrong.

The resolution of this dilemma produces a given allocation of risk and direct control between the two organizations. At one extreme the vendor supplies a turnkey project in which it absorbs all of the risk but maintains a great deal of control over what happens. At the other extreme, the customer decides to do the project on its own. Its substantial risk is mitigated by having the necessary infrastructure or receiving financial resources from the government, while control is retained in its own hands. Between these two poles are varying degrees of joint participation. Thus, the critical variable which describes the nature of the customer-vendor relationship is the degree to which a turnkey project exists.

Determinants of the Relationship

It is assumed here that the customer's or vendor's perceived risk is inversely related to its desire for direct control. As risk goes up the desire for control decreases. Situations that are risky for the customer (vendor) and not to the vendor (customer) will therefore be handled by the latter. Situations that are risky to neither party will undoubtedly be shared. The most interesting situation is characterized by risk to both parties so that neither wants complete control.

Technical complexity contributes to a relationship being technically, financially, and socially risky for both parties. Of the variables influencing complexity perhaps the system's degree of customization is the most significant. The vendor, for example, minimizes its technical and financial risk if it is able to pull standardized designs off the shelf. A second factor influencing complexity is whether or not the customer is developing a new product in conjunction with the radical technology. Interactions between parts and manufacturing process are then more problematical. The system's size, state of the art, degree of automation, and extent of integration are also influential.

One would expect that as a system's technical complexity increases the vendor's desire for a participative relationship and the customer's wish for a turnkey project would also rise. Interviews suggested the veracity of this assertion with respect to vendors and supplied no information with respect to customers. A robot system vendor has sold about 2000 systems costing in the hundreds of thousands of dollars each and is willing to take on all responsibility except installation. An FMC and FMS vendor has sold around forty systems each costing in the millions of dollars and is an ardent believer in customer participation.

When a customer's and a vendor's infrastructures lack sufficient experience, mutual risk is high. This condition, to be expected for the former, may also affect the latter when it is just breaking into the market for integrated systems. Sometimes users choose vendors with which they have had satisfactory prior relationships even if the latter do not yet know much about FMSs (Graham and Rosenthal 1986).

Certain activities performed during the innovation process are more risk prone than others for the two parties. Due to the small number of firms interviewed, it was not possible to determine whether these activities vary from one relationship to another or remain the same across relationships. Consider parts processing, that is, the design of tooling, fixtures, and programs. Because each customer has different parts a vendor may not be able to transfer the knowledge gained in working with one customer to others. A customer may not be able to transfer its knowledge of its own parts from the relatively unsophisticated environment in which they were originally manufactured to the new computer-controlled environment. Installation of the equipment is a second example. A vendor may find that installation depends on the particular way in which a given factory is constructed and that the customer's blueprints are of low quality. A customer usually does not have the necessary knowledge of the machine tools to facilitate their installation.

The interorganizational relations literature helps in identifying power-related variables influencing the extent to which a mutually risky situation winds up as a turnkey project. Undoubtedly, if one organization depends less on the resources it desires than the other organization, it will have the power to shift risk. If one party has more alternative sources from which to obtain the desired resources, it will have the power to shift risk. The robot system manufacturer does a large share of its business with one of the Big Three auto makers. This customer, however, has several vendors from which to choose. Traditionally, customers have been responsible for installation of the system, but now the auto company is expecting the vendor to assume responsibility. The vendor, which considers installation a headache, is reluctant to say no because it cannot afford to lose this customer's business.

Tradition also plays a powerful role in influencing the degree of turnkey status in mutually risky situations. The manner of securing resources tends to persist over time and cannot be altered at will (Zeitz 1980). Customers have grown accustomed to turnkey relationships in their purchases of stand-alone equipment. Vendors are willing to assume all the responsibility in providing individual machines due to the lack of risk. When seeking to introduce novel practices such as the sharing of responsibility for radical manufacturing systems, vendors must contend with customers' established beliefs and expectations based on previous ways of doing business. Most of the FMS users studied by Graham and Rosenthal (1986) began their relationship by assuming the vendor would be responsible for everything. They did not perceive the transaction as differing significantly from a traditional purchase.

When a party is able to shift risk, it attempts to compensate for the resulting loss of direct control. Introducing more formalization into the relationship is an indirect means of putting limits on the control ceded to the other party (Van de Ven and Walker 1984). As a relationship moves towards the turnkey end of the spectrum, a user tries to protect itself by making the contract more specific and rule bound. Consider performance standards, the basic parameters to which the manufacturing system must conform if the user is to accept it. The user will want language in the contract that strictly holds the vendor to the standards and that makes the vendor responsible for changes at its own expense if the standards are not met initially.

RADICAL MANUFACTURING INNOVATION AS AN INTERORGANIZATIONAL EXCHANGE **299**

The vendor, however, is reluctant to accept. Due to the prevailing uncertainty, cycle times, tolerances, and uptime are difficult to estimate. If the standards are set too high then problems may arise in attaining them. If they are set too low, then it may not make the sale. The vendor's solution is to press for joint responsibility in achieving the standards.

Other strategies exist for a party left with more risk than it wants. It can negotiate with a third party to mitigate the risk. If a vendor or a customer is reluctant to perform a certain activity, it can hire a firm that specializes in dealing with it. This solution increases the integration problem. Another alternative is for the party to hide behind the existing uncertainty. It is sometimes questionable who is at fault if performance standards are not met even if one side has most of the responsibility. Are the problems due to the vendor's technology, the customer's design choices, or both? Even a vendor performing a turnkey project may find an opportunity to shift blame back onto the customer. Often the result is mutual recriminations and perhaps disintegration of the relationship.

Activity Analyses

One way of studying the customer-vendor relationship, implied in the previous discussion, is to consider it as the unit of analysis. Then characteristics of this unit, such as the degree of participation, are conceptualized and measured. A more process-oriented alternative is to identify the sequence of activities that occur during innovation and consider them as the units of analysis. Then one can study the degree of participation and other characteristics for each activity. This second approach allows a more refined determination of where problems arise in the relationship.

The interviews identified a set of eight activities that describe the innovation process for a CAM system.

1. After contact is made between the two parties, the initial step is identification of a specific problem faced by the customer.
2. Based on this information a preliminary design is created.
3. A detailed design of the system's components is prepared. It includes parts processing, machine tools, material handling (MHS), computer controls, and a floor plan.
4. The equipment is manufactured and the software is written.
5. The system is installed on the customer's factory floor.
6. Integration—getting the components to work together smoothly—is undertaken.
7. A demonstration follows during which the performance standards must be achieved.
8. Training and the necessary adjustments in the customer's organization go on simultaneously with the last few activities.

Table 12.1 shows the results of an activity analysis conducted with a project manager supervisor for a leading vendor of FMSs and FMCs, and with the project

TABLE 12.1 A Vendor's and a Customer's Activity Analyses

Activities	Vendor	Customer	Joint
Problem definition		V, C	
Preliminary design	V, C	v	
Parts processing design			V, C
Machine tools design	V, C		
MHS design	V, C		
Computer controls	V, C		
Floor plan	c	C	V
Manufacture	V, C		
Installation			V, C
Integration			V, C
Demonstration			V, C
Training			C
Prepare organization		C	

manager for a firm which purchased an FMC from the same vendor. The former was asked to identify with a V which activities the vendor wants performed by itself, the customer, or jointly. As an example the vendor wants problem definition done by the customer. The latter individual was asked to identify with a C which steps the customer wants done by the vendor, itself, or jointly. For two activities, preliminary design and floor plan, a respondent indicated that one party should have primary responsibility (capital letter) and the other should have secondary responsibility (small letter). A more comprehensive analysis could also ask each representative for his perceptions of what the other organization wants. Then it would be possible to compare one unit's perceptions with the other unit's actual desires.

There is a high degree of agreement between the two individuals. Both concur that the vendor's responsibility lies in preliminary design and the standardized aspects of detailed design, that the customer's responsibility lies in problem definition, and that both should share responsibility for customized activities. There is little disagreement concerning the floor plan as both individuals believe in some degree of cooperation. No information was collected from the vendor's representative on training and preparing the organization.

Both respondents believed that the relationship was a good one and that the project was a success. One can point to the high degree of agreement on the allocation of responsibilities as a contributing factor. This specific allocation distinguished by the customer's willingness to accept significant responsibility should also be taken into account. It was not possible, however, to make comparisons with a poor relationship and a failed project. Neither vendors nor customers wish to talk about them.

The respondents' verbal explanations for their choices illuminated problems that arise in the relationship although all problems did not necessarily occur in this particular association.

1. The vendor's representative maintained that customers run the gamut from wanting to do problem definition to wanting the vendor to do it. He considered the latter situation a signal of future problems in the relationship.
2. The customer's representative pointed out that sometimes a CAM system does not work properly because the vendor, rather than trying to meet the customer's needs, prepares a standardized solution.
3. Both individuals saw the design of computer controls as a potential source of friction. Customers may want to meddle with the vendor's standardized solutions or prepare the software themselves, but a relative lack of experience often defeats them.
4. The vendor's respondent believed that installation and integration should be joint activities so that customers can become familiar with the new system. Customers, however, do not always want to get involved due to a lack of manpower. The customer's respondent indicated that some vendors don't want to take the time and effort to communicate with and educate users.
5. The customer's representative pointed out that users must provide enough time for qualified employees to be trained and then give them opportunities to use their new knowledge. He also indicated that the vendor can help the user identify needed organizational changes but sometimes is reluctant presumably because it may affect the transaction.

More (1988) developed an activity analysis for studying customer-vendor relationships for new hardware/software systems such as computer assisted learning. The supplier's development process and the user's adoption process were each characterized by the same nine activities. They included problem recognition, need analysis, product concept, technology choice, financial analysis, product design, production (for the vendor)/sourcing (for the customer), commitment by the vendor to produce and by the customer to purchase, and implementation.

The analysis involves constructing a matrix in which the rows represent the user's steps and the columns the developer's steps, and then identifying which are done separately and which jointly. Using case studies More identified a supplier-driven process in which the vendor develops the product by itself and then interfaces with a potential customer, and a codevelopment process in which most of the nine activities are done jointly. He believed that the second approach was more likely to produce success for both parties.

Future Research Possibilities

Explaining the customer-vendor relationship from an interorganizational perspective starts from the innovation's technical complexity and the sophistication of the participants' infrastructures. With high complexity and low sophistication, a large degree of perceived risk exists. The vendor aims for joint decision making, whereas the customer wants a turnkey project. Giving up direct control is preferable to being stuck with the blame for a failure. The degree to which a turnkey project emerges is affected by power-related variables such as how critical to each side are the resources it desires and the presence of alternative sources of the resources. The manner of purchasing more conventional manufacturing equipment, which emphasizes a turnkey relationship, is also influential. Customers have grown accustomed to this way of doing business.

If one party is successful in shifting risk to the other, it tries to compensate for the resulting loss of control by introducing more formalization into the relationship. It tries, for example, to pin down the contract's provisions. The party that absorbs more risk has the option of subcontracting or using the existing uncertainty to avoid blame if something goes wrong.

Future research may want to consider that vendor and user lie at the hub of a wider interorganizational network. Mohr's (1987) interorganizational processes by which a firm learns about CAM are relevant here. A firm may imitate what comparable organizations are doing or take the advice of consultants. Newly hired people from other enterprises may push for consideration of the technology or a key customer may exert influence to have it used. Not identified by Mohr is the possibility that employees have contacts with noncompeting enterprises with CAM which provide useful information. Finally, governmental bodies may provide financial resources and each of the two main parties may work with subcontractors.

In many situations these interorganizational networks have long run staying power (Weiss and Birnbaum 1989). A continuing stable relationship among all the relevant parties reduces the uncertainties and costs of frequent innovation. The network can be used as a barrier to the entry of new firms, but it discourages radical innovation and increases the difficulty of protecting first-mover advantages. North America is likely to see this type of linkage grow in significance as more and more user firms establish long run relationships with equipment vendors.

Research into the customer-vendor relationship will have true value to decision makers if it can indicate how to make innovation more successful. The major issue is whether a participative relationship leads to more widespread adoption and more effective implementation than a turnkey situation. Available evidence from case studies suggests this may be true (Gerwin 1982, Graham and Rosenthal 1986, More 1988), but no large-scale empirical tests have been conducted. What if a customer is forced by a vendor's superior power to take on activities it is not prepared to handle? Future work will undoubtedly discover that the degree of customer participation for successful innovation is contingent on factors peculiar to each party, factors in the relationship, and factors in the environment surrounding the relationship.

13
LESSONS FOR THE FUTURE

By challenging traditional wisdom, research into advanced manufacturing technology has had a significant impact on the production management and management of technology literatures. The research has:

- Challenged the assumption that technology alone can solve a firm's competitive problems
- Contributed to the recognition of a link between a company's strategy and its process technology
- Called into question the ironclad inverse relationship between flexibility and productivity
- Led to new methods for evaluating capital projects with significant intangible benefits and costs
- Cast further doubt on the suitability of traditional, hierarchical modes of work organization
- Called on systems designers to give more credence to the knowledge and experience of workers
- Helped in understanding that a good deal of the work in innovating a new technology is done after adoption

In this final chapter we review in detail these and other lessons learned.

UNCERTAINTY AND ITS IMPLICATIONS

Uncertainty is a defining characteristic of discrete parts manufacturing. Long-run market and manufacturing factors combined with short run technical, financial, and social factors are persistent realities, although their impact declines in the progression from custom to batch to mass production. In continuous manufacturing, uncertainties also exist but they are less influential particularly in the market.

Hirschhorn (1984) and others have explained why uncertainty has greater significance for discrete versus continuous production.

1. A firm makes independent units that have a larger range of variation than liquids and gases with commodity characteristics.
2. With some exceptions, a poorer understanding of the manufacturing process means less ability to define and control its key variables.
3. More human input is required, which introduces additional errors and delays, although workers' experiences are useful in reducing uncertainties.
4. Uncertainties, particularly in the market, not only persist over time they are constantly changing—sometimes faster than people can cope with them.

The *technological* approach, discussed in Chapter 1, seeks through AMT, including expert systems, to make discrete parts manufacturing more like continuous processing. For example, CIM is used to achieve the type of integration that already exists in many process industries (Browne, Harhan, and Shivnan 1988). The ideal manufacturing operation is not only fully integrated, it is continuous, automated, without direct labor, dependent only on planned maintenance, and monitored perfectly. Manufacturing innovations are selected to close the gap between this ideal and the actual situation (Bessant and Grunt 1985). These innovations are judged as sufficient in themselves to solve competitive problems, provided the most sophisticated technology obtainable is used. The ultimate expression of this view is the integrated, automated, unmanned factory of the future.

Although the omnipresence of uncertainty makes discrete manufacturing different than continuous production, a technological approach is insensitive to this difference. Uncertainty is handled through a combination of reduction and adaptation, but the various methods for reduction cannot lower uncertainty enough and AMT, the principal method for adaptation, does not yield sufficient flexibility. Moreover, a factory designed according to technological assumptions generates new uncertainties. These points are illustrated by the experience of General Motors with its Hamtramck plant. A second example is the vulnerability of some CIM factories to unexpected downward shifts in demand. Narrow, rigid dimensional envelopes inhibit their changeover flexibility. Their volume flexibility is hampered by close integration; one segment of a plant cannot be shut down because it affects other segments.

One can predict that the following uncertainty-related managerial problems will arise in a completely automated factory.

1. Due to high interdependencies, unanticipated market and process changes reverberate throughout the plant in inexplicable ways.

2. Since a plant's design complexity is inversely related to its overall reliability, a large number of process uncertainties will exist.
3. Unanticipated variations in the composition and dimensions of raw materials, formerly handled by humans, create new process uncertainties.
4. The need for in-process inventories to keep production going when equipment breaks down works against installing a JIT philosophy.
5. The more that automation—even computerized automation—replaces humans, the less flexible the factory becomes.
6. Integration and automation also hinder flexibility responsiveness, the altering of a plant's flexibility capabilities through reconfiguration of the manufacturing process. Flexibility responsiveness is needed to meet changes in strategic priorities as the nature and intensity of uncertainties shift over time.
7. Design choices reflect the ideal of continuous operations rather than the nature of discrete production. As one example, human monitors of automated equipment are stationed in control rooms to interpret computer outputs. They should also be on the shop floor using their senses to understand what is going on, provided safety is not a factor.

The *integrative* approach, also presented in Chapter 1, does a better job of recognizing and handling uncertainty's pervasiveness in discrete operations. It is based on an interacting sociotechnical system in which each component complements the other. Decision support systems feed data and suggested alternatives to operators with the ability to respond to uncertainties. People therefore have ultimate responsibility for dealing with unexpected events particularly when the flexible automation cannot. Strategic needs dictate the types of hardware and software; AMT is not an end in itself. Unnecessary operations have been eliminated so that they are not automated. Existing technology is utilized, if it can do the job, rather than investing in the "bleeding edge."

In the integrative approach decisions on strategy, organization, and technological innovation require human activity at all levels of the factory. Unless novel strategies are developed by top managers who understand the technology's capabilities and limitations it will not be possible to seize the competitive initiative by redefining uncertainties. A sophisticated human support system must exist to operate, maintain, and control the technology. In particular, workers trained in problem solving and with a broad range of skills must participate in groups to handle unexpected events. To back up the groups, supervisors act as consultants and interact with other units to obtain needed resources. During innovation, the adoption decision must be based on considerable human judgment, because the benefits and costs of AMT have major intangible aspects. There exists a need for a champion who believes enough in the technology to accept the risks associated with getting it approved, and who is sensitive enough to the organizational implications that the risks are minimized. In short, although less human resources exist, those that remain have a critical role in the factory's success.

STRATEGIC CONSIDERATIONS

Any linkage between strategy and advanced manufacturing technology depends upon a company's need for flexibility and AMT's ability to deliver it. There are however numerous ways of providing manufacturing flexibility, through work organization, materials management, and product design, for example. In the past many companies tended to concentrate on just one or another, technology in particular. Their unsatisfactory experience led to the realization that it is necessary to balance and integrate these various approaches. As one example effective implementation of a flexible manufacturing system calls for an adaptable work organization and Just-in-Time production.

Effective selection of AMT requires a top-down planning process in which the technology is used as a means of realizing a firm's objectives. Company and manufacturing strategy influence requirements for flexibility, which in turn determine the technology's specifications. Low-level involvement is also necessary if commitment to new AMT is to occur. In practice however a bottom-up approach is often used. Low-level decisions to increase automation or imitate competitors lead to technologies with flexibility characteristics that do not necessarily support strategic objectives.

Most researchers and managers only see flexibility as a means of adapting to market and manufacturing process uncertainties. This is a dangerously narrow view; a firm also needs to be proactive. Managers have the option of directly reducing uncertainties through long-term contracts or Total Quality Management for example, provided their firm is large relative to customers and competitors, and has a manufacturing process susceptible to comprehension and control.

A firm may also take actions that lead customers to rethink their market needs as when it uses superior manufacturing flexibility to help introduce new products faster than competitors. Honda's successful strategy against Yamaha during the H-Y War is one example (Stalk 1988). In this manner a firm redefines an industry's competitive uncertainties to its own advantage provided it is capable of creative strategy development.

Finally, it is possible to bank flexibility, that is, hold it in reserve for a future need or opportunity. The Department of Defense, for example, compels military contractors to be able to increase production volume quickly in case of a crisis. Developing a multiskilled workforce is another example. A voluntary depository of flexibility maintained by enterprises in a mature stage of development would help insure that sudden augmentations of uncertainty originating from outside the industry can be handled. Creating a reserve of flexibility requires that investment alternatives are not unduly penalized for their long-term intangible advantages.

Achieving an appropriate balance between adapting to and reducing uncertainties is currently a salient issue. Many North American firms are pursuing both approaches without taking note of their three major interconnections. First, in the early fluid stages of development for a new product and process, adaptation is more prevalent, but in the later mature stages it gives way to reduction (Abernathy 1978). Second, adaptation can be pursued in the short run while attempting to reduce uncertainties in the long run. Third, the more one has of adaptation or reduction the less is the need for the other. Too much adaptive capability however lessens the motivation to eliminate uncertainties, and too much control over uncertainties prevents an adaptive response to crises.

This last type of interconnection is illustrated by the relationships between our six flexibility dimensions (adaptive mechanisms) and certain marketing, design, and manufacturing practices (reduction devices).

1. Uncertainty for the kinds of products offered leads to mix flexibility which allows a plant to handle a range of existing products with setups that are not time consuming and expensive. Excess product variety, however, produces complexity and confusion which drives up overhead costs. Where feasible, long-term contracts with customers should also be pursued.
2. Uncertainty as to the length of product life cycles creates a need for changeover flexibility, which facilitates a quick substitution of a variety of new products for those currently being manufactured. It is, however, difficult to specify in advance an appropriate product envelope within which substitutions can occur. Managers at IBM's Lexington plant found they had specified the dimensional envelope for the assembly of printers too narrowly when a Japanese manufacturer subsequently marketed a larger model than they could produce. Furthermore, by the time current products are out of date, so may be their flexible manufacturing processes. Consequently, traditional product life extension practices are also necessary.
3. There is often uncertainty as to the nature of the appropriate characteristics for a new product or a customized design. Modification flexibility is an ability to speedily manufacture a range of design changes on the basis of market feedback. Too much of this type of flexibility however reduces pressures on product engineers to get designs right the first time. Cross-functional design teams that maintain close contacts with customers help reduce the uncertainty.
4. Unpredictable variation in the amount of demand makes for volume flexibility, which permits quick increases or decreases in the production level, but requires heavy investments in excess capacity, floor space, and slack time in the production schedule. Leveling demand, an aspect of JIT production, is also needed.
5. Machine downtime creates a need for rerouting flexibility, which provides an opportunity to speedily adjust a part's routing when a breakdown occurs. Alternate routings, however, discourage efforts to eliminate breakdowns so preventive maintenance is also essential.
6. Uncertainties exist in the dimensional and compositional characteristics of the raw materials being processed. Material flexibility permits a manufacturing process to quickly adjust to a wide variety of these problems, but it also reduces pressures on upstream activities to stop quality problems from occurring in the first place. Total Quality Management helps to restore a balance.

Under what circumstances is it appropriate to use each of the four strategic implications that is, adaptation, reduction, redefinition, and banking? One can begin to specify conditions by using a discrepancy analysis. Having required greater than potential for a given flexibility dimension often arises because the state of the art in AMT may not provide as much flexibility as needed. Closing the gap is achieved through uncertainty reduction, which decreases the amount of flexibility required, or by investing in more adaptive potential through methods of providing flexibility other than technology. When potential is greater than actual one can augment adaptability by removing flexibility bottlenecks in the value added chain such as slow feedback from the marketing department about changes in customers' preferences. A firm can also operate AMT to realize its strategic potential rather than as another example of mass production technology (Jaikumar 1986).

Suppose potential is greater than required implying there is more capability for flexibility than is needed. Creeping flexibility is one reason. Over time, product lines might have proliferated beyond what is necessary to meet most customers' needs. Under these circumstances a firm can employ the excess flexibility to its advantage by redefining market uncertainties, thus raising required levels. Instead, it can decide to bank the excess flexibility for later use, which implies that the gap remains. Another option is to reduce adaptive potential, say through rationalizing the product structure. An American electric motor producer pared the number of different horsepower motors offered but allowed a customer desiring an unobtainable size to have the next largest available at no extra cost.

A discrepancy analysis also puts the strategic choice between flexibility and focus into context. Moving toward a flexible factory is appropriate when the factory's potential for mix flexibility is less than its market requires. In a flexible factory, more product variety and shorter lead times occur without raising overhead costs substantially as the result of implementing JIT and/or CIM. When all of its individual plants can make an expanded range of items, a business unit enjoys the benefits of a relatively broad product line. This approach works best for large business units with abundant resources and is susceptible to other firms in the industry specializing in different niches.

Moving toward a focused factory is appropriate when the factory's potential for mix flexibility is greater than its market requires. If all the plants in a business unit are given more limited but different tasks, it is still possible to achieve a broad product line at the business unit level. If however the plants are given the same limited task, often high volume production of a few standardized items, the business unit will necessarily also have narrow product variety. The former variation is illustrated by the motor division of Reliance Electric Corporation which makes a broad size range by having each factory specialize on a narrow size range (Hayes and Wheelwright 1984). The latter is exemplified by many Japanese firms when they broke into international competition.

In the first version of the focused factory there is a tendency to inefficiently utilize duplicated human and capital resources, and associated pressures to make some resources less specialized. Sharing expensive resources, using the plant within a plant concept for example, is a compromise between the benefits and limitations of providing each unit with everything it needs to accomplish its tasks. In the second version concentration on high volume production of a few items fosters an efficient use of resources, but cripples market responsiveness if competitive conditions change radically. For example, during the recession of the early 1980s many factories dedicated to oil rig parts production were forced out of operation.

The concepts already discussed are useful for establishing a procedure that will change a factory's flexibility.

1. Phase I, at the business unit level, identifies flexibility dimensions relevant for competition in the industry.
2. In Phase II, task forces at the plant level determine the level of required, potential and actual flexibility for each relevant dimension. Any gaps are prioritized to determine the order in which they will be handled given limited financial resources.
3. In Phase III, the task forces identify ways of closing the gaps.
4. Phase IV involves continuous monitoring of what has been accomplished. Adjustments are made to changing conditions by shifting the values of required flexibilities. At Wang Labs each unit in the workflow has performance measures based on the requirements of its downstream neighbor. A shift in customers' needs therefore reverberates backwards through the workflow (Cross and Lynch 1988/89).

Any procedure of this nature must come to grips with a number of problems.

1. Changing flexibility may have consequences for other important objectives such as cost, productivity, and quality. Tradeoffs must be considered. A related point is that limitations on financial resources may prevent closing (or creating) gaps entirely. General Motors' assembly plant at Oshawa, Ontario found that the capital and maintenance costs of flexible assembly systems to make the desired level of production modifications were so great it had to reduce the number of options it wanted to offer.
2. Conflict may exist among the business functions in determining the magnitudes and priorities of the gaps, especially when much of the information is subjective.
3. Functional managers may resent the transfer of their influence to the task forces.
4. Measuring gaps, which will be discussed next, is not easy to do.
5. Justifying any necessary capital investments, which will be discussed subsequently, may flounder when intangible factors are significant.

Incorporating nonfinancial quantities into performance measurement helps managers in computerized manufacturing environments. It permits control over activities that significantly affect costs, adoption of a philosophy of continuous improvement, an improved balance between short-run and long-run considerations, and reestablishment of a correspondence between strategic manufacturing objectives and performance indicators, which has been broken in many firms. A better allocation of overhead costs can also be achieved. An experimental unmanned FMS recently developed by a large military contractor with the U.S. Air Force will not be charged with any overhead costs because the government requires that overhead be allocated using direct labor hours.

In order to develop scales that express the flexibility dimensions in physical units the following questions need to be asked:

1. Which organizational levels (business unit, factory, manufacturing system, machine) are relevant?
2. Which sources of data are available, and will objective or subjective information be collected?
3. Which alternative measure for a given flexibility dimension will be selected? Criteria for making this determination include cost-effectiveness, validity, and reliability.
4. Will customers' needs, competitors' capabilities, or the firm's past levels serve as baselines to determine how much improvement is necessary or has been accomplished?
5. Will the measures be used for evaluation, self-improvement, or both?

Most efforts to develop flexibility scales have concentrated on the mix and changeover dimensions. The former's range aspect is often expressed in terms of the number of different outputs but should also include consideration of the extent to which these outputs differ from each other. Measures of mix flexibility's time aspect are often based on the reaction time concept, say the interval between receipt of an internal order to produce and shipment. The range of changeover flexibility is measured in different ways including the number of new outputs introduced per unit time or the turnover of outputs in a given time period. The time aspect represents the interval devoted to manufacturing's activities during new product development.

Due to its novelty and radicalism in most firms, AMT is faced with significant obstacles during the capital appropriation process. Concept justification, determining whether the basic idea is worthwhile, is a necessary initial step. Although it contains elements of an informal advocacy process, concept justification serves as a valuable educational experience for top management. Financial justification should interact with equipment design. Information from one should feed the other as plans become more concrete. Once more, any new techniques for selecting investments need to be accompanied by new organizational arrangements to support them.

The major stumbling block is the intangible nature of many of AMTs benefits and costs such as flexibility. A popular way to handle intangible factors is to modify the traditional discounted cash flow analysis. Factors not normally considered, but which are expressible in financial terms with some effort, are included. Some tendency exists however to include hard to express revenues as opposed to costs. Lowering discount rates to more accurately reflect a capital alternative's degree of risk is a second modification. This, however, removes a protection against advocacy bias, the tendency of requestors to make a favored project appear more beneficial than it actually is. Simulation models, which provide an explicit account of risk and dynamic considerations, represent a third modification.

All three of the major financial investment methodologies—modified discounted cash flow, multiattribute analysis, and the strategic approach—are subject to advocacy bias. Choosing which intangibles to consider, estimating financial impacts, selecting attributes, or providing information on weights and values are opportunities for advancing preconceived notions. The first two methods also make unreasonable impositions on an enterprise's information processing capabilities. They require knowledge of all significant benefits and costs in situations where unanticipated consequences are more the rule than the exception. They also force individuals to be precise about uncertain factors including financial implications, probabilities, values, and weights.

Although the strategic approach requires less heroic information processing it offers hardly any precision at all. Consistent use might spell financial disaster for an enterprise. It has been successful for firms in life-threatening crises where maximizing profits over some time frame is not the paramount issue. The Peerless Saw Company invested in a unique CAD/CAM system in order to save itself even though the system did not pass financial tests (Meredith and Hill 1987).

One possible solution to these problems occurs when a firm implements AMT using a gradual incremental approach successfully applied by companies such as Ingersoll Milling Machine (Kirton 1986). Information processing requirements are reduced at each step. Advocacy bias is moderated since requestors are not proposing one large system for a go or no-go decision, and feedback on the results of previous steps acts as a reality check on biased estimates.

What is the flexibility of a significant aspect of AMT—CAM technology—in practice? There is surprisingly good agreement between the results of the pilot study and the mail survey. In those two investigations, mix, modification, and volume flexibility increased while changeover, rerouting, and material flexibility decreased. Introducing CAM into a manufacturing process will not necessarily improve all the dimensions. Manufacturing managers must therefore be aware of which types they need and monitor the design and selection process to see that they are delivered.

The mixed pattern of increases and decreases in individual flexibility dimensions is influenced by the automation level of the original manufacturing process (OAL) that was replaced. As the original process moves from being labor intensive to rigidly automated, the values of three combined measures of change in flexibility tend to increase. The change in market flexibility increases with OAL and is positive over the ranges of observation in both studies. Given the market turbulence of the recent past this is an encouraging result. The change in process flexibility remains negative over a wide range in both studies until finally turning positive. Companies that start with labor-intensive processes must be careful that introducing CAM does not prevent them from making adjustments to process uncertainties which cannot be eliminated. The change in global flexibility, which is negative at low OAL levels, turns positive as OAL increases. In the mail survey turning positive occurred at a lower OAL level than in the pilot study.

Whether one considers the individual or combined measures, the changes in flexibilities are rather small in magnitude. CAM does not appear to have a very significant influence on this manufacturing characteristic.

Decline is not inevitable for an industry in a mature stage of development. Three alternative theories exist for how to check a headlong plunge towards economic rigor mortis. First, dematurity may occur, that is, a reversal in the evolutionary process. Individual firms in a mature industry can move backward to an earlier, more fluid stage of development if environmental uncertainty reasserts itself and the manufacturing process is flexible enough. Changing consumer tastes pushed Ford in the late 1920s, after it reached a highly mature state of development, to shift back to more fluid conditions in order to introduce new models. The tardiness and cost of this reversal were in part due to the rigid nature of its original production process. One problem with de-maturity is that it substitutes product innovation for cost efficiency when both are currently needed.

Second, under certain conditions, introducing CAM into a plant will raise the levels of product innovation and cost efficiency simultaneously causing a trajectory to be traced through upward shifts in the life cycle curve. This occurs when the original manufacturing process was rigidly automated as opposed to labor intensive.

Third, renewal may occur at the industry level rather than at the firm level. Old, sluggish firms can revitalize themselves with NC and CNC technology but may not seize the opportunity. They are replaced by new dynamic entries with NC and CNC. Implicit in this view is that organizational and individual characteristics, which over time have been fine tuned to take advantage of relative certainty, are a critical stumbling block to the survival of older firms when environmental turbulence appears.

ORGANIZATIONAL CONSIDERATIONS

What we have learned with confidence in the past ten years is what does not work. We have, as Donald Schon (1971) so eloquently stated, lost the stable state, and with its demise many of our organizational pillars have collapsed. We are too slowly beginning to appreciate that our organizations are open systems, open to an increasingly turbulent and uncertain environment that can no longer be buffered at the boundaries of the organization. That environment enters from the top and from the bottom, via suppliers and through customers, on the back of technology and on the wings of unpredictability, and when it comes face to face with the stable, bureaucratic structures that provided us with so much affluence in the 1950s and 1960s, it cannot find a way to be accepted. The gap between the demands of the turbulent outside world and the slow to change organizational ways becomes increasingly large, but it is not the environment that makes the adjustment.

The stable structures that started to shake in the 1970s began to crumble in the 1980s. We learned that long hierarchies are poor at communications and particularly poor at communicating the changes emanating from an uncertain environment. We learned that the rigid relationships and controlling behavior of stable structures alienates people from their work, particularly when their own environments and their own experiences tell them they can expect better. We learned that even in times of recession people will no longer put up with unsatisfying work in order to find their satisfaction outside the factory floor. They consider themselves entitled to dignity as well as a paycheck from their places of employment.

Technological change is a major component of the turbulence in the environment. Major technological changes induce significant organizational changes. They have immediate effects on the content of jobs and on the skills and qualifications of people. Almost everyone has to learn to "keyboard" and interact with a terminal; operators have to understand what lies behind the black boxes that control their machines and equipment; and supervisors are confronted with a plethora of manufacturing processes they never even heard mentioned three years earlier—JIT, CAD, CAM, and TQM for example. At first, the technological changes do not modify the existing organization (Liu et al. 1990). In time, however, the old organization design becomes increasingly suboptimal with respect to the new technology's potential. Suddenly, there is a qualitative difference in the relationship of structure to technology from that of the last few decades. When technology and the environment were stable, there was a one-to-one relationship that almost mandated a traditional organizational structure. However, what structural form is appropriate for uncertain environmental conditions and changing technology? There is no quick answer.

We have continued to hear calls for a kind of return to the certainty of the past through advocacy of total technological solutions that would account for all uncertainties by offering us infinitely flexible technology. However, we have also learned to be wary of a technological approach. It has rarely delivered what it has promised and the evidence is now accumulating to explain why (Womack et al. 1990).

From the Total Quality Management (TQM) and Just-in-Time (JIT) perspectives, we have heard advocacy of an alternative set of solutions that would eliminate uncertainties at the input to the organization and at every stage of the processes of production. The gains achieved by those who followed this perspective have been significant. But they have not successfully bridged the boundary to the organizational side of the firm, to the people, to their social systems, and to their management structures. We have not been able to ensure that people who enter the organization's doors will make no errors or that their relationships with each other will be smooth or that their communications to each other will be free of transmission errors or that they will exercise no more influence or authority over each other than what is specifically required for each situation. Other than offering a mild bow to the importance of people, TQM has largely ignored the organizational aspects of the workplace.

From Japan and the automobile industry we have seen advocacy of "lean" manufacturing that has begun to unmask many of the false assumptions we have had about how to manufacture goods. Building from a quality management base, the "lean" approach has begun to revolutionize our manufacturing methods and taught us much about how to move up the scale of increased flexibility. Yet it too appears to fall short when the Western world's dream of transferring the democracy of its political structures to its workplaces is invoked. It falls no shorter than our own experiences, but we have already found them wanting. We are aware of exciting new manufacturing technologies and advanced thinking about manufacturing methods, but the beacon that should guide our way to how to organize ourselves to capture the potential of these possibilities is still a dim one.

This beacon does shine, however, in an increasingly wide variety of applications. A Western model of organization is evolving that promises to facilitate innovation in both the social and the technical subsystems of the organization and, more importantly, in their joint interaction. It has theory and concepts, design principles and processes, methodology, and experience associated with it but no prescribed solutions. It is local in its application because it is driven by the idea that there is organizational choice in every application. The parameters that inform that choice are "soft" factors such as values, culture, vision statements, and organizational philosophies. They underlie a new paradigm of work that assumes that commitment only arises out of involvement, open communications, and trust; that power equalization is needed for openness and collaboration to develop; and that all organizational members can grow and develop (Beer and Spector 1985). The changes to the workplace that result from this new paradigm are revolutionary, not evolutionary. That is why adoption has been slow. However, new technologies, particularly new manufacturing technologies, are accelerating the pace of adoption. We almost have no option but to move in this direction. Technology is becoming increasingly complex. Its very nature is changing: the interdependencies it is based on are so great that its outcomes are increasingly unpredictable; it is abstract to most who must interact with it; and the distinction we once understood between discrete technologies and continuous ones has begun to blur. What we have learned, particularly in the last decade, is that the more complex technology becomes, the more dependent organizations become on their people to make the technology function effectively. That effectiveness will not be realized without a revolutionary change in how we manage and organize our relationships at work.

When GE Canada's airfoil blade and vane plant in Bromont, Quebec was being designed, the proprietary pinch-and-roll process that gives the correct shape to the jet engine blades was implemented manually because the engineers did not have the confidence that the proposals they had received for automating the process could deliver the required shape or quality. However, the plant was designed to function participatively. That culture soon evolved a joint problem solving process between the engineers and the workers that resulted in their own proposals for automating the pinch-and-roll process. Those proposals turned into successful applications.

The approach or philosophy was often repeated in a plant that was successful enough to grow to several times its originally proposed size. The senior management team knew they would be continually automating their manufacturing processes, but they chose to do so with their employees. Two robots were installed in the center of the plant to be used for training and as demonstration devices. Employees were encouraged to become familiar with their functioning. Nine original robots grew slowly to forty-three as the plant encouraged its employees to make use of their accumulating understanding of the technology to select the new automated equipment that they would themselves eventually operate. The plant was designed to a sociotechnical system philosophy and that perspective never let the technology get ahead of the accompanying social system, except once. That one purchase remains in the warehouse, unused.

From the start the Bromont plant's employees were instructed in the social system alternatives that were arising as well as in a thorough appreciation of what the technology available could do for the task they were to perform. Only then, when they had a basis for being innovative, were the employees asked to create and design the kind of organization in which they could best imagine themselves wanting to work. The outcome of that process has been a great success, but it is only one of a burgeoning number of models of new ways of working that are arising in every sector of the North American manufacturing environment. Bromont's model was based on sociotechnical or high commitment principles, but the application was uniquely their own. Every other successful application has that same local uniqueness.

The earliest of these innovative organizational arrangements arose, some twenty years ago in North America, in continuous process petrochemical factories and in food processing and consumer goods plants where at least part of the production process is continuous process. Investment per employee in petrochemical plants ranged from $1,000,000 to $4,000,000. Until recently, investment per employee in discrete manufacturing was much smaller but could reach $500,000 in the more automated plants. Now with AMT, investment per employee is approaching the numbers of the petrochemical sector and the logic that drove that industry to invest in its people is being replicated in discrete manufacturing. That logic recognizes the good and the harm that people can do to and with the technology. With it comes an appreciation that an investment in the social system, for example, in training, is small compared with the cost of the factory. The corollary is that the cost of an uncommitted and untrained employee can be incalculable when the total plant investment has become so high.

The economics of high technology plants that makes social system costs small compared with the costs of the technology, the complexity of technology that makes it increasingly dependent on the organization's members if it is to function, the competitors who appear to have discovered alternative ways to organize and produce that are consistently more effective, the changing value systems stimulated by increased educational levels and a growing psychology of entitlement—all of these are putting enormous pressure on manufacturing organizations to reform the ways in which they organize themselves. It is the confluence of all these factors that is changing the paradigm of work and with it the structural arrangements that will reinforce and sustain paradigmatic change.

There are proven design processes to accompany the design principles. Steering committees of people who can provide sanction and support and resources and direction can ensure that innovative efforts will have the opportunity to be diffused across the organization and will not become isolated and encapsulated. Design committees composed of a vertical slice of people from the unit under design can ensure that the resources of the organization are used and ownership of the processes is assumed by those who have a vested interest in the new design or redesign. Study committees and task forces can provide forums for everyone interested in influencing the outcomes to do so.

Pratt and Whitney Canada (PWC) designed their CIM plant in Halifax, Nova Scotia following exactly such a process (Betcherman et al. 1990). The steering committee was made up of the top functional managers of the plant, corporate representatives, and the chairperson of the design committee. They wrote the initial mission and philosophy statements for the plant, capturing the company's strategic vision, its organizational identity, the prevailing values, and the principles that guided its activities. The design committee was formed before employees were hired for the greenfield site. Initially the committee was composed of the plant manager, representatives from all key departments, and members of the Manufacturing Modernization Program and was assisted by design consultants. As employees were hired they joined the committee until they constituted half the membership. For the first few months, all employees also served one day each week on one of the different subcommittees set up to design different aspects of the plant.

The design process needs active attention. Participation in the process requires training. At PWC the design committee went through a team building process and an organizational design course to learn the design process. Creative design requires building of awareness and stimulation of ideas. Extensive support is needed, both from champions and supporters of champions. But it has been done now and is being done now in plants across North America (Walton 1985). The organizational lessons of the last decade have created a strong platform from which advanced manufacturing technologies can approach the year 2000.

ISSUES IN INNOVATION

Large firms are most likely to innovate CAM, particularly integrated manufacturing systems. The first phase, adoption, may commence with the perception of performance gaps resulting from needs or opportunities. An American firm that responded to a decline in sales and profits, a British firm that took advantage of government funding, and a German company with a new contract that doubled its capacity illustrate the possibilities (Gerwin 1984). A search is conducted for ways of eliminating the gaps in which informal contacts with users of CAM are instrumental in learning about it.

The technology is also found through less rational means (Mohr 1987). Individuals in a firm may look for new challenges, an increase in social status, or new toys with which to play. An organization may take the advice of consultants, slavishly imitate competitors, or respond to the request of a key customer. The last alternative is not significant in the United States except perhaps in the auto and defense industries. These two industries, however, are instrumental in stimulating the adoption of AMT in other sectors of the economy.

Having discovered CAM a firm must decide on its originality level, the extent to which invention, adaption, or borrowing will take place (Pelz and Munson 1982). A high degree of originality, that is, invention, occurs when an enterprise intends to become a supplier of the technology, has a need for specialized equipment or can obtain governmental financial aid. In the United States invention is infrequently employed by firms interested in FMSs.

The selection decision takes place under a good deal of financial uncertainty. If strategic management has a short-term orientation, reflecting uncertainty avoidance, CAM is hindered from getting a fair hearing in the capital appropriation process. A short-term orientation occurs when the firm's stock price is low, rewards are based on immediate profits, a high cost of capital exists, strategic management's background and experience are in financial control, and the proposed equipment is technically complex. It is exacerbated in large firms where bureaucratic specialization keeps production and finance people separated from each other. A short-term orientation creates a self-fulfilling prophecy in which the task force and vendor emphasize an alternative's concrete advantages and downplay its intangible benefits. The chances of adopting CAM are diminished, unless the vendor offers a low acquisition cost. One reason a leading vendor of FMSs and FMCs sold its first system was that the low cost compensated the customer for the uncertain financial returns (Gerwin 1981a).

Since strategic management operates with a lack of reliable information it depends on social and political criteria in choosing an alternative. It is concerned with the past reliability of the task force's leader, whether a coalition of key functions supports a proposal and if the task force appears confident in its choice. The task force attempts to develop a coalition, appear confident, avoid excessive biasing of its proposal, and reduce the proposal's visibility. One example of limiting visibility is to bury a request for a computerized manufacturing system in a much larger capital improvement program.

Given the high degree of uncertainty, strategic management often exercises the option of hesitation instead of coming to a definitive decision on whether or not to purchase. In one American firm strategic management sent back the task force's proposals for an FMS over thirty times for revision before a decision was made (Gerwin and Tarondeau 1982).

During preparation it is necessary to develop an infrastructure sophisticated enough to operate, maintain, and control a manufacturing innovation. Many companies fail to put together the required ingredients because of uncertainty. A lack of knowledge exists on what specific changes to make. Once more, conditions necessary to facilitate redesign of the infrastructure are not in place.

Due to the technology's radicalism and recency, information on specific changes is not available from the company's history, comparable organizations, or the vendor. The vendor, for example, may find the technology too complex to install in its own test facilities prior to shipment. Once more, the existence of knowledge transfer points leads to distortion or leakage of valuable information. Some of the information loss is controlled by keeping the number of transfer points to a minimum and maintaining overlapping memberships in the groups that hand over and receive the technology.

Certain conditions must be altered in order to facilitate revision of the infrastructure. Factors susceptible to change in the short run include excessive dependence on the vendor, pressures for immediate full-scale production, and a lack of participation in the task force by functional units affected by the new technology. Other factors are possible to change only in the long run. Management, impressed by the technology's sophistication, may believe that human support is unnecessary. The infrastructure's sophistication level at the beginning of preparation may be too low to attain the required level in time. Due to its lack of NC and CNC experience the British motor producer studied by Gerwin and Tarondeau (1982) realized that integration of its new computerized equipment would be achieved slowly. The German aircraft producer, with an abundance of this experience, could progress much more rapidly. Finally, certain institutional constraints, such as work rules imposed by trade unions and cost accounting practices imposed by governments on the firms with which they do business, also hinder revision of the infrastructure.

Organizational lag is the result. The rate of technical innovation outpaces the rate of administrative innovation although both are required. Since the infrastructure is not ready at the start of implementation, further adjustments must be made in it during that stage.

Implementation is a control process in which the first attempts are made to collect performance data, compare to expectations, and align discrepancies. Even at this stage uncertainty is a powerful force. Control systems, which are part of the infrastructure, are imperfectly developed. As an organization tries to eliminate apparent discrepancies it must also improve the reliability and validity of these systems. Once more, based in part on information from the imperfect control systems, individuals estimate the innovation's consequences for themselves. Those who believe they are adversely affected will stubbornly resist.

When operationalizing expectations it is unclear which types of standards to use and whether correct initial levels have been chosen. A firm relies on information supplied by the vendor, which is undoubtedly biased, and on the intuitive judgments of its experts. Measures of the innovation's performance are ambiguous due to simultaneously occurring changes elsewhere in the factory, indirect impacts that are hard to operationalize, the technology's complexity, and the possible existence of new parts to manufacture.

Performance levels are heavily influenced by the degree of compatibility between the innovation's complexity and the infrastructure's sophistication. Technical complexity is reduced if the innovation can be implemented in stages, and if it is not required to manufacture new parts. General Motors' Hamtramck, Michigan plant ran into costly debugging problems because its new state-of-the-art manufacturing process assembled new car models (Keller 1989). The infrastructure's level of sophistication is determined by some of the same factors that existed during preparation.

Organizational conflict over and resistance to the technology is not automatically dangerous. It allows inappropriate innovations to be modified or disregarded. The extent of conflict and resistance depends upon the amount of change taking place in the infrastructure. New people, occupational specialties, and procedures create changes in the roles, powers, and statuses attached to organizational positions (Hage 1980). For example, Gerwin (1984) found that role conflict occurred in an American firm with an FMS where operating personnel resisted close tolerances set for them by product engineers. Power conflict also occurred as shop floor supervisors became more dependent on the accounting and maintenance departments. In a British bearing manufacturer with a rudimentary direct numerical control system, status conflict occurred when young process planners who had volunteered to do programming were paid more than older process planners who had not volunteered.

Subunits that control the new technology deal with important uncertainties affecting the activities of others. They stand to gain power, whereas subunits that handled uncertainties through obsolete means will lose power, especially if the former are linked with many other subunits and no alternative exists that can perform their services (Hickson et al. 1971). However, short-run uncertainties introduced with the innovation threaten, at least temporarily, to impair the ability to handle other units' long-run requirements.

All the methods that exist for implementing a radical manufacturing technology have problems (Hage 1980). Revolution demands abundant financial resources and leads to sabotage unless a crisis exists, the organization is centralized, and there are few technical specialists. Evolution produces meaningful results only over a long period of time, and then only if the innovation has not been watered down. It presupposes that the technology is divisible, the organization is decentralized, and there are numerous technical experts. Isolation is often infeasible due to financial constraints and the need for a sophisticated well-knit infrastructure. A German aircraft manufacturer, which was doubling its capacity, augmented its existing facility rather than build a new plant in part because the former already had a well-developed support system (Gerwin 1984). Isolation is used primarily to avoid significant resistance from the existing labor force. An American manufacturer of automative parts could not successfully implement a robotic welding line in an existing factory because workers feared losing their jobs. When the line became part of a new factory with new workers it was running smoothly in a few months (Beatty and Gordon 1988).

Strategic management's support for a new technology is generally believed to influence the rank and file to help in implementation or at least not to resist. It has the greatest impact on people who have little need for the technology or who display low competence on the job. Use by individuals with a high need or high competence depends upon the innovation's accessibility (Leonard-Barton and Deschamps 1988).

Initial feedback from the control systems should indicate sizeable gaps. Expectations are high, spurred on by the innovation's advocates, and performance is low, due to the mismatch between technical complexity and the infrastructure's sophistication. Reducing the gaps is an exercise in organizational learning. The prevailing uncertainty and other factors cause learning to occur very slowly and wastefully. Consider that certain groups are not allowed to meaningfully participate in the production process, it is difficult to determine the impacts of changes because so many occur simultaneously, and environmental fluctuations constantly alter what is expected.

At least three coping strategies exist for reducing the gaps. First, the technology is altered to better fit the organization. This type of change usually boosts performance directly, but it also augments technical complexity producing a negative indirect impact on performance. Second, the organization is adjusted to better match the technology. Any change in the infrastructure, however, while improving performance directly also leads to more conflict and resistance which negatively impacts on performance. Third, the nature and levels of expectations are adjusted. The organization learns on which dimensions the innovation does well, and how much can reasonably be expected of it. Winch (1989) cited the example of a British firm that stopped measuring the productivity of a new CAD/CAM system when it became apparent the expected gains would not materialize. By then management had decided the real purpose of the system was to reduce lead times.

In the final analysis, although the innovation is utilized, conditions may be too uncertain to know whether implementation is a success. The firm studied by Gerwin (1981a) had not found a completely reliable basis for computing accounting standards even after seven years of FMS operations. Cancellation tends not to occur due to the heavy financial and psychological costs already incurred, even though they are sunk. An exception occurs when large discrepancies exist that require additional large financial commitments to remove.

Suggestions for Redesign

How can a firm better decide whether or not to purchase a radical manufacturing technology, and then whether or not to permanently use it? One overall suggestion is to establish a new role of (manufacturing) process manager analogous to a product manager. The role would cope with uncertainties from the identification of a need until the technology is phased out, with different individuals occupying it over time. A second overall suggestion is for the process manager to incorporate significant time allowances for unforeseen events into any plans.

The process manager would function as a champion who assembles and leads the project team, sells the project within the company, gathers needed resources, deals with organizational and technical obstacles, and interacts with the vendor. The champion's intense commitment to the proposed innovation must be tempered by the realization that excessive zeal may lead a firm into the wrong choice. In large organizations a division of labor exists between the manufacturing process champion who is frequently technically oriented and hierarchically inferior, a midlevel executive champion who acts as a liaison with strategic management, and a senior level sponsor who protects the project (Dean 1987).

In the adoption phase it is paramount to eliminate strategic management's short-term orientation, but that is amenable to change mainly in the long run. Some specific recommendations include utilizing planning systems such as Texas Instruments' Objectives, Strategies, and Tactics (OST) System, which balances short- and long-term considerations (Jelinek and Schoonhoven 1990), introducing more manufacturing people into strategic management and establishing a special fund outside of the normal capital appropriation process to finance experimental projects. The Ford Motor Company established a five-year $75 million fund to encourage the development of new manufacturing technologies. Purchasing new manufacturing technology that is not so technically complex and building up the infrastructure's sophistication are also helpful.

There are a few useful suggestions for revamping preparation and implementation. Little can be done, due to the innovation's radicalism, about increasing information on how to change the infrastructure. However, conditions should exist that facilitate designing a new support system. Encourage participation of affected units in the task force, avoid undue pressures for immediate full-scale production, and eschew turnkey projects. Installing a new technology in stages and limiting the number of new parts to be manufactured reduces technical complexity which in turn stimulates higher performance. Learning is stimulated during implementation by allowing the task force to remain with a new integrated system long after installation, ensuring that operators are engineers with multifunctional responsibilities, and establishing a shop floor reporting system that includes soft data gathered from experience (Jaikumar 1986, Jonsson 1987). Only the most necessary changes should occur in control systems at the same time that the new technology is implemented. In some new auto assembly plants operators cannot verify whether a signaled problem indicates defective auto bodies or a deficiency in the computer-aided inspection equipment (Goodman et al. 1990).

INTEGRATION OF THE KEY FACTORS

The significance of strategy, organization, and innovation for AMTs fulfillment is more accepted today but by no means universally so. In the future, it will also be necessary to learn the ways in which these three factors interact. If decisions in one arena foreclose options in another, one must consider the interrelated system of the three variables to successfully manage AMT. Unfortunately, very little research has inquired into interrelationships.

One exception is a conceptual model developed by Sorge and Streeck (1988). Their model traces out some interconnections among product strategy, work organization, and technical change for AMT from the viewpoints of different factions in a company. Managers want the interactions to adhere to the following sequence. Market signals influence product strategy and design. Market signals and product design affect decisions on process technology. Process technology impacts on work organization, which in turn influences the quantity and quality of labor.

From a trade union's perspective the ideal sequence is initiated in the labor market. By assuming control over a firm's labor supply the union prescribes wage levels, skill structures, and conditions of employment. Management must then adjust work organization to reflect these outcomes. Process technology decisions are based on the labor supply and organization of work. Product strategy and design follow from the nature of process technology.

Although they did not explain why, Sorge and Streeck believed that in reality relationships among the model's variables are more characterized by mutual dependence than by one way causation. At least three possible reasons come to mind. First, management and union each pushes for its own opposing linear sequence. The result is some degree of mutual interaction among the variables depending on the balance of power between the two factions. Second, the current nature of competition is forcing a company to integrate its internal activities leading to more mutual interaction among them. Third, new approaches for solving competitive problems emphasize mutual interaction. Sociotechnical systems experiments lead to more joint consideration of process technology and work organization. Simultaneous engineering creates more mutuality in choices on product design and process technology. Computer-based technologies lend themselves to more intensive integration.

The Flexibility Cycle

Are there certain values for each of the three key variables—strategy, organization, and innovation—which when linked together facilitate the successful management of AMT? Necessary linkages among any two of them have been discussed in the literature but have not been put together into a three-way relationship. Doing so represents one way to summarize in a concise fashion some of this book's major themes.

Consider two way interactions:

1. A strategy based on flexibility leads to a work organization in which employees with broad general skills participate in groups that have the capacity to adapt

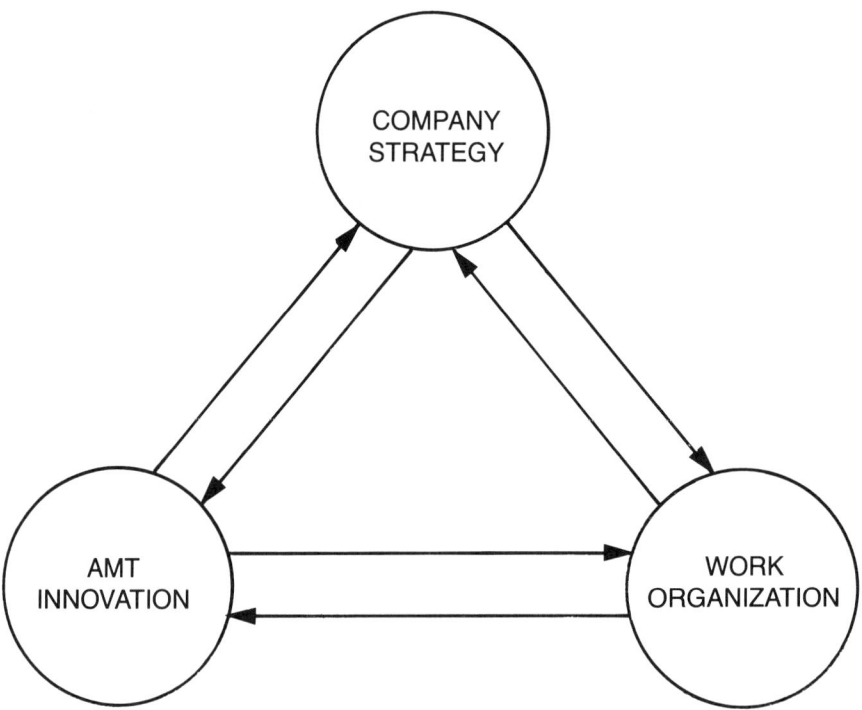

FIGURE 13.1. The flexibility cycle.

to unanticipated circumstances. Simultaneously, this type of work organization encourages a strategy oriented toward flexibility.
2. An adaptable work organization lessens the resistance of employees to AMT innovations. If displaced, people with broad general skills can readily adjust to a job transfer. Management is therefore more willing to provide employment security which further diminishes resistance to change (Sorge and Streeck 1988). At the same time management can use the occasion of AMT innovation to devise a more adaptable work organization.
3. AMT innovation gives management the opportunity to develop strategies that emphasize flexibility, and these same strategies influence process technology choices towards AMT. Note that a strategy based on flexibility implies a long-term orientation, which impacts directly on the chances of adopting AMT in the innovation theory.

When the two-way interactions are linked together they form a mutually reinforcing cycle of flexibility. As indicted in Figure 13.1:

1. A strategy emphasizing flexibility makes for an adaptable work organization that lessens resistance to AMT innovations, which in turn supports the strategy.
2. A strategy based on flexibility influences process technology choices toward AMT. These choices support an adaptable work organization which in turn facilitates the strategy.

Companies that establish a well-functioning cycle of this nature will have a sound basis for competing in the 1990s and beyond.

APPENDIX 1
CHARACTERISTICS OF SOME EMPIRICAL STUDIES

Study	Characteristics
Aguren and Edgren (1980)	A synthesis of case studies to examine organizational innovation in a variety of technologies in over 20 manufacturing plants, with the organizational emphasis on product focused forms
Bennett et al. (1987)	Case studies in seven American firms with NC, CAD/CAM, FMS, and AMH
Betcherman, Newton, and Godin (1990)	Nine case studies in Canada that combine innovation in AMT in a range of technologies with best practices in human resource management
Boer and During (1987), and Boer, Hill, and Krabbendam (1988)	Longitudinal case studies of three Dutch and three British firms with FMSs
Collins, Hull, and Hage (1989)	Statistical analyses from a random sample of 54 New Jersey manufacturing plants in 1981, some of which had CAM technology—mostly robotics and CNC
Dean (1987)	Real time and retrospective case studies in five firms considering the adoption of some form of AMT including CAD, robotics, and MRP II
Ettlie (1988b)	Statistical analyses from interviews in 39 U.S. firms that planned or implemented CAM systems such as FMS and FAS

CHARACTERISTICS OF SOME EMPIRICAL STUDIES

Study	Characteristics
Gerwin and Tarondeau (1989)	Statistical analyses from a mail survey of 81 French firms using some type of CAM technology such as NC, CNC, PCs, and robotics
Gerwin and Tarondeau (1986)	Case studies in seven French and American auto fabrication and assembly activities possessing some type of CAM technology
Graham and Rosenthal (1986a)	Case studies in eight American firms that had recently adopted FMSs
Halpern (1984)	A comprehensive case description by the organizational designer of a greenfield continuous process plant
Hicks (1986)	Statistical analyses from a mail survey of 1200 U.S. metal working firms with <250 employees and possessing ≥1 NC or CNC machine
Howell et al. (1987)	Descriptive statistics from a mail survey of about 350 U.S. firms with some form of AMT such as CAD, robotics, NC, and FMS
Jaikumar (1986)	Descriptive statistics from interviews in 35 American and 60 Japanese firms, all with FMSs
Kelley and Brooks (1988)	Statistical analyses from a 1987 mail survey of a stratified random sample of over 1000 U.S. firms of which 44 percent had at least one example of NC, CNC, or FMS
Klein (1991)	An extended case study of the impact on autonomy of imposing JIT in a high commitment work system with considerable CAM technology
Krafcik and MacDuffie (1989)	Statistical analysis of interviews and quantitative data from 52 automobile assembly plants on five continents with a range of high and low robot utilization
Leonard-Barton (1988)	Case studies (mostly retrospective) of the implementation in large corporations of 12 new technologies including expert systems, MRP II, and CAE
Painter (1990)	A synthesis of 16 case studies in office and factory settings where new technology

(*Table continues on p. 326.*)

Study	Characteristics
Painter (1990)(*Continued*)	has been used to innovate in the redesign of jobs
Rosenthal (1984)	Descriptive statistics from mail surveys of 57 users of, 38 suppliers of, and 64 experts on various forms of CAM technology including CNC, DNC, PCs, AS/RS, robotics, CAD/CAM, and FMS
Swamidass and Newell (1987)	Statistical analyses based on interviews with corporate and manufacturing executives in 35 U.S. (Seattle, Washington area) machinery and machine tool firms engaged in small batch production
Wall, Corbett, Martin, Clegg, and Jackson (1990)	A statistical analysis of a longitudinal change intervention in a CNC printed circuit assembly plant

APPENDIX 2
AMERICAN MANUFACTURING ACTIVITIES IN THE PILOT STUDY

In the pilot study the American body framing activities are labeled BA1 and BA2. When the interviews began BA1 had been assembling a new small car model for about six months using a computerized system for body framing. Framing of unitized bodies commences with a "toy tabbing" operation in which body sides and underbodies are loosely fitted together by hand. Essentially, metal tabs on one component are inserted through slots in another component. The body is then conveyed to an automated body framing unit, which is a transfer line with eight stations. An electric eye identifies the particular body style so that the appropriate fixturing mechanism, consisting of gates on either side of the body, will move in to clamp it rigidly in place. Critical welds are placed by six robots and some fixed automation in order to establish the correct dimensions. The body is then conveyed to another unit where the roof is placed on by hand and welded with fixed automation. Finally, it is conveyed to an automated respotting line with thirteen stations where fifteen robots and some fixed automation place additional welds. The two transfer lines are each controlled by a programmable controller (PC) that handles operations such as positioning the body in the fixturing mechanism and transferring bodies from one station to another.

BA1 was also assembling another small and a medium size car model using conventional framing techniques. The body on chassis process begins in a shell buck with the hand clamping of roof and body sides together and the manual placing of critical welds. Next, in a decking station critical welds are made by manual and fixed means in order to attach this subassembly to the chassis. Three sets of shell bucks and decking stations operate in parallel. Finally, the entire body frame is sent to a manual respotting operation for additional welds.

When the interviews began in BA2 it had been producing two market brands of small cars for almost a year. The unitized body framing operation begins with toy tabbing where body sides and underbody are loosely fit together using a special conveyor system whose carts have individual drive mechanisms for more or less independent operation. Then an overhead conveyor transports the subassembly to either of two automated framing units depending on body style. The gates of the appropriate fixturing mechanism clamp the body and critical welds are placed in a build station containing eight robots. Next, the body is conveyed to another special conveyor for loading the roof manually, welding the roof using robotics and fixed automation, respotting the entire body using thirty robots, and some minor operations.

The whole system is under hierarchical control using PCs. A master controller keeps track of each body being produced to ensure that the proper fixtures and welding programs are used. It also supervises three local PCs for underbody assembly, automated framing, and roofing and respotting respectively.

Conventional assembly in BA2 was stopped in 1980 when it was decided to refurbish the plant to assemble small cars. It had been producing two full-size car models of the same market brand. The body on chassis framing process was different in concept from that in BA1. A completed underbody was hand clamped onto a cart serving as a fixture on a floor conveyor. Left-hand body sides were hand clamped onto fixtures mounted on a conveyor. Another conveyor handled right-hand body sides. Each body fixture was suspended off the ground and oriented as if the body side was on the car. These conveyors approached the chassis conveyor from either side. The body framing process commenced with its most critical step, the "marriage" of body sides and underbody. The parts were hand clamped together as the three lines met. The fixtured body continued along the floor conveyor where the roof was manually placed and manual welders placed critical spots. A manual respotting line completed the allotment of welds.

The American engine fabrication activity in the pilot study is called EF1. The computerized process was selected to be a new four-cylinder engine line. During the interviews it was moving from pilot to mass production of components for new model vehicles. The main computerized activity is the use of programmable controllers (PCs) for directing machining and transfer operations on the engine lines. Typically, a PC is used for either a transfer line or a large single operation machine. PCs have a stored program consisting of a sequence of IF . . . THEN . . . statements in the form of a numeric code. Inputs to the program are signals that certain events have occurred, for example, a component has been clamped. Outputs are instructions to the machinery such as to transfer a component. The stored program is easily modified by keypunching into the PCs memory.

A six-cylinder line was chosen for the conventional operation. It is closely matched in tooling, design, and methods of operation to the computerized system. Some operations for minor engine components are in common. Its operations are mainly controlled by hard-wired relay panels that require time-consuming rewiring by electricians in order to modify.

In general, the components made in an engine line are machined in parallel. Large parts such as engine blocks are machined mainly by sequences of transfer lines. Non programmable "pick and place" robots perform material handling between the stations. Small parts such as crank shafts are fabricated mainly by sequences of single-operation machines. The equipment is essentially dedicated to particular components, although the single operation machines can be more readily converted to other uses.

APPENDIX 3
QUESTIONNAIRE

As part of a research project being jointly conducted by the University of Wisconsin–Milwaukee and ESSEC, Cergy, France, we are studying the relationships between flexible automation and various aspects of factory operations. These include auto or truck components, innovation, the production process, capacity, material inputs, and work organization. By the term "flexible automation" we mean process equipment that can be relatively easily adapted to a variety of uses through computer programming and related means.

All the information collected will be treated confidentially. Only the researchers will have access to your responses. A report summarizing the results in this and other factories will be made available to you.

When answering the questions please keep the following in mind:

1. Try to indicate what actually is the situation and what could be the situation. For example, we want to know about what is actually produced and what could be produced.
2. Most of the questions involve making comparisons between the situation with flexible equipment and the situation with conventional equipment that manufactures the same or similar automobile or truck components.
3. If you are not sure of an answer or do not know the answer, please tell the interviewer.

PRELIMINARY INFORMATION

1. Current job and previous jobs of respondent, how long in each job
2. Nature of the factory studied
   A. Final products produced

B. Major raw materials and components received as inputs
   C. Major kinds of manufacturing activities (e.g., engine assembly, body building)
   D. Number of employees
   E. When factory was built
3. Nature of the equipment
   A. Flexible equipment
      (a) Identify the major kinds of equipment, the function performed, and how many of each kind there are. Characterize the process (one of a kind, small batch, medium batch, large batch, mass)
      (b) In what years was the equipment introduced? Estimate the percent of the equipment introduced in each of those years.
      (c) What percent of the equipment was developed inside the company versus outside the company?
      (d) Kinds of employees and their numbers
      (e) Capacity of the equipment
   B. Conventional equipment
      (a) Identify the major kinds of equipment, the function performed, how many of each kind there are. Characterize the process (one of a kind, small batch, medium batch, large batch, mass)
      (b) Kinds of employees and their numbers
      (c) Capacity of the equipment
4. Nature of the components produced
   A. By the flexible equipment
      (a) How many different components typically use the equipment? Identify the most important components typically produced. What final products of the factory is each a component of?
      (b) What percent of the production dollar volume that requires these operations:
         • Typically utilizes only this equipment?
         • Typically utilizes this equipment and other equipment?
      (c) What percent of the production dollar volume typically utilizing this equipment:
         • Had been produced in this factory on conventional equipment before?
         • Has been transferred from another factory?
      (d) In what years were the components typically produced on the equipment now first manufactured on it? What is the percent of production dollar volume in each year?
   B. By the conventional equipment
      (a) How many different components typically use the equipment? Identify

the most important components typically produced. What final products of the factory is each a component of?
   (b) What percent of the production dollar volume that requires these operations:
     • Typically utilizes only this equipment?
     • Typically utilizes this equipment and other equipment?
5. Sketch out a flowchart of the production process indicating where the flexible equipment and conventional equipment are located.

I. Component Characteristics

Consider each of the following characteristics first for the flexible equipment and then for the conventional equipment. Indicate how much of a change has occurred in the characteristic when the flexible equipment is compared to the conventional equipment as a baseline.

1. (a) The number of different components manufactured at any one time

Decreased a lot	Decreased	No change	Increased	Increased a lot
−2	−1	0	+1	+2

   (b) The range of difference in the components

Decreased a lot	Decreased	No change	Increased	Increased a lot
−2	−1	0	+1	+2

2. The annual production volume of a component

Decreased a lot	Decreased	No change	Increased	Increased a lot
−2	−1	0	+1	+2

3. The lot size of a component

Decreased a lot	Decreased	No change	Increased	Increased a lot
−2	−1	0	+1	+2

4. The complexity of component design

Decreased a lot	Decreased	No change	Increased	Increased a lot
−2	−1	0	+1	+2

5. The degree to which components are modified to meet customer specifications

Decreased a lot	Decreased	No change	Increased	Increased a lot
−2	−1	0	+1	+2

6. The tolerances used in manufacturing components

Decreased a lot	Decreased	No change	Increased	Increased a lot
−2	−1	0	+1	+2

7. What characteristics of a component make it suitable for flexible automation?

II. Component Design and New Components

Consider each of the following characteristics first for the flexible equipment and then for the conventional equipment. Indicate how much of a change has occurred in the characteristic when the flexible equipment is compared to the conventional equipment as a baseline.

1. (a) The frequency of manufacture of new components (a new component is one that cannot be obtained by a design change in an existing component)

Decreased a lot	Decreased	No change	Increased	Increased a lot
−2	−1	0	+1	+2

(b) The frequency of manufacture of major design changes in existing components

Decreased a lot	Decreased	No change	Increased	Increased a lot
−2	−1	0	+1	+2

(c) The frequency of manufacture of minor design changes in existing components

Decreased a lot	Decreased	No change	Increased	Increased a lot
−2	−1	0	+1	+2

2. (a) The frequency of manufacture of new component introductions relative to design changes in existing components

Decreased a lot	Decreased	No change	Increased	Increased a lot
−2	−1	0	+1	+2

(b) The frequency of manufacture of major design changes relative to minor design changes

Decreased a lot	Decreased	No change	Increased	Increased a lot
−2	−1	0	+1	+2

3. The time needed to adapt the production process to
(a) A new component

Decreased a lot	Decreased	No change	Increased	Increased a lot
−2	−1	0	+1	+2

(b) A major design change in an existing component

Decreased a lot	Decreased	No change	Increased	Increased a lot
−2	−1	0	+1	+2

(c) A minor design change in an existing component

Decreased a lot	Decreased	No change	Increased	Increased a lot
−2	−1	0	+1	+2

4. The constraints exercised by the production equipment on
   (a) The design of a new component

Decreased a lot	Decreased	No change	Increased	Increased a lot
−2	−1	0	+1	+2

   (b) A major design change in an existing component

Decreased a lot	Decreased	No change	Increased	Increased a lot
−2	−1	0	+1	+2

   (c) A minor design change in an existing component

Decreased a lot	Decreased	No change	Increased	Increased a lot
−2	−1	0	+1	+2

5. The time between conception of a new idea and full-scale manufacture for a
   (a) New component

Decreased a lot	Decreased	No change	Increased	Increased a lot
−2	−1	0	+1	+2

   (b) Major design change in an existing component

Decreased a lot	Decreased	No change	Increased	Increased a lot
−2	−1	0	+1	+2

   (c) Minor design change in an existing component

Decreased a lot	Decreased	No change	Increased	Increased a lot
−2	−1	0	+1	+2

APPENDIX 3

6. The time between initial full-scale manufacture and phasing out for a
   (a) New component

Decreased a lot	Decreased	No change	Increased	Increased a lot
−2	−1	0	+1	+2

   (b) Major design change in an existing component

Decreased a lot	Decreased	No change	Increased	Increased a lot
−2	−1	0	+1	+2

   (c) Minor design change in an existing component

Decreased a lot	Decreased	No change	Increased	Increased a lot
−2	−1	0	+1	+2

7. What problems are encountered when:
   (a) A new component is to be manufactured on flexible automation?
   (b) A major design change in an existing component is to be manufactured using flexible automation?

III. Characteristics of the Production Process

1. Maximum automation level is an estimate of the highest level of automation achieved in any machine tool according to Bright's scale:

Type of Machine Response	Level Number	Level of Automation
Responds with action	17	Anticipates action required and adjusts to provide it
Modifies own action over a wide range of variation	16	Corrects performance while operating
	15	Corrects performance after operating
Selects from a limited range of possible prefixed actions	14	Identifies and selects appropriate set of actions
	13	Segregates or rejects according to measurement

QUESTIONNAIRE 337

Type of Machine Response	Level Number	Level of Automation
	12	Changes speed, position, and direction according to measurement signal
Responds with signal	11	Records performance
	10	Signal preselected values of measurement (includes error detection)
	9	Measures characteristic of work
Fixed within the machine	8	Actuated by introduction of work piece or material
	7	Power tool system, remote controlled
	6	Power tool, program control (sequence of fixed functions)
	5	Power tool, fixed cycles (single function)
Variable response initiated by man	4	Power tool, hand control
	3	Powered hand tool
	2	Hand tool
	1	Hand

(a) What is the automation level for the flexible equipment?

(b) What is the automation level for the comparison equipment?

Consider each of the following characteristics first for the flexible equipment and then for the comparison equipment. Indicate how much of a change has occurred in the characteristic when the flexible equipment is compared to the conventional equipment as a baseline.

2. The automation of material handling between machines

Decreased a lot	Decreased	No change	Increased	Increased a lot
−2	−1	0	+1	+2

3. The grouping of production operations into single machines (e.g., the number of different welds performed at the same location)

```
Decreased     Decreased     No change     Increased     Increased
  a lot                                                   a lot
    |             |             |             |             |
   -2            -1             0            +1            +2
```

4. The interdependence between a work station and the preceding and following work stations (Interdependence means that the ability of a work station to carry out its activities depends upon whether the preceding and following work stations are carrying out their activities)

```
Decreased     Decreased     No change     Increased     Increased
  a lot                                                   a lot
    |             |             |             |             |
   -2            -1             0            +1            +2
```

5. The ability to predict daily production volumes

```
Decreased     Decreased     No change     Increased     Increased
  a lot                                                   a lot
    |             |             |             |             |
   -2            -1             0            +1            +2
```

6. The levels of in-process inventories

```
Decreased     Decreased     No change     Increased     Increased
  a lot                                                   a lot
    |             |             |             |             |
   -2            -1             0            +1            +2
```

7. The amount of time that those managers responsible for production devote to problems with the production process

```
Decreased     Decreased     No change     Increased     Increased
  a lot                                                   a lot
    |             |             |             |             |
   -2            -1             0            +1            +2
```

8. The degree of flexibility of the production process, in particular
   (a) Ease in changing the annual production volume for a component

```
Decreased     Decreased     No change     Increased     Increased
  a lot                                                   a lot
    |             |             |             |             |
   -2            -1             0            +1            +2
```

(b) Ease in changing the lot size of a given component

Decreased a lot	Decreased	No change	Increased	Increased a lot
−2	−1	0	+1	+2

(c) Ease in changing the number of different components being manufactured over time

Decreased a lot	Decreased	No change	Increased	Increased a lot
−2	−1	0	+1	+2

(d) Ease in changing the organization of work (the allocation of tasks among the operators)

Decreased a lot	Decreased	No change	Increased	Increased a lot
−2	−1	0	+1	+2

(e) Ease in rerouting components when a machine breakdown occurs

Decreased a lot	Decreased	No change	Increased	Increased a lot
−2	−1	0	+1	+2

(f) Ease in making design changes in existing components

Decreased a lot	Decreased	No change	Increased	Increased a lot
−2	−1	0	+1	+2

(g) Ease in changing the output rate (number of components per hour)

Decreased a lot	Decreased	No change	Increased	Increased a lot
−2	−1	0	+1	+2

9. The amount of time devoted to setting up for the next lot

Decreased a lot	Decreased	No change	Increased	Increased a lot
−2	−1	0	+1	+2

10. The degree of mechanization of quality control operations

Decreased a lot	Decreased	No change	Increased	Increased a lot
−2	−1	0	+1	+2

11. The degree to which an equipment improvement in one spot affects an integrated system of different machines (rather than a single type of machine)

Decreased a lot	Decreased	No change	Increased	Increased a lot
−2	−1	0	+1	+2

12. How often during a typical month you break into scheduled production for a special order

Decreased a lot	Decreased	No change	Increased	Increased a lot
−2	−1	0	+1	+2

13. The output rate (number of components per hour)

Decreased a lot	Decreased	No change	Increased	Increased a lot
−2	−1	0	+1	+2

14. The ability to adhere to tolerances

Decreased a lot	Decreased	No change	Increased	Increased a lot
−2	−1	0	+1	+2

15. What are the major limitations of the flexible equipment?

16. What are your biggest problems with the production equipment in:
    (a) The flexible situation?
    (b) The conventional situation?

IV. Production Capacity

Consider each of the following characteristics first for the flexible equipment and then for the conventional equipment. Indicate how much of a change has occurred in the characteristic when the flexible equipment is compared to the conventional equipment as a baseline.

1. The ability to accurately estimate the capacity limits of your equipment

Decreased a lot	Decreased	No change	Increased	Increased a lot
-2	-1	0	$+1$	$+2$

2. The ability to vary capacity limits (versus being constrained by a rigid limit)

Decreased a lot	Decreased	No change	Increased	Increased a lot
-2	-1	0	$+1$	$+2$

3. By how much has flexible automation:
    (a) Increased capacity? _____ %
    (b) Replaced existing capacity? _____ %
4. How many other locations in your company produce:
    (a) Exactly the same components as the flexible equipment? _____
    (b) Variations of the same components? _____
5. What are your biggest problems with production capacity for:
    (a) The flexible equipment?
    (b) The conventional equipment?

V. Material Inputs

Consider each of the following characteristics first for the flexible equipment and then for the conventional equipment. Indicate how much of a change has occurred in the characteristic when the flexible equipment is compared to the conventional equipment as a baseline.

1. The number of different kinds of materials used in a component

Decreased a lot	Decreased	No change	Increased	Increased a lot
−2	−1	0	+1	+2

2. The number of different sources of material inputs for a component

Decreased a lot	Decreased	No change	Increased	Increased a lot
−2	−1	0	+1	+2

3. The fraction of suppliers outside your company with whom you have contracts lasting for a year or more

Decreased a lot	Decreased	No change	Increased	Increased a lot
−2	−1	0	+1	+2

4. The fraction of your suppliers' production (in $) that is devoted to you

Decreased a lot	Decreased	No change	Increased	Increased a lot
−2	−1	0	+1	+2

5. The fraction of your material inputs (in $) supplied by single sources

Decreased a lot	Decreased	No change	Increased	Increased a lot
−2	−1	0	+1	+2

6. The proportion of material inputs (in $) from inside the company as opposed to outside the company

Decreased a lot	Decreased	No change	Increased	Increased a lot
−2	−1	0	+1	+2

7. The level of material input inventories (in $)

```
Decreased     Decreased    No change    Increased    Increased
a lot                                                 a lot
|_____|_____|_____|_____|
-2            -1            0           +1           +2
```

8. The frequency at which material input orders are placed

```
Decreased     Decreased    No change    Increased    Increased
a lot                                                 a lot
|_____|_____|_____|_____|
-2            -1            0           +1           +2
```

9. The amount of an order for material inputs (in $)

```
Decreased     Decreased    No change    Increased    Increased
a lot                                                 a lot
|_____|_____|_____|_____|
-2            -1            0           +1           +2
```

10. The tolerances defining the quality of material inputs

```
Decreased     Decreased    No change    Increased    Increased
a lot                                                 a lot
|_____|_____|_____|_____|
-2            -1            0           +1           +2
```

11. Explain any changes that have been made in purchasing policies.
12. Discuss any changes in the nature of material inputs.
13. What are your biggest problems with material inputs for:
    (a) The flexible equipment?
    (b) The conventional equipment?

VI. Organization of Work

Consider each of the following characteristics first for the flexible equipment and then for the conventional equipment. Indicate how much of a change has occurred in the characteristic when the flexible equipment is compared to the conventional equipment as a baseline.

1. The skill level of operators

```
Decreased     Decreased    No change    Increased    Increased
a lot                                                 a lot
|_____|_____|_____|_____|
-2            -1            0           +1           +2
```

2. The range of skills used by an operator

Decreased a lot	Decreased	No change	Increased	Increased a lot
−2	−1	0	+1	+2

3. The qualification level of operators in terms of diplomas or certificates

Decreased a lot	Decreased	No change	Increased	Increased a lot
−2	−1	0	+1	+2

4. The amount of inside the company training for operators

Decreased a lot	Decreased	No change	Increased	Increased a lot
−2	−1	0	+1	+2

5. The amount of outside the company training for operators

Decreased a lot	Decreased	No change	Increased	Increased a lot
−2	−1	0	+1	+2

6. The amount of pay for operators

Decreased a lot	Decreased	No change	Increased	Increased a lot
−2	−1	0	+1	+2

7. The amount of time it takes to perform a complete cycle of the tasks of an operator

Decreased a lot	Decreased	No change	Increased	Increased a lot
−2	−1	0	+1	+2

8. The degree of discretion operators have over:
   (a) The execution of their work

Decreased a lot	Decreased	No change	Increased	Increased a lot
−2	−1	0	+1	+2

   (b) The planning of their work

Decreased a lot	Decreased	No change	Increased	Increased a lot
−2	−1	0	+1	+2

   (c) Their work pace

Decreased a lot	Decreased	No change	Increased	Increased a lot
−2	−1	0	+1	+2

9. The amount of the following activities performed by operators:
   (a) Maintenance

Decreased a lot	Decreased	No change	Increased	Increased a lot
−2	−1	0	+1	+2

   (b) Inspection

Decreased a lot	Decreased	No change	Increased	Increased a lot
−2	−1	0	+1	+2

   (c) Computer programming

Decreased a lot	Decreased	No change	Increased	Increased a lot
−2	−1	0	+1	+2

(d) Tool setting

Decreased a lot	Decreased	No change	Increased	Increased a lot
−2	−1	0	+1	+2

(e) Others (please specify)

Decreased a lot	Decreased	No change	Increased	Increased a lot
−2	−1	0	+1	+2

10. The degree to which operators do the same tasks in the same way every day (repetition of tasks)

Decreased a lot	Decreased	No change	Increased	Increased a lot
−2	−1	0	+1	+2

11. The degree to which operators' tasks are well defined and structured

Decreased a lot	Decreased	No change	Increased	Increased a lot
−2	−1	0	+1	+2

12. The amount of direct vs. indirect labor performed

Decreased a lot	Decreased	No change	Increased	Increased a lot
−2	−1	0	+1	+2

13. What changes have occurred in the activities of the following persons?
    (a) Supervisors
    (b) Operators
14. What changes have occurred in the skills of the following persons?
    (a) Supervisors
    (b) Operators
15. What changes have occurred in the training of the following persons?
    (a) Supervisors
    (b) Operators

16. What changes have occurred in the policies and procedures of the following?
    (a) Production scheduling
    (b) Maintenance
    (c) Quality control
    (d) Inventory control
17. What changes have occurred in the reporting of operating information?
18. What are your biggest problems with the organization of work for:
    (a) The flexible equipment?
    (b) The conventional equipment?

VII. Criteria of Effectiveness

Consider each of the following characteristics first for the flexible equipment and then for the conventional equipment. Indicate how much of a change has occurred in the characteristic when the flexible equipment is compared to the conventional equipment as a baseline.

1. (a) The productivity of labor (output/man-hour)

Decreased a lot	Decreased	No change	Increased	Increased a lot
-2	-1	0	+1	+2

(b) Machine utilization

Decreased a lot	Decreased	No change	Increased	Increased a lot
-2	-1	0	+1	+2

(c) Quality and reliability of the components produced

Decreased a lot	Decreased	No change	Increased	Increased a lot
-2	-1	0	+1	+2

(d) The length of time it takes to manufacture a component (including machining time, waiting time, etc.)

Decreased a lot	Decreased	No change	Increased	Increased a lot
-2	-1	0	+1	+2

(e) Costs of manufacturing a component

Decreased a lot	Decreased	No change	Increased	Increased a lot
−2	−1	0	+1	+2

(f) In-process inventory levels

Decreased a lot	Decreased	No change	Increased	Increased a lot
−2	−1	0	+1	+2

(g) Flexibility for volume changes

Decreased a lot	Decreased	No change	Increased	Increased a lot
−2	−1	0	+1	+2

(h) Flexibility for component and component design changes

Decreased a lot	Decreased	No change	Increased	Increased a lot
−2	−1	0	+1	+2

(i) Others (please specify)

Decreased a lot	Decreased	No change	Increased	Increased a lot
−2	−1	0	+1	+2

2. Indicate the priority order for your most important manufacturing criteria (up to five):

For the Conventional Equipment
1. _____
2. _____
3. _____
4. _____
5. _____

For the Flexible Equipment
1. _____
2. _____
3. _____
4. _____
5. _____

VIII. Strategy

1. (a) What capabilities must your unit have in order to satisfy strategic, financial, and marketing requirements over the *next three years*?
   (b) What will be the most difficult capability to acquire—the particular challenge?
   (c) Which capabilities and requirements in the above are different from the past?
2. How does flexible automation help you to or prevent you from developing the necessary capabilities?
3. What major changes are anticipated in flexible automation in the next three years?
4. A. Relative to the degree of flexibility in the flexible process equipment what is the degree of flexibility for:
   (a) Component design changes and new components

```
Much          Less          No           More          Much
less                        difference                 more
  |             |             |             |             |
 -2            -1             0            +1            +2
```

   (b) Production capacity

```
Much          Less          No           More          Much
less                        difference                 more
  |             |             |             |             |
 -2            -1             0            +1            +2
```

   (c) Material inputs

```
Much          Less          No           More          Much
less                        difference                 more
  |             |             |             |             |
 -2            -1             0            +1            +2
```

   (d) Organization of work

```
Much          Less          No           More          Much
less                        difference                 more
  |             |             |             |             |
 -2            -1             0            +1            +2
```

(e) The rest of your process equipment

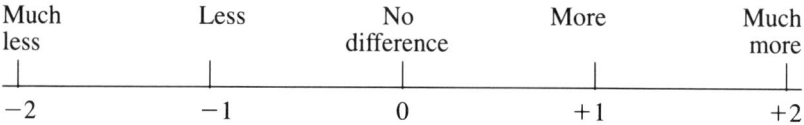

B. What problems arise as the result of any mismatches in flexibility?

C. What actions are being taken to overcome the problems?

5. In what ways is the flexible equipment not living up to its potential? Why?

APPENDIX 4
REASONS FOR THE CHANGES IN FLEXIBILITY IN THE AMERICAN ACTIVITIES OF THE PILOT STUDY

MIX FLEXIBILITY

Mix flexibility was defined as the range of components currently produced. In body framing, changes along this dimension were measured through comparison of the number of components processed on the computerized and conventional systems in each plant. The figures for the actual situation are not very meaningful as they are heavily influenced by short-term fluctuations in market demand. More stable values on the systems' potential are required. Potential is considered in terms of what can be produced with the existing design. It does not include what could be processed with equipment modifications and additions.

In BA1 the potential for the number of different components that can be produced has increased. The computerized system can manufacture four different components, whereas the conventional system handles three after taking into consideration components that are virtually identical.

The number of different components that can be produced by the new system is a function of the fixturing mechanisms on the framing line in conjunction with the geometry of the body being considered for manufacture. It is a question of finding room on a gate to place more clamps without getting in the way of existing clamps. Strictly speaking there is no fixed number for potential. For practical purposes we can consider that each of the two gates of the framing unit can accommodate two body styles. Consequently, the system was operating at its potential. Two- and four-door hatchback styles of a new small car had been in production for about six months. Pilot production for three- and five-door hatchback styles was underway.

The conventional equipment has the potential for a four-door and wagon style of a medium-sized car and a three-door style of a small car. Each of these models can be produced in a regular or deluxe version but with very little difference in bodies. When the interviews began, the two medium-sized models were being produced, but soon after this part of the plant was temporarily closed.

In BA2 there has been a large increase in potential due to a change from three to six different components. The new system has two automated framing units each with three sets of gates and for all practical purposes allows one body style per gate. Currently four gates are being used for a two-door, three-door, four-door, and wagon style of a small car. One of the company's brands uses all four gates, whereas a second brand uses only the four-door one.

The conventional equipment had the potential to produce a two-door, four-door, and wagon style of a full-sized car. This was done for two models of the same brand whose bodies were very similar. One of the brands had a regular and a deluxe wagon the bodies of which were also very similar.

In BA1 the potential range of differences among components has decreased. The new equipment can manufacture from a small two-door body to a small five-door body, whereas the old can handle from a small three-door to a medium wagon. Similarly, in BA2 there has been a decrease in potential. The range of differences embedded in moving from a small two-door body to a small wagon is less than the corresponding move from a full-sized two door to a full-sized wagon. In addition, each of the full-sized models used many unique parts, whereas now the emphasis is on modular construction with interchangeable parts.

In EF1, the engine fabrication activity, the components actually produced were studied. Since all the different parts manufactured by a line are ultimately assembled into a single engine short-term market fluctuations cannot lead to some of them not being produced. Either all are made or none are made.

There has been virtually no change in the number of different kinds of parts manufactured. The four-cylinder line machines essentially the same components as the six-cylinder line. Neither has there been a shift in the range of differences among components. The main difference is that the products of the computerized process are smaller than those of the conventional system.

Changeover Flexibility

Changeover flexibility refers to the ease with which different components are added to the mix and existing components are removed over time. In body framing there has not been much experience with product mix changes during the short period the computerized process has been operational in both plants. The responses from BA1 are based on the initial implementation of the two- and four-door styles, the subsequent addition of the three- and five-door cars, and expectations. The responses from BA2 are based on initial implementation and expectations. As with mix flexibility it is necessary to consider that there can be differences from one manufacturing system to another in the variation among components that can be handled. An attempt was made to control for this factor by focusing only on major design changes in bodies.

In BA1 the ability to process major design changes has decreased because the new equipment has a great deal more hard tooling. The fixturing mechanisms in the framing unit are particularly difficult to retool. Moreover, the new system is considerably more interdependent and complex. Changes made in one part may produce unanticipated side effects elsewhere in the system. Since these problems are difficult to correct there may be serious consequences for system functioning. Adjustments in a fixturing mechanism will lead to a need to redesign the control system, but the latter's complexity makes troubleshooting difficult thus increasing the chances of inadvertently smashing up a machine. There is, however, less of a problem in proving out robotic changes as their control systems do not involve a lot of interrelated functions.

Due to these problems the time to design, build, and implement process changes has increased. It was estimated that implementation takes about twice as long as for the conventional system.

Correspondingly, process constraints on design changes have greatly increased at least as far as structural questions are concerned. There is a good deal more give and take between process engineers and product designers. For example, alternative ways of making a joint exist and one particular approach might have advantages from the process equipment's perspective.

In BA2 the situation is different. There has been a large increase in the ability to accommodate major design changes. This increase was attributed to a change in the philosophy of equipment design. The computerized process is based on a single tooling concept, whereas the conventional process was based on a multiple tooling approach. Now, only one set of gates is used to fixture all the bodies of a particular body style. Before, bodies of a certain style could be mounted on up to forty-eight sets of body side fixtures. Consequently, a great deal less hard tooling is affected by a major design change.

There has been no change in the time needed to handle a major design change. The conventional system required more time for retooling, but the new one needs more time for electrical adjustments including reprogramming of the robots.

Process constraints on structural design changes have increased, but it is the changing nature of the constraints that was of particular concern. Product engineers have traditionally designed bodies to be put together by people. They face an entirely new situation in designing bodies to be assembled by automation and are experiencing some difficulty in adjusting to it. The body is designed so that clamps on the fixturing mechanisms cannot be moved to allow more access for the robots. Due to the design it is not possible to automatically load a part such as the roof which must interlock with other subcomponents. The need for a new way of thinking has been complicated by the fact that the original small-car designs were not made to be assembled by automation. When the decision was made to automate the plant the designs were modified insofar as possible. For example, the toy tabs were added. Now a toy-tabbed body may not sit completely upright because there is not quite enough play in the fit.

In EF1 changeover flexibility was also investigated by inquiring into major design changes. The four-cylinder line has no experience with this type of flexibility as it is just entering mass production. It was estimated that there will be no change in the ease of handling major design changes. It remains at a very low level in both situations due to the dedicated nature of the equipment. However, the PCs can be removed and used to control different machining operations.

The time needed to accommodate a major design change is likely to be shorter due to the ease in changing the sequence of operations when using PCs. They also facilitate experimenting with alternative sequences of operations to discover which one is best. Furthermore, they help to get new equipment operational faster due to their diagnostic capabilities.

MODIFICATION FLEXIBILITY

Modification flexibility refers to the degree to which engineering improvements can be made in a given component and was investigated by inquiring into the ability to make minor design changes. In BA1 this ability has increased because minor design changes are less likely to involve the hard tooling and more likely to involve reprogramming of the robots. Furthermore, minor process adjustments can be taken care of on third shift and weekends so that production is not likely to be affected. With the conventional process it was necessary to change the positions of welding guns, provide new fixturing and retrain workers. The respondent in BA2 perceived a large increase for the same reasons. In BA1 no difference was perceived in the time it takes to make minor design changes. In BA2 a decrease in time was reported.

In EF1 the ability to accommodate minor design changes has increased. With programmable controllers it is easier to make changes in the sequence of operations than with hard-wired automation. This was especially useful in starting up the new four-cylinder line. The time to accommodate minor design changes has also decreased for the same reason.

VOLUME FLEXIBILITY

Volume flexibility refers to the ease with which changes in the aggregate amount of production can be achieved. In BA1 there has been somewhat of an increase in volume flexibility. If the computerized system is operating below capacity then it is simply a matter of changing line speed in order to change the amount of production. Furthermore, running the robots below their maximum speed makes for less repairs and more accuracy. There is no comparable situation for the manual system since it is assumed to be operating at capacity given the existing resources. Rebalancing is necessary whether production is to be increased or decreased.

The new system has somewhat more flexibility than the old in adjusting the capacity limits of welding operations. The robot line can be rebalanced by shifting some spot weld operations to manual welders who are located after the body framing process. With fewer operations to perform the line's speed can be increased. The manual system's capacity limits can be raised in an analogous manner.

The capacity limit of the computerized system's transfer equipment is quite rigid being governed by the maximum line speed obtainable. It is not desirable to invest in a major equipment change in order to augment capacity as production would be stopped for from three to six months. It would be prohibitively expensive to add to capacity by purchasing another system. This potential source of uncertainty has been handled by incorporating a great deal more capacity than that in the conventional system. There is no limitation for all practical purposes. The conventional system's capacity is determined by the fact that four people can work at a shell buck at most. This capacity could be augmented without affecting production on the other shell bucks. It is also much less expensive to add another shell buck-decking station combination. In BA2 the respondents perceived an increase in volume flexibility for similar reasons. Their new system has the ability to quickly increase production if it is not at capacity. It is somewhat easier to rebalance the robots than the manual line. The transfer equipment's capacity has been designed to be very high.

In EF1 volume flexibility remains unchanged, but this is the result of two opposing tendencies cancelling out. The capacity of the six-cylinder line is greater than that of the four-cylinder line, which was judged to produce a slight decrease in the ability to adjust aggregate production.

The ability to vary capacity limits has increased slightly although in both situations it is limited due to the fixed machine cycles. If a machine operating at its capacity is discovered to be a bottleneck, another would be added to operate in parallel. With PCs it is easier to establish the controls that will allow the following machine in the operating sequence to receive dual inputs.

REROUTING FLEXIBILITY

This is defined as the degree to which the sequence of machines through which a component flows can be changed. It arises from the necessity to cope with machine shutdowns due to equipment or quality problems. Ease in rerouting in BA1 has greatly decreased. The increased complexity of the new equipment leads to more frequent shutdowns which take longer to correct. If the shutdown affects the transfer line equipment there is no possibility of rerouting and all production ceases. Due to the three parallel lines of the conventional system there is an option to reroute using overtime if necessary. If one of the framer's robots is shut down, there is a backup manual system or backup robots to take over. If a respotting robot is affected then work can be transferred to manual welders who are stationed after the respot line. The ability to reallocate spot welding also exists for the manual system.

In BA2 ease in rerouting has decreased as the frequency and duration of shutdowns has gone up. If the shutdown affects the transfer equipment rerouting is not possible on such short notice, but stopping one framer would not affect the operation of the other. For the conventional equipment no rerouting existed and all production ceased. The work of only some of the robots can be shifted to human workers because the positioning of the clamps on the fixtures may prevent access to manual welding. If a welding gun broke down in the conventional situation a worker at the next station could make one or two welds to hold everything together and a pickup gun at the end of the line could finish the job.

No change has occurred in rerouting flexibility in EF1. Most critical operations have a backup machine so it is just a matter of transferring the operators. However, PCs, being more reliable than relay controls, break down less. They also possess superior diagnostic capabilities, which helps reduce the amount of downtime.

MATERIAL FLEXIBILITY

This is the ability of a manufacturing system to handle unexpected variations in the dimensional or material quality of the metal being processed. Without material flexibility, interdependence is heightened. Performance at one step affects performance at subsequent ones. In BA1 there has been almost a large decrease in material flexibility and a corresponding increase in interdependence. For example, if the car is not put together correctly during toy tabbing there can be a smashup in the framing unit or the robots will not put welds in the proper position. Since the thinner metal is not as resistant to improperly placed welds it will bend causing further problems down the line. In the conventional system the operators could make the necessary adjustments to properly position the metal or could call for help to fix it.

The situation in BA2 is similar. There has been a large decrease in material flexibility and a large increase in interdependence. Now the emphasis is on doing everything as correctly as possible at each station and if necessary shutting down machines in order to find the causes of quality problems. Previously, human intervention handled many unexpected variations as in BA1.

No change was reported in the low material flexibility of EF1s two lines. The ability to adjust to dimensional and metallurgical variations is about the same since it lies in the similar mechanical and hydraulic features rather than the different control systems. However, a PCs diagnostic capabilities can aid in identifying problems caused by material variations. As would be expected, no change has occurred in the high degree of interdependence within or between transfer lines.

REFERENCES

Abegglen, James C. and George Stalk, Jr. *Kaisha, the Japanese Corporation.* New York: Basic Books, 1985.

Abernathy, William J. *The Productivity Dilemma: Roadblock to Innovation in the Automobile Industry.* Baltimore: The Johns Hopkins University Press, 1978.

Abernathy, William J., Kim B. Clark, and Alan M. Kantrow. *Industrial Renaissance: Producing a Competitive Future for America.* New York: Basic Books, 1983.

Abernathy, William J., Kim B. Clark, and Alan M. Kantrow. "The New Industrial Competition." *Harvard Business Review*, 1981, 68–81.

Abernathy, William, Chair, and participants of the Automotive Panel, Committee on Technology and International Economic and Trade Issues. *The Competitive Status of the U.S. Auto Industry.* Washington, DC: National Academy Press, 1982.

Ackoff, Russell L. and Fred Emery. *On Purposeful Systems.* Chicago: Aldine, 1972.

Adler, Paul. "Workers and Flexible Manufacturing Systems: Three Installations Compared." Working Paper, Stanford University, January 1989a.

Adler, Paul. "CAD/CAM: Managerial Challenges and Research Issues." *IEEE Transactions on Engineering Management*, 36(3), 1989b, 202–215.

Adler, Paul. "Managing Flexible Automation." *California Management Review*, 30, 1988, 34–56.

Adler, Paul. "New Technologies, New Skills." *California Management Review*, 29, 1986, 9–28.

Adler, Paul and Robert Howard. *Technology and the Future of Work.* Palo Alto, CA: Stanford University, Department of Industrial Engineering and Engineering Management, 1990.

Agervald, Olaf. *Exempel pa Monteringssystem for Mindre Montage (Examples of Assembly Systems)* (IVR-resultat 81607). Sweden: Mekanforbund, 1981.

Aguren, Stefan and Jan Edgren. *New Factories.* Stockholm: Swedish Employers' Confederation, 1980.

Aguren, Stefan, Christer Bredbacka, Reine Hansson, Kurt Ihregren, and K. G. Karlsson. *Volvo Kalmar Revisited: Ten Years of Experience*. Stockholm: Efficiency and Participation Development Council, SAF-LO-PTK, 1984.

Aguren, Stefan, Reine Hansson, and K. G. Karlsson. *The Volvo Kalmar Plant: The Impact of New Design on Work Organization*. Stockholm: The Rationalization Council, SAF-LO, 1976

Aldrich, Howard and David A. Whetten. "Organization-sets, Action Sets and Networks: Making the Most of Simplicity." In P. C. Nystrom and W. H. Starbuck, Eds., *Handbook of Organizational Design, Vol. 1*. New York: Oxford University Press, 1981, pp. 385–408.

Argote, Linda, Sara L. Beckman, and Denis Epple. "The Persistence and Transfer of Learning in Industrial Settings." *Management Science*, 36(2), 1990, 140–154.

Argyris, Chris. *The Applicability of Organizational Sociology*. Cambridge, England: Cambridge University Press, 1972.

Arnopoulos, Shiela. *Participative Management in an Advanced Technology Plant: Canadian General Electric in Bromont*. Montreal: McGill Human Resource Associates Inc., 1985.

Ashby, W. Ross. *An Introduction to Cybernetics*. New York: Wiley, 1956.

Ayres, Robert V. *The Next Industrial Revolution: Reviving Industry Through Innovation*. Cambridge, MA: Ballinger, 1984.

Azzone, Giovanni and Umberto Bertele. "Measuring the Economic Effectiveness of Flexible Automation." *International Journal of Production Research*, 27(5), 1989, 735–746.

Baily, Martin Neil and Alok K. Chakrabarti. *Innovation and the Productivity Crisis*. Washington, DC: The Brookings Institution, 1988.

Baldridge, J. Victor and Robert A. Burnham. "Organizational Innovation: Individual, Organizational, and Environmental Impacts." *Administrative Science Quarterly*, 20, 1975, 165–176.

Barley, Stephen R. "Technology as an Occasion for Structuring: Evidence from Observation of CT Scanners and the Social Order of Radiology Departments." *Administrative Science Quarterly*, 31(1), 1986, 78–108.

Beatty, Carol A. and John R. M. Gordon. "Barriers to the Implementation of CAD/CAM Systems." *Sloan Management Review*, 29(4), Summer 1988, 25–33.

Beer, Michael and Bert Spector. "Corporate Wide Transformations in Human Resource Management." In R. E. Walton and P. R. Lawrence, Eds., *Human Resource Management: Trends and Challenges*. Boston, MA: Harvard Business School Press, 1985, pp. 219–254.

Beer, Michael, Bert Spector, Paul R. Lawrence, D. Quinn Mills, and Richard E. Walton. *Human Resource Management: A General Manager's Perspective*. New York: The Free Press, 1985.

Bennett, Robert E., James A. Hendricks, David E. Keys, and Edward J. Rudnicki. *Cost Accounting for Factory Automation*. Montvale, NJ: National Association of Accountants, 1987.

Benson, J. Kenneth. "The Interorganizational Network as a Political Economy." *Administrative Science Quarterly*, 20, 1975, 229–249.

Berger, Peter L. and Thomas Luckmann. *The Social Construction of Reality*. New York: Doubleday, 1966.

Berliner, Collie and James A. Brimson, Eds. *Cost Management for Today's Advanced*

Manufacturing: The CAM-I Conceptual Design. Boston, MA: Harvard Business School Press, 1988.

Bessant, John R. and K. E. Dickson. *Issues in the Adoption of Microelectronics*. London: Pinter, 1982.

Bessant, John, E. Braun, and R. Moseley. "Microelectronics in Manufacturing Industry: The Rate of Diffusion." In T. Forester, Ed., *The Microelectronics Revolution*. Cambridge, MA: The MIT Press, 1981, pp. 198–218.

Bessant, John and Manfred Grunt. *Management and Manufacturing Innovation in the United Kingdom and West Germany*. Aldershot, England: Gower Publishing Co., 1985.

Bessant, John and Bill Haywood. "Flexibility in Manufacturing Systems." *Omega*, 14(6), 1986, 465–473.

Bessant, John and Peter Senker. "Societal Implications of Advanced Manufacturing Technology." In T. D. Wall, C. W. Clegg, and N. J. Kemp, Eds., *The Human Side of Advanced Manufacturing Technology*. London: Wiley, 1987, pp. 153–171.

Betcherman, Gordon, Keith Newton, and Joanne Godin. "Systems and People: Managing Socio-technical Change at Pratt and Whitney Canada," *Two Steps Forward: Human Resource Management in a High-Tech World*. Ottawa: Canadian Government Publishing Centre, 1990, pp. 27–34.

Bjorn-Anderson, Niels and Paul H. Pedersen. "Computer Facilitated Changes in the Management Power Structure," *Accounting, Organization, and Society*, 5, 1980, 203–216.

Blumberg, Melvin and Donald Gerwin. "Coping with Advanced Manufacturing Technology," *Journal of Occupational Behavior*, 5(2), 1984, 113–140.

Blundell, William R. C. "Prescription for the 90s: The Boundaryless Company." *Business Quarterly*, 55(2), Autumn 1990, 71–73.

Boaden, Ruth J. *Justification of Computer-Integrated Manufacturing: Some Insights into Practice*. Manchester, England: Manchester School of Management, UMIST, 1989.

Boddy, David and David A. Buchannan. *Managing New Technology*. Oxford, England: Basil Blackwell, 1986.

Boer, Harry and Willem E. During. "Management of Process Innovation—The Case of FMS: A Systems Approach." *International Journal of Production Research*, 25(11), 1987, 1671–1682.

Boer, Harry, Malcolm R. Hill, and Koos Krabbendam. "It Is One Thing to Promise and Another to Perform." Enschede, Holland: School of Management Studies, University of Twente, 1988.

Bolwijn, P. T., J. Boorsma, Q. H. van Brevkelen, S. Brinkman, and T. Kumpe. *Flexible Manufacturing: Integrating Technological and Social Innovation*. Amsterdam: Elsevier, 1986.

Bomba, Steven J. "FMS Organizing for Success." In C. J. Travora, Ed., *Flexible Assembly Systems Round Table*. Dearborn, MI: Society of Manufacturing Engineers, 1989.

Bower, Joseph. *Managing the Resource Allocation Process*. Boston, MA: Division of Research, Graduate School of Business, Harvard, 1970.

Braverman, Harry. *Labor and Monopoly Capital*. New York: Monthly Review Press, 1974.

Bright, J. C. *Automation and Management*. Boston: Division of Research, Graduate School of Business Administration, Harvard University, 1958.

Browne, Jim, Didier Dubois, Keith Rathmill, Suresh P. Sethi, and Kathryn E. Stecke. "Classification of FMS." *FMS Magazine*, April 1984, 114–117.

Browne, Jim, John Harhen, and James Shivnan. *Production Management Systems: A CIM Perspective*. Reading, MA: Addison-Wesley, 1988.

Buckley, William. *Sociology and Modern Systems Theory*. Englewood Cliffs, NJ: Prentice-Hall, 1967.

Buitendam, Arend. "The Horizontal Perspective of Organization Design and New Technology." In J. M. Pennings and A. Buitendam, Eds., *New Technology as Organizational Innovation: The Development and Diffusion of Microelectronics*. Cambridge, MA: Ballinger, 1987, 59–86.

Bullock, R. J. and Edward E. Lawler. "Gainsharing: A Few Questions and Fewer Answers." *Human Resource Management*, Spring 1984, 23–40.

Burbridge, J. L. *The Introduction of Group Technology*. London: Heinemann, 1975.

Burnes, B. and M. Fitter. "Control of Advanced Manufacturing Technology: Supervision Without Supervisors?" In T. D. Wall, C. W. Clegg, and N. J. Kemp, Eds., *The Human Side of Advanced Manufacturing Technology*. London: Wiley, 1987, pp. 83–99.

Bushe, Gervase R. and A. B. Shani. *Parallel Learning Structures: Increasing Innovation in Bureaucracies*. Reading, MA: Addison-Wesley, 1991.

Butera, Federico. "Environmental Factors in Job and Organization Design: The Case of Olivetti." In E. Davis, and A. B. Cherns, Eds., *The Quality of Working Life*. New York: The Free Press, 1975, pp. 166–200.

Buzacott, John A. "The Fundamental Principles of Flexibility in Manufacturing Systems." *Proceedings of the First International Congress on Flexible Manufacturing Systems*, Brighton, England, 1982, pp. 13–22.

Campbell, Adrian and Malcolm Warner. "Managing Advanced Manufacturing Technology." In M. Warner, W. Wobbe, and P. Broadner, Eds., *New Technology and Manufacturing Management*. Chichester, England: Wiley, 1990, pp. 253–264.

Canada, John R. and William G. Sullivan. *Economic and Multiattribute Evaluation of Advanced Manufacturing Systems*. Englewood Cliffs, NJ: Prentice-Hall, 1989.

Cardozo, R. N. and J. W. Cagley. "Experimental Study of Industrial Buyer Behavior." *Journal of Marketing Research*, 8, 1971, 329–334.

Carlsson, Matts and Lars Trygg. "Assembly Strategies—Some Implications of Engineering Organization, Design Work, Manufacturing System and the Product Itself." Goteborg, Sweden: Department of Industrial Management, Chalmers University, 1987.

Carnall, C. A. "Semi-autonomous Work Groups and the Social Structure of the Organization." *Journal of Management Studies*, 19(3), 1982, 279–294.

Carpenter, A. J. "Computer Aided Manufacture: Overcoming Organizational Inertia for Competitive Advantage." Tewkesbury, England: Dowty Mining Equipment Co., 1988.

Carter, E. E. "The Behavioral Theory of the Firm and Top Level Corporate Decision." *Administrative Science Quarterly*, 16(4), 1971, 413–429.

Carter, Joseph R., Ram Narsimhan, and Shawnee K. Vickery. *International Sourcing for Manufacturing Operations*. Waco, TX: Operations Management Association, 1988.

Carter, Michael F. "Designing Flexibility into Automated Manufacturing Systems." *Proceedings of the Second ORSA/TIMS Conference on Flexible Manufacturing Systems*, Ann Arbor, Michigan: 1986, 107–118.

Cavestro, William. "Beyond the Deskilling Controversy." *Computer Integrated Manufacturing Systems*, 1990, 38–46.

Chandler, Alfred D. J. *Strategy and Structure: Chapters in the History of Industrial Enterprise.* Cambridge: The MIT Press, 1962.

Cherns, Albert B. "The Principles of Sociotechnical Design." *Human Relations*, 29(8), August 1976, 783–792.

Child, John. "Organizational Structure, Environment and Performance: The Role of Strategic Choice." *Sociology*, 6, 1972, 1–22.

Child, John, Hans-Dieter Ganter, and Alfred Kierser. "Technological Innovation and Organizational Conservatism." In J. M. Pennings and A. Buitendam, Eds., *New Technology as Organizational Innovation: The Development and Diffusion of Microelectronics.* Cambridge, MA: Ballinger, 1987, 87–115.

Choate, Pat and J. K. Linger. *The High-Flex Society: Shaping America's Economic Future.* New York: Knopf, 1986.

Clarkson, Max. *Searching.* Paper presented at QWL and the 80s Conference. Toronto, Sept. 1981.

Cohen, Stephen S. and John Zysman. *Manufacturing Matters: The Myth of the Post-Industrial Economy.* New York: Basic Books, 1987.

Cole, Robert E. "The Macropolitics of Organizational Change: A Comparative Analysis of the Spread of Small Group Activities." *Administrative Science Quarterly*, December 1985, 510–585.

Collins, Paul D., Jerald Hage, and Frank Hull. "A Framework for Analyzing Technical Systems in Complex Organizations." *Research in the Sociology of Organizations*, 6, 1988, 81–100.

Collins, Paul D., Frank Hull, and Jerald Hage. "Programmable Automation, Technology Strategy, and Organization: A Profile of Adopters." Lafayette, IN: Krannert School, Purdue University, 1989.

Cooley, Michael. *Problems of Automation.* Amsterdam: North Holland, 1984.

Corbett, J. Martin. "Strategic Options for CIM: Technology-Centered versus Human-Centered Systems Design." *Computer-Integrated Manufacturing Systems*, 1(2), May 1988, 75–81.

Corbett, Martin. "Design for Human-Machine Interfaces." In M. Warner, W. Wobbe, and P. Brodner (Eds.), *New Technology and Manufacturing Management: Strategic Choices for Flexible Production Systems.* Chichester: Wiley, 1990, pp. 113–114.

Cross, Kelvin F. and Richard L. Lynch. "The 'Smart' Way to Define and Sustain Success." *National Productivity Review*, 8(1), 1988/89, 23–33.

Cummings, Thomas. "Self-Regulating Work Groups: A Socio-Technical Synthesis." *Academy of Management Review*, 1978, 625–634.

Cummings, Thomas and Blumberg, Melvin. "Advanced Manufacturing Technology and Work Design." In T. D. Wall, C. W. Clegg, and N. J. Kemp (Eds.), *The Human Side of Manufacturing Technology.* Chichester, England: Wiley, 1987, pp. 37–60.

Cummings, Thomas and Molloy, Edmond S. *Improving Productivity and the Quality of Work Life.* New York: Praeger, 1977.

Cunningham, M. T. and J. G. White. "The Behavior of Industrial Buyers in Their Search for Suppliers of Machine Tools." *Journal of Management Studies*, 1974, 115–128.

Cyert, Richard M. and James G. March. *A Behavioral Theory of the Firm.* Englewood Cliffs, NJ: Prentice-Hall, 1963.

Daft, Richard L. "A Dual-Core Model of Organizational Innovation." *Academy of Management Journal*, 21(2), 1978, 193–210.

Daft, Richard L. *Organization Theory and Design*, St. Paul, MN: West, 1989.

Damanpour, Fariborz, and William M. Evan. "Organizational Innovation and Performance: The Problem of 'Organizational Lag'." *Administrative Science Quarterly*, 29, 1984, 392–409.

Darrow, William P. "A Survey of Flexible Manufacturing Systems Implementation." Gaithersburg, MD: U.S. Department of Commerce, National Bureau of Standards, 1986.

Davis, Louis E. "Workers and Technology: The Necessary Joint Basis for Organizational Effectiveness." *National Productivity Review*, Winter, 1983–84, 7–14.

Davis, Louis E. "Learnings for the Design of New Organizations." In H. Kolodny and H. van Beinum (Eds.), *The Quality of Working Life and the 1980s*. New York: Praeger, 1983, pp. 65–86.

Davis, Louis E. "Organization Design." In G. Salvendy (Ed.), *Handbook of Industrial Engineering*, Atlanta, GA: Industrial Engineering and Management Press, 1982, Ch. 2.1.

Davis, Louis E. "Evolving Alternative Organization Designs: Their Sociotechnical Bases." *Human Relations*, 39(3), 1977, 261–273.

Davis, Louis E., R. R. Canter, and J. Hoffman. "Current Job Design Criteria." In L. E. Davis, and J. C. Taylor, Eds., *Design of Jobs*. Middlesex, England: Penguin Books, 1972.

Davis, Louis E. and Albert B. Cherns. *The Quality of Working Life*. New York: The Free Press, 1975.

Davis, Louis E. and James C. Taylor. *Design of Jobs*. Santa Monica, CA: Goodyear Publishing, 1979.

Davis, Louis E. and James C. Taylor. "Technology, Organization, and Job Structure." In R. Dublin, Ed., *Handbook of Work, Organization and Society*. Skokie, IL: Rand-McNally, 1976.

Davis, Stanley M. and Paul R. Lawrence. *Matrix*. Reading, MA: Addison-Wesley, 1977.

Dean, James W. Jr. *Deciding to Innovate: How Firms Justify Advanced Technology*. Cambridge, MA: Ballinger, 1987.

Dean, James W. and Gerald I. Susman. "Organizing for Manufacturable Design." *Harvard Business Review*, January-February 1989, 28–30, 31, 36.

De Meyer, Arnaud, Jinichiro Nakane, Jeffrey G. Miller, and Kasra Ferdows. "Flexibility: The Next Competitive Battle." Manufacturing Roundtable Research Report Series. Boston University, School of Management, 1987.

De Pietro, Rocco A. and Gina Massaro Schremser. "The Introduction of Advanced Manufacturing Technology (AMT) and Its Impact on Skilled Workers' Perceptions of Communication, Interaction, and Other Job Outcomes at a Large Manufacturing Plant." *IEEE Transactions on Engineering Management*, EM-34(1), 1987, 4–11.

Dertouzos, Michael L., Richard K. Lester, and Robert M. Solow. *Made in America: Regaining the Productive Edge*. Cambridge, MA: The MIT Press, 1989.

D'Iribarne, Alain. "Qualification and Skill Requirements." In M. Warner, W. Wobbe, and P. Brodner (Eds.), *New Technology and Manufacturing Management*, Chichester, England: Wiley, 1990, pp. 61–71.

Dewar, Robert D. and Jerald Hage. "Size, Technology, Complexity and Structural Differentiation: Toward a Theoretical Synthesis." *Administrative Science Quarterly*, 23(1), March, 1978, 111–136.

Dooner, Michael. "Flexibility: A Review and Proposal for Its Expression in a Modelling Environment." Research Report No. 88/03. Milton Keynes, England: The Open University, 1988.

Downs, George W. and Lawrence B. Mohr. "Conceptual Issues in the Study of Innovation." *Administrative Science Quarterly*, 21, 1976, 700–714.

Duncan, Robert B. Characteristics of Organizational Environments and Perceived Environmental Uncertainty." *Administrative Science Quarterly*, 17(3), 1972, 313–327.

Dyer, James S. "Remarks on the Analytic Hierarchy Process." *Management Science*, 36(3), 1990, 249–258.

Edgren, Jan. *Production and Work Organization Design in Sweden: The Developmental Lines*. The Swedish Management Group: PA Council, 1981.

Emery, Fred E. "Characteristics of Sociotechnical Systems." In F. Emery, Ed., *The Emergence of a New Paradigm of Work*, Canberra, Australia: Centre for Continuing Education, Australian National University, 1978a.

Emery, Fred E. *The Emergence of a New Paradigm of Work*. Canberra, Australia: Centre for Continuing Education, Australian National University, 1978b.

Emery, Fred E. "Some Ways in Which Tasks May Be More Effectively Put Together to Make Jobs." In F. Emery, Ed., *The Emergence of a New Paradigm of Work*, Canberra, Australia: Centre for Continuing Education, Australian National University, 1978c.

Emery, Fred E. *Futures We Are In*. Leiden: Martinus Nihoff, 1976.

Emery, Fred E. and Emery, Merrilyn. *Participative Design*. Canberra: Australia: Centre for Continuing Education, Australian National University, 1973.

Emery, Fred and Einar Thorsrud. *Democracy at Work*. Leiden: Martinus Nihoff, 1976.

Emery, Fred E. and Einar Thorsrud. *Form and Content in Industrial Democracy*. London: Tavistock, 1969.

Emery, Fred and Eric L. Trist. The Causal Texture of Organizational Environments. *Human Relations*, 18(21), 1965, 21–32.

Emery, Fred, and Eric L. Trist. "Socio-Technical Systems." In C. W. Churchman and M. Verhurst, Eds., *Management Science Models and Techniques*. London: Pergamon Press, 1960, pp. 83–97.

Ernst, Dieter. *The Global Race in Microelectronics: Innovation and Corporate Strategies in a Period of Crisis*. Frankfurt, Germany: Campus Verlag, 1983.

Ettlie, John E. "What Makes a Manufacturing Firm Innovative?" *Academy of Management Executive*, 4(4), 1990, 7–20.

Ettlie, John E. "Implementation Strategies for Discrete Parts Manufacturing Technologies." University of Michigan: Business School, 1988a.

Ettlie, John E. *Taking Charge of Manufacturing*. San Francisco, CA: Jossey-Bass, 1988b.

Ettlie, John E. "Implementing Manufacturing Technologies: Lessons from Experience." In Donald D. Davies, Ed., *Managing Technological Innovation*. San Francisco, CA: Jossey-Bass, 1986, pp. 72–104.

Ettlie, John E. and Henry W. Stoll. *Managing the Design-Manufacturing Process*. New York: McGraw-Hill, 1990.

Eurojobs. *A Special Issue on the Uddevalla Experience-Exchange Meeting.* S-103 30 Stockholm, Sweden: Eurojobs, 1990.

Farley, John V., Barbara Kahn, Donald R. Lehmann, and William L. Moore. "Modeling the Choice to Automate." *Sloan Management Review*, 28(2), 1987, 5–15.

Faux, Victor and L. Greiner. *Donnelley Mirrors, Inc.* (9-473-088). Cambridge, MA: Harvard Business School Case Services, 1973.

Fennell, Mary L. "Synergy, Influence and Information in the Adoption of Administrative Innovation." *Academy of Management Journal*, 27(1), 1984, 113–129.

Fidler, Lori A. and J. David Johnson. "Communication and Innovation Implementation." *Academy of Management Review*, 9(4), 1984, 704–711.

Fine, Charles, H. "Developments in Manufacturing Technology and Economic Evaluation Models." Cambridge, MA: Sloan School of Management, Massachusetts Institute of Technology, 1989.

Fligstein, Neil. "The Spread of the Multidivisional Form Among Large Firms." *American Sociological Review*, 50, June 1985, 377–391.

Forslin, Jan, Britt-Marie Thulestedt, and Sven Andersson. "Computer-Aided Design: A Case of Strategy in Implementing a New Technology." *IEEE Transactions on Engineering Management*, 36(3), August 1989, 191–201.

Foster, George and Charles T. Horngren. "Flexible Manufacturing Systems: Cost Management and Cost Accounting Implications." Palo Alto, CA: Stanford University, Graduate School of Business, 1988.

Gagnon, Roger J. and Samuel J. Mantel, Jr. "Strategies and Performance Improvement for Computer-Assisted Design." *IEEE Transactions on Engineering Management*, Em-34(4), 1987, 223–235.

Galbraith, Jay. *Designing Complex Organizations.* Reading, MA: Addison-Wesley, 1973.

Galbraith, Jay. "Organization Design: An Information Processing View." In J. Lorsch, and P. Lawrence, Eds., *Organization Planning: Cases and Concepts.* Homewood, IL: Irwin-Dorsey, 1972, pp. 49–74.

Gardell, Bertil and Bjorn Gustavsen. "Work Environment Research and Social Change—Current Developments in Scandinavia." *Journal of Occupational Behaviour*, 1(1), 1980, 3–18.

Gerwin, Donald. "Manufacturing Flexibility in the CAM Era." *Business Horizons*, 32(1), 1989, 78–84.

Gerwin, Donald. "A Theory of Innovation Processes for Computer-Aided Manufacturing Technology." *IEEE Transactions on Engineering Management*, 35(2), 1988, 90–100.

Gerwin, Donald. "An Agenda for Research on the Flexibility of Manufacturing Processes." *International Journal of Operations and Production Management*, 7(1), 1987, 38–49.

Gerwin, Donald. "Organizational Implications of CAM." *Omega*, 13(5), 1985, 443–451.

Gerwin, Donald. "Innovation, Microelectronics and Manufacturing Technology." In Malcom Warner, Ed., *Microprocessors, Manpower and Society.* Aldershot, England: Gower Press, 1984, pp. 66–83.

Gerwin, Donald. "The Do's and Don'ts of Computerized Manufacturing." *Harvard Business Review*, 60(2), 1982, 107–116.

Gerwin, Donald. "Control and Evaluation in the Innovation Process: The Case of Flexible Manufacturing Systems." *IEEE Transactions on Engineering Management*, Em-28(3), 1981a, 62–70.

Gerwin, Donald. "Relationships Between Structure and Technology." In P. C. Nystrom and W. H. Starbuck, Eds. *Handbook of Organizational Design*, Vol. II. New York: Oxford University Press, 1981b, pp. 3–38.

Gerwin, Donald and Jean-Claude Tarondeau. "International Comparisons of Manufacturing Flexibility." In Kasra Ferdows, Ed., *Managing International Manufacturing*. Amsterdam: Elsevier-North Holland, 1989, pp. 169–185.

Gerwin, Donald and Jean-Claude Tarondeau. "Consequences of Programmable Automation of French and American Auto Factories: An International Case Study." In Benjamin Lev, Ed., *Production Management: Methods and Studies*. Amsterdam: Elsevier-North Holland, 1986, pp. 85–98.

Gerwin, Donald and Jean-Claude Tarondeau. "Case Studies of Computer Integrated Manufacturing Systems: A View of Uncertainty and Innovation Processes." *Journal of Operations Management*, 2(2), 1982, 87–99.

Giles, Harold. *Learning for Change in G.E.* Toronto, Ontario: NTL Institute and Organization Development Network, K1A-8A, 1989.

Gold, Bela. "CAM Sets New Rules for Production." *Harvard Business Review*, 60(6), 1982, 88–94.

Gold, Bela. "Revising Managerial Evaluations of Computer-Aided Manufacturing Systems." *Proceedings of the Autofact Western Conference*. Dearborn, MI: Society of Manufacturing Engineers, 1980.

Goldhar, Joel D. and Mariann Jelinek. "Plan for Economics of Scope." *Harvard Business Review*, 61(6), 1983, 141–148.

Goodman, Paul S., Gerri L. Griffith, and Deborah B. Fenner. "Understanding Technology and the Individual in an Organizational Context." In Paul S. Goodman, Lee S. Sproull, and Associates, *Technology and Organizations*. San Francisco: Jossey-Bass, 1990, pp. 45–86.

Gould, Lawrence. "How Flexible Is FMS?" *Managing Automation*, 1986, 57–63.

Graham, Margaret B. W. and Stephen R. Rosenthal. "Institutional Aspects of Process Procurement for Flexible Machining Systems." Boston: Boston University School of Management, 1986a.

Graham, Margaret B. W. and Stephen R. Rosenthal. "Flexible Manufacturing Systems Require Flexible People." *Human Systems Management*, 6(3), 1986b, 211–222.

Greenhalgh, Len, Robert B. McKersie, and R. W. Gilkey. *Rebalancing the Workforce at IBM: A Case Study of Redeployment and Revitalization*. Cambridge, MA: Sloan School of Management, Massachusetts Institute of Technology, 1985.

Grondahl, Peter. *Problem vid overgang till produktverstader (Difficulties in dividing into product shops)* IVF-resultat 79618. Stockholm: Mekanforbund, Nov. 1979, 31 pages.

Groover, Mikell P., Mitchell Weiss, Roger Nagel, and Nicholas Odrey. *Industrial Robots: Technology, Programming and Applications*. New York: McGraw-Hill, 1986.

Gulowson, Jan A. "Measure of Work Group Autonomy." In L. E. Davis and T. J. Taylor (Eds.), *Design of Jobs*, 2nd Ed. Santa Monica: Goodyear Publishing, 1979.

Gunn, Thomas G. *Manufacturing for Competitive Advantage: Becoming a World Class Manufacturer*. Cambridge, MA: Ballinger, 1987.

Gunzberg, Doron. *Industrial Democracy Approaches in Sweden: An Australian View*. Melbourne Productivity Promotion Council of Australia, 1978.

Gupta, Nina, G. Douglas Jenkins, and William P. Curington. "Paying for Knowledge: Myths and Realities." *National Productivity Review*, Spring 1986, 107–123.

Gustavsen, Bjorn and Per H. Engelstad. "The Design of Conferences and the Evolving Role of Democratic Dialogue in Changing Working Life." *Human Relations*, 39(2), 1986, 101–116.

Gustavsson, Sten-Olaf. "Flexibility and Productivity in Complex Production Processes." *International Journal of Production Research*, 22(5), 1984, 801–808.

Gwynne, S. C. "The Right Stuff." *Time*, October 29, 1990, 42–44, 46, 48, 49, 51.

Gyllenhammar, Pehr G. *People at Work*. Reading, MA: Addison-Wesley, 1977.

Haas, Ain. "The Aftermath of Sweden's Codetermination Law: Workers' Experiences in Gothenburg 1977–1980." *Economic and Industrial Democracy*, 4(1), February 1983, 19–46.

Hackman, J. Richard. *Groups that Work (and Those that Don't)*. San Francisco: Jossey-Bass, 1990.

Hackman, J. Richard. "The Design of Work Teams." In J. W. Lorsch, Ed., *Handbook of Organizational Behavior*, 1987, pp. 315–342.

Hackman, J. Richard. "Work Design." In J. R. Hackman and J. L. Suttle, Eds., *Improving Life at Work: Behavioral Science Approaches to Organizational Change*. Santa Monica, CA: Goodyear, 1977, pp. 96–162.

Hackman, J. Richard and Edward E. Lawler. "Employee Reactions to Job Characteristics." *Journal of Applied Psychology*, Monograph 55, 1971, 259–286.

Hackman, J. Richard and G. R. Oldham. *Work Redesign*. Reading, MA: Addison-Wesley, 1980.

Hackman, J. Richard and G. R. Oldham. Motivation Through the Design of Work: Test of a Theory." *Organizational Behavior and Human Performance*, 16, 1976, 250–279.

Hage, Jerald. "Reflections on New Technology and Organizational Change." In J. M. Pennings and A. Buitendam, Eds., *New Technology as Organizational Innovation: The Development and Diffusion of Microelectronics*. Cambridge, MA: Ballinger, 1987, pp. 261–276.

Hage, Jerald. *Theories of Organizations*. New York: Wiley, 1980.

Hage, Jerald. "Routine Technology, Social Structure and Organizational Goals." *Administrative Science Quarterly*, 14(3), September 1969, 366–377.

Hall, Richard H. *Organizations: Structures, Processes and Outcomes*, 4th Ed. Englewood Cliffs, NJ: Prentice-Hall, 1987.

Hall, Robert W. *Attaining Manufacturing Excellence*. Homewood, IL: Dow Jones-Irwin, 1987.

Halpern, Norman. "Sociotechnical Systems Design: The Shell-Sarnia Experience." In B. Cunningham and T. White, Eds. *Quality of Working Life: Contemporary Cases*. Ottawa: Labour Canada, 1984, pp. 31–75.

Harris, Louis. "Bush Isn't a Shoo-In for '92." *The New York Times*, Wednesday, May 23, 1990, p. A29.

Hatvany, Joseph, O. Bjorke, M. E. Merchant, O. I. Semenkov, and H. Yoshikowa. "Advanced Manufacturing Systems in Modern Society." In T. M. R. Ellis and O. I. Semenkov, Eds., *Advances in CAD/CAM*. Amsterdam: North Holland Publishing Co., 1983, pp. 3–26.

Havn, Erling. "Work Organization, Integrated Manufacturing, and CIM System Design." *Computer Integrated Manufacturing Systems*, 3(4), 1990, 230–235.

Hayes, Robert H. "Why Japanese Factories Work." *Harvard Business Review*, July-August 1981, 57–66.

Hayes, Robert H. and William J. Abernathy. "Managing Our Way to Economic Decline." *Harvard Business Review*, July-August 1980, 67–77.

Hayes, Robert H. and Steven C. Wheelwright. *Restoring Our Competitive Edge: Competing Through Manufacturing*. New York: Wiley, 1984.

Hayes, Robert H., Steven C. Wheelwright, and Kim B. Clark. *Dynamic Manufacturing: Creating the Learning Organization*. New York: The Free Press, 1988.

Haywood, Bill and J. Bessant. "Organization and Integration of Production Systems." In M. Warner, W. Wobbe, and P. Brodner, Eds., *New Technology and Manufacturing Management*. Chichester, England: Wiley, 1990.

Hazelhurst, R. J., R. J. Bradbury, and E. N. Corlett. "A Comparison of the Skills of Machinists on Numerically Controlled and Conventional Machines." *Occupational Psychology*, 43, 1969, 169–182.

Hedberg, Bo. "How Organizations Learn and Unlearn." In P. Nystrom, and W. H. Starbuck, Eds., *Handbook of Organizational Design*, Vol. I. Oxford, England: Oxford University Press, 1981, 3-27.

Herbst, Philip G. *Autonomous Group Functioning*. London: Tavistock, 1962.

Hicks, Donald A. *Automation Technology and Industrial Renewal: Adjustment Dynamics in the U.S. Metalworking Sector*. Washington, DC: American Enterprise Institute for Public Policy Research, 1986.

Hickson, David J., C. R. Hinings, C. A. Lee, Rodney E. Schneck, and Johannes M. Pennings. "A Strategic Contingencies Theory of Intraorganizational Power." *Administrative Science Quarterly*, 16(2), 1971, 216–229.

Hickson, David J., Derek S. Pugh, and Dianne C. Pheysey. "Operations Technology and Organizational Structure: An Empirical Reappraisal." *Administrative Science Quarterly*, 14, September 1969, 378–397.

Hill, Paul. *Toward a New Philosophy of Management*. New York: Barnes and Noble, 1972.

Hinings, C. R., David J. Hickson, Johannes M. Pennings, and Rodney E. Schneck. "Structural Conditions of Intraorganizational Power." *Administrative Science Quarterly*, 19(1), 1974, 22–24.

Hirschhorn, Larry. *Beyond Mechanization: Work and Technology in a Postindustrial Age*. Cambridge, MA: The MIT Press, 1984.

Hoerr, John. "Sharpening Minds for a Competitive Edge." *Business Week*, December 17, 1990, 72, 74, 78.

Hottenstein, Michael. "Managing the Design/Manufacturing Interface in Selected High Technology Firms" Working Paper 88-2. Center for the Management of Technological and Organizational Change. Pennsylvania State University, 1988, 18 pages.

Howell, Robert A., James D. Brown, Stephen R. Soucy, and Allen H. Seed. "Management Accounting in the New Manufacturing Environment." Montvale, NJ: National Assocation of Accountants, 1987.

Hull, Frank M., and Paul D. Collins. "High-Technology Batch Production Systems: Woodward's Missing Type." *Academy of Management Journal*, 30(4), 1987, 786–797.

Hundy, B. B. and D. J. Hamblin. "Risk and Assessment of Investment in New Technology." *International Journal of Production Research*, 26(11), 1988, 1799–1810.

Hutchinson, George K., and John R. Holland. "The Economic Value of Flexible Automation." *Journal of Manufacturing Systems*, 1(2), 1982, 215–227.

Hyer, Nancy L. "The Potential of Group Technology for U.S. Manufacturing." *Journal of Operations Management*, 4(3), 1984, 140–149.

Hyer, Nancy L. and Urban Wemmerlov. "Group Technology in the US Manufacturing Industry: A Survey of Current Practices." *International Journal of Production Research*, 27(8), 1989, 1287–1304.

Industrial Management Roundtable. "Concurrent Engineering, Global Competitiveness, and Staying Alive: An Industrial Management Roundtable." *Industrial Management*, July-August 1990, 6–10.

Jaikumar, Ramchandran. "Postindustrial Manufacturing." *Harvard Business Review*, November-December 1986, 69–76.

Jelinek, Mariann. *Institutionalizing Innovation*. New York: Praeger, 1979.

Jelinek, Mariann and Claudia Bird Schoonhoven. *The Innovation Marathon: Lessons from High Technology Firms*. Oxford, England: Basil Blackwell, 1990.

Jenkins, David. *The Age of Job Design: National Patterns of QWL Activities in Western Europe*. Stockholm: Eurojobs, 1982.

Jenkins, David. *Job Power*. New York: Penguin Books, 1973.

Johnson, H. Thomas and Robert S. Kaplan. *Relevance Lost: The Rise and Fall of Management Accounting*. Boston, MA: Harvard Business School Press, 1987.

Johnston, Russell and P. R. Lawrence. "Beyond Vertical Integration: The Rise of the Value-Adding Partnership." *Harvard Business Review*, July-August 1988, 94–101.

Jonsson, Berth. "Corporate Strategy for People at Work: The Volvo Experience." Paper presented at QWL and 1980s Conference. Toronto: University of Toronto, Canada, Sept. 1981.

Jonsson, Sten. "Limits of Information Technology for Facilitating Organizational Learning." In J. M. Pennings, and A. Buitendam, Eds., *New Technology as Organizational Innovation: The Development and Diffusion of Microelectronics*. Cambridge, MA: Ballinger, 1987, pp. 217–234.

Jordan, Nehemiah. "Allocation of Functions Between Man and Machines in Automated Systems." In L. E. Davis and J. C. Taylor, Eds., *Design of Jobs*, 2nd Ed. Santa Monica, CA: Goodyear Publishing, 1979, pp. 6–11.

Kaimann, Richard A. and Barbara A. Bechler. "Emerging Concepts in Production: Part II, Manufacturing Cells." *Industrial Management*, January-February 1983, 7–10.

Kanter, Rosabeth Moss. *The Change Masters*. New York: Simon and Schuster, 1983.

Kaplan, Robert S. "Limitations of Cost Accounting in Advanced Manufacturing Environments." In Robert S. Kaplan, Ed., *Measures for Manufacturing Excellence*. Boston, MA: Harvard Business School Press, 1990, pp. 15–38.

Kaplan, Robert S. "Must CIM Be Justified by Faith Alone?" *Harvard Business Review*, March-April, 1986, 87–95.

Kaplan, Robert S. "Measuring Manufacturing Performance: A New Challenge for Management Accounting Research." *The Accounting Review*, LVIII(4), 1983, 686–705.

Kaplinsky, Raphael. *Automation: The Technology Society*. Harlow, Essex, England: Longman Group Ltd., 1984.

Kapstein, J. and John Hoerr. "Volvo's Radical New Plant: "The Death of the Assembly Line?" *Business Week*, 1989, 92–93.

Karasek, Robert A. "Job Demands, Job Decision Latitude and Mental Strain." *Administrative Science Quarterly*, 24, 1979, 285–308.

Karmarkar, Uday S., Phillip J. Lederer, and Jerold L. Zimmerman. "Choosing Manufacturing Production Control and Cost Accounting Systems." In Robert S. Kaplan, Ed., *Measures for Manufacturing Excellence*. Boston, MA: Harvard Business School Press, 1990, pp. 353–396.

Kearney and Trecker Corporation, Milwaukee, WI. "Applying the Technology of Flexible Production."

Keen, Peter G. W. "Value Analysis: Justifying Decision Support Systems." *MIS Quarterly*, March 1981, 183–197.

Kekre, Sunder and Kannan Srinivasan. "Broader Product Line: A Necessity to Achieve Success?" *Management Science*, 36(10), 1990, 1216–1231.

Keller, Maryann. *Rude Awakening: The Rise, Fall and Struggle for Recovery of General Motors*. New York: William Morrow, 1989.

Kelley, John E. "A Reappraisal of Sociotechnical System Theory." *Human Relations*, 31(12), 1978, 1069–1099.

Kelly, Maryellen R. and Harvey Brooks. "Patterns of Adoption of Programmable Automation Technologies in the U.S. Industrial Base." Cambridge, MA: Harvard University, John F. Kennedy School of Government, 1988.

Khandwalla, Pradip N. "Mass Output Orientation of Operations Technology and Organizational Structure." *Administrative Science Quarterly*, 19, March 1974, 74–97.

Kiggundu, Moses. "Task Interdependence and Job Design." *Organizational Behavior and Human Performance*, 31, 1983, 145–172.

Kimberly, John R. "Organizational and Contextual Influences on the Diffusion of Technological Innovation." In J. M. Pennings, and A. Buitendam, Eds., *New Technology as Organizational Innovation: The Development and Diffusion of Microelectronics*. Cambridge, MA: Ballinger, 1987, pp. 237–259.

Kimberly, John R. "Managerial Innovation." In P. Nystrom, and W. H. Starbuck, Eds., *Handbook of Organizational Design*. Oxford, England: Oxford University Press, 1981.

Kimberly, John R., and M. J. Evanisko. "Organizational Innovation: The Influence of Individual, Organizational, and Contextual Factors on Hospital Adoption of Technical and Administrative Innovations." *Academy of Management Journal*, 24(4), 1981, 689–713.

King, Resa W., James R. Treece and Keith H. Hammonds. "Engineering Made Easy, Thanks to Igor and Friends." *Business Week*, Sept. 3, 1990, 103–104.

Kirby, Philip. A Socio-Technical Approach to Reducing Manufacturing Cycle-Time: A Case Study. Technical Paper MM85-721. Dearborn, MI: Society of Manufacturing Engineers, 1985, 24 pages.

Kirton, J. A. "Implementing Advanced Manufacturing Technology: The Way It Is." In C. A. Voss, Ed., *Managing Advanced Manufacturing Technology*. London, England: IFS Publications, 1986, pp. 57–67.

Klein, Janice A. "A Reexamination of Autonomy in Light of New Manufacturing Practices." *Human Relations*, January 1991, 44(1), 21–38.

Klein, Janice A. "The Human Cost of Manufacturing Reform." *Harvard Business Review*, March-April, 1989, 60, 61, 64–66.

Kochan, Thomas A. "On the Human Side of Technology." *ICL Technical Journal*, November 1988, 391–400.

Kochan, Thomas A., Harry C. Katz, and Robert B. McKersie. *The Transformation of American Industrial Relations*. New York: Basic Books, 1988.

Koelsch, James R. "Flexible Cells Take Control." *Manufacturing Engineering*, Sept. 1990, 75–77.

Kolodny, Harvey F. "Some Characteristics of Organizational Designs in New/High Technology Firms." In L. R. Gomez-Meija and M. W. Lawless, Eds., *Organizational Issues in High Technology Management*. Greenwich, CT: JAI Press, 1990, pp. 165–176.

Kolodny, Harvey. Product Focussed Forms. Toronto Canada: Society of Manufacturing Engineers, CASA, 1986a.

Kolodny, Harvey. "Assembly Cells and Parallelization: Two Swedish Cases." In O. J. Brown and H. W. Hendrick, Eds., *Human Factors in Organization Design and Management-II*. North Holland, Michigan: Elsevier Science Publishers, 1986b.

Kolodny, Harvey F. "Work Organizations in Sweden: Some Impressions from 1982–83." *Human Systems Management*, 1985, 207–219.

Kolodny, Harvey. "Evolution to a Matrix Organization." *Academy of Management Review*, 4(4), 1979, 543–553.

Kolodny, Harvey and Barbara Dresner. "Linking Arrangements and New Work Designs." *Organizational Dynamics*, 14(3), 1986, 33–51.

Kolodny, Harvey and Torbjorn Stjernberg. "The Change Process in Innovative Work Designs: New Design and Redesign in Sweden, Canada and the USA." *Journal of Applied Behavioral Science*, 1986, 287–301.

Kolodny, Harvey and Hans van Beinum. *The Quality of Working Life and the 1980s*. New York: Praeger Publishers, 1983.

Krafcik, John F. and John Paul MacDuffie. *Explaining High Performance Manufacturing: The International Motor Vehicle Assembly Plant Study*. Massachusetts Institute of Technology, International Motor Vehicle Program, 1989.

Kuhn, Thomas S. *The Structure of Scientific Revolutions*. Chicago: The University of Chicago Press, 1962.

Kulatilaka, Nalin. "Valuing the Flexibility of Flexible Manufacturing Systems." *IEEE Transactions on Engineering Management*, 35(4), 1988, 250–257.

Kumar, Vinod. "Entropic Measures of Manufacturing Flexibility." *International Journal of Production Research*, 25(7), 1987, 957–966.

Kumpe, Ted and Piet T. Bolwijn. "Manufacturing: The New Case for Vertical Integration." *Harvard Business Review*, March-April 1988, 75–81.

Lawler, Edward E. *High-Involvement Management*. San Francisco: Jossey-Bass, 1986.

Lawler, Edward E. *Pay and Organization Development*. Reading, MA: Addison-Wesley, 1981.

Lawler, Edward E. and Gerald E. Ledford. "Skill-based Pay: A Concept That's Catching On." *Personnel*, 62(9), September 1985, 30–37.

Lawrence, Paul R. and Davis Dyer. *Renewing American Industry*. New York: The Free Press, 1983.

Lawrence, Paul R. and Jay W. Lorsch. *Organization and Environment: Managing Differentiation and Integration*. Boston: Harvard University, Graduate School of Business Administration, 1967.

Lee, Gloria. "Managing Change with CAD and CAD/CAM." *IEEE Transactions on Engineering Management*, 36(3), August 1989, 227–233.

Leonard-Barton, Dorothy. "Implementation as Mutual Adaptation of Technology and Organization." *Research Policy*, 17, 1988, 251–267.

Leonard-Barton, Dorothy and Isabelle Deschamps. "Managerial Influence in the Implementation of New Technology." *Management Science*, 34(10), 1988, 1252–1265.

Levi, Daniel J., Charles M. Slem, and Andrew Young. "Technological Versus Team Driven Approaches to Implementing Advanced Manufacturing Technology." San Luis Obispo, CA: Department of Psychology, California Polytechnic Institute, 1989.

Levitt, Barbara and James G. March. "Organizational Learning." *Annual Review of Sociology*, 14, 1988, 319–340.

Lim, S. H. "Flexible Manufacturing Systems and Manufacturing Flexibility in the United Kingdom." *International Journal of Operations and Production Management*, 7(6), 1987, 44–54.

Lindholm, Rolf. *Job Reform in Sweden*. Stockholm: Swedish Employers' Confederation, 1975.

Lindholm, Rolf and Sven Flykt. "The Design of Production Systems: New Thinking and New Lines of Development." In G. Kanawaty, Ed., *Managing and Developing New Forms of Work Organization*. Geneva: International Labour Office, 1981, 33–76.

Lindholm, Rolf and Jan-Peder Norstedt. *The Volvo Report*. Stockholm: Swedish Employers' Confederation, 1975.

Liu, Michel, Hélène Denis, Harvey Kolodny, and Bengt Stymne. "Organization Design for Technological Change." *Human Relations*, 43(1), 1990, 7–22.

Lund, Robert T. and John A. Hansen. *Keeping America at Work: Strategies for Employing the New Technologies*. New York: Wiley, 1986.

MacCrimmon, Kenneth R. and D. A. Wehrung. *Taking Risks: The Management of Uncertainty*. New York: The Free Press, 1986.

Madique, Modesto A. "Entrepreneurs, Champions, and Technological Innovation." *Sloan Management Review*, 21(2), 1980, 59–76.

Majchrzak, Ann. "A National Probability Survey on Education and Training for CAD/CAM." *IEEE Transactions on Engineering Management*, EM-33(4), November 1986, 197–206.

Majchrzak, Ann. *The Human Side of Factory Automation*. San Francisco: Jossey-Bass, 1988.

Majchrzak, Ann and Harold Salzman. "Introduction to the Special Issue: Social and Organizational Dimensions of Computer-Aided Design." *IEEE Transactions on Engineering Management*, EM-36(3), August 1989, 174–179.

Mansell, Jacquie and Tom Rankin. *Changing Organizations: The Quality of Working Life Process*. Toronto, Ontario: Ontario Quality of Working Life Centre, 1983.

Mansfield, Edwin. "The Speed and Cost of Industrial Innovation in Japan and the United States: External vs. Internal Technology." *Management Science*, 34(10), 1988, 1157–1168.

Manufacturing Studies Board. *Human Resource Practices for Implementing Advanced Manufacturing Technology*. Washington, D.C.: Manufacturing Studies Board, 1986.

March, James G. and J. P. Olsen. *Ambiguity and Choice in Organizations*. Bergen, Norway: Universitetsforlaget, 1976.

Margirier, Gilles. "Flexible Automation in Machining in France: Results of a Survey."

Grenoble, France: Institute of Economic Research, University of Social Sciences of Grenoble, 1986.

Markus, M. Lynne. *Systems of Organizations: Bugs and Features*. Marshfield, MA: Pitman Publishing, 1984.

Martin, John M. "Cells Drive Manufacturing Strategy." *Manufacturing Engineering*, January 1989, 49–54.

Maton, Bob. "Socio-technical Systems: Conceptual and Implementation Problems." *Relations Industrielles*, 43(4), 1988, 868–888.

Merchant, M. Eugene. "The Inexorable Push for Automated Production." *Production Engineering*, January 1977, 44–49.

Meredith, Jack R. *Managerial Lessons in Factory Automation: Three Case Studies in Flexible Manufacturing Systems*. Monograph 4, *Operations Management Association*. Texas: Waco, 1989.

Meredith, Jack R. "Strategic Planning for Factory Automation by the Championing Process." *IEEE Transactions on Engineering Management*, EM-33(4), 1986, 229–232.

Meredith, Jack R. and Marianne M. Hill. "Justifying New Manufacturing Systems: A Managerial Approach." *Sloan Management Review*, Summer 1987, 49–61.

Meredith, Jack R. and Nallan C. Suresh. "Justification Techniques for Advanced Manufacturing Technologies." *International Journal of Production Research*, 24(5), 1986, 1043–1057.

Meyer, Alan D. and James B. Goes. "Organizational Assimilation of Innovations: A Multilevel Contextual Analysis." *Academy of Management Journal*, 31(4), 1988, 897–923.

Miles, Robert and Charles Snow. "Network Organizations." *California Management Review*, 1986.

Miller, Danny. "Toward a New Contingency Approach." *Journal of Management Studies*, 18, 1981, 1–26.

Miller, Jeffrey G. and Aleda V. Roth. "Manufacturing Strategies: Executive Summary of the 1988 North American Manufacturing Futures Survey." Boston: Boston University School of Management, 1988.

Miller, Jeffrey G. and Aleda V. Roth. "Manufacturing Strategies: Executive Summary of the 1987 North American Manufacturing Futures Survey." Boston: Boston University School of Management, 1987.

Mintzberg, Henry. *Structures in Fives: Designing Effective Organizations*. Englewood Cliffs, NJ: Prentice-Hall, 1983.

Mishne, Patricia P. "A Passion for Perfection." *Manufacturing Engineering*, November 1988, 46–58.

Mohr, Lawrence B. "Innovation Theory: An Assessment from the Vantage Point of New Electronic Technology in Organizations." In J. M. Pennings and A. Buitendam, Eds., *New Technology as Organizational Innovation: The Development and Diffusion of Microelectronics*. Cambridge, MA: Ballinger, 1987, pp. 13–31.

Mohr, Lawrence B. "Organizational Technology and Organizational Structure." *Administrative Science Quarterly*, 16(4), December 1976, 661–674.

Moore, Brian E. and Timothy L. Ross. *The Scanlon Way to Improved Productivity: A Practical Guide*. New York: Wiley, 1978.

More, Roger A. "Supplier/User Interfacing in the Development and Adoption of New Hardware/Software Systems: A Framework for Research." *IEEE Transactions on Engineering Management*, 35(3), 1988, 190–196.

Mumford, Enid and M. Weir. *Computer Systems in Work Design—The ETHICS Method.* New York: Wiley, 1979.

Munro, H. and Hamid Noori. "Measuring Commitment to New Manufacturing Technology: Integrating Technological Push and Marketing Pull Concepts." *IEEE Transactions on Engineering Management*, 35(2), 1988, 63–70.

Nag, Amal. "Tricky Technology." *Wall Street Journal*, 1, May 13, 1986.

Nevins, James L., Daniel E. Whitney, and Thomas L. De Fazio. *Concurrent Design of Products and Processes: A Strategy for the Next Generation in Manufacturing.* New York: McGraw-Hill, 1989.

Nightingale, Donald V. and Richard L. Long. *Gain and Equity Sharing.* Ottawa, Ontario: Labour Canada, 1984.

Noble, David. "Social Choice in Machine Design: The Case of Automatically Controlled Machine Tools." In A. Zimbalist, Ed., *Case Studies in the Labor Process.* New York: Monthly Review Press, 1979, pp. 18–50.

Nohara, Hikari. "Reconsidering the Japanese Production Systems Model." Paper presented at the Workplace Australia Conference, Monash University, Melbourne, March 1, 1991.

Noori, Hamid. *Managing the Dynamics of New Technology: Issues in Manufacturing Management.* Englewood Cliffs, NJ: Prentice-Hall, 1990.

Norstedt, Jan-Peder and Stefan Aguren. *The Saab-Scania Report.* Stockholm: Swedish Employers' Confederation, 1973.

Nunnally, J. C. *Psychometric Theory.* New York: McGraw-Hill, 1978.

Nutt, Paul C. "Tactics of Implementation." *Academy of Management Journal*, 29(2), 1986, 230–261.

Oney, Mark K. "Encouraging New Disciplines in a Flexible Manufacturing System." In C. J. Travora (Ed.), *SME Blue Book Series.* Dearborn, MI: Society of Manufacturing Engineers, 1989, pp. 19–22.

Orne, Daniel L. and Leo E. Hanifin. "International Manufacturing Strategies and Computer Integrated Manufacturing (CIM): A Review of the Emerging Interactive Effects." In B. Lev, Ed., *Production Management: Methods and Studies.* B. V. North Holland, Michigan: Elsevier Science Publishers, 1986, pp. 61–83.

Painter, Bert. *Good Jobs with New Technology.* Vancouver, BC: B.C. Research, 1991.

Painter, Bert. *The Real Possibility of Good Jobs with New Technology.* Vancouver, BC: B.C. Research, 1990.

Palframan, Diane. "FMS: Too Much, Too Soon." *Manufacturing Engineering*, March 1987, 34–38.

Pasmore, William A. and Paul D. Tolchinsky. "Doing It Right from the Start." *The Journal for Quality and Participation*, December 1989, 56–60.

Pasmore, William A. *Designing Effective Organizations: The Sociotechnical Systems Perspective.* New York: Wiley, 1988.

Patterson, W. P. "Where Is Technology Taking Us?" *Industry Week*, May 30, 1983, 30–40.

Pava, Cal H. P. *Managing New Office Technology: An Organizational Strategy.* New York: The Free Press, 1983.

Peganoff, Joseph. *Employment Security.* Jamestown, NY: Cummins Engine Company, 1984.

Pelz, Donald C. and F. C. Munson. "Originality Level and the Innovating Process in Organizations." *Human Systems Management*, 3, 1982, 173–187.

Peters, Tom J. and Robert H. Waterman. *In Search of Excellence*. New York: Warner Books, 1982.

Pfeffer, Jeffrey. "Bringing the Environment Back In: The Social Context of Business Strategy." In David J. Teece, Ed., *The Competitive Challenge: Strategies for Industrial Innovation and Renewal*. Cambridge, MA: Ballinger, 1987, pp. 119–135.

Pfeffer, Jeffrey. *Power in Organization*. Boston, MA: Pitman, 1981.

Pfeffer, Jeffrey and Gerald R. Salancik. *The External Control of Organizations: A Resource Dependence Perspective*. New York: Harper & Row, 1978.

Porter, Michael E. *Competitive Advantage: Creating and Sustaining Superior Performance*. New York: The Free Press, 1985.

Porter, Michael E. "The Technological Dimension of Competitive Strategy." In Richard L. Rosenbloom, Ed., *Research on Technological Innovation, Management and Policy*, Vol. 1. JAI Press, 1983, pp. 1–33.

Poza, Ernesto. "Twelve Actions for Strong U.S. Factories." *Sloan Management Review*, 1983, 27–38.

Primrose, P. L. and R. Leonard. "Predicting Future Developments in Flexible Manufacturing Technology." *International Journal of Production Research*, 26(6), 1988, 1065–1072.

Primrose, P. L. and R. Leonard. "Identifying the Flexibility of FMS Systems." *Proceedings of the 26th International Machine Tool Design and Research Conference*, Manchester, England, September 1986, pp. 117–127.

Qualls, William, Richard W. Olshavsky, and Ronald E. Michaels. "Shortening of the PLC—An Empirical Test." *Journal of Marketing*, 45(4), 1981, 76–80.

Rahimi, M., P. A. Hancock, and Ann Majchrzak. "On Managing the Human Factors Engineering of Hybrid Production Systems." *IEEE Transactions on Engineering Management*, 35(4), 1988, 238–249.

Rankin, Tom. *New Forms of Work Organization: The Challenge for North American Unions*. Toronto: University of Toronto Press, 1990.

Ranney, Joyce. "Bringing Sociotechnical Systems from the Factory to the Office." *National Productivity Review*, Spring 1986, 124–133.

Ranson, Stewart, Bob Hinings, and Royston Greenwood. "The Structuring of Organizational Structures." *Administrative Science Quarterly*, 25(2), March 1980, 1–17.

Reich, Robert B. *The Next American Frontier*. New York: Times Books, 1983.

Rhenman, Eric. *Organization Theory for Long-Range Planning*. New York: Wiley, 1973.

Rice, A. K. *Productivity and Social Organization: The Ahmedabad Experiment*. London: Tavistock Publications, 1958.

Roberts, Edward B. "What We've Learned: Managing Invention and Innovation." *Research/Technology Management*, 31(1), January/February, 1988, 11–30.

Roberts, Karlene H. and William Glick. "The Job Characteristics Approach to Task Design: A Critical Review." *Journal of Applied Psychology*, 66(2), 1981, 193–217.

Roethlisberger, Fritz J. *The Elusive Phenomenon: An Autobiographical Account of My Work in the Field of Organizational Behavior at the Harvard Business School*. Cambridge, MA: Harvard University Graduate School of Business Administration, 1977.

Rohan, Thomas. "Tooling Up For Flexibility." *Industry Week*, September 6, 1982, 34–38.

Rohan, Thomas. "Thinking Small Helps Spur Production." *Industry Week*, September 21, 1981, 98–100.

Rosenthal, L. "Progress Toward the 'Factory of the Future.'" *Journal of Operations Management*, 4(3), 1984.

Rosow, Jerome and Robert Zager. *New Roles for Managers, Part I: Employee Involvement and the Supervisor's Job.* Scarsdale, N.Y.: Work in America Institute, 1989.

Rothwell, Shiela. "Selection and Training for Advanced Manufacturing Technology." In T. D. Wall, C. W. Clegg, and N. J. Kemp, Eds., *The Human Side of Advanced Manufacturing Technology*. Chichester, England: Wiley, 1987, pp. 61–82.

Rubenowitz, Sigvard, Flemming Norggren, and Arnold S. Tannenbaum. "Some Social Psychological Effects of Direct and Indirect Participation in Ten Swedish Companies." *Organization Studies*, 4(3), 1983, 243–259.

Rumelt, Richard. *Strategy, Structure, and Economic Performance*. Boston: Harvard University Graduate School of Business Administration, 1974.

Saaty, Thomas L. *The Analytic Hierarchy Process*. New York: McGraw-Hill, 1980.

Sakurai, Michiharu. "The Influence of Factory Automation on Management Accounting Practices: A Study of Japanese Companies." In Robert S. Kaplan, Ed., *Measures for Manufacturing Excellence*. Boston, MA: Harvard Business School Press, 1990, pp. 39–62.

Salancik, Gerald R. and Jeffrey Pfeffer. "Who Gets Power—and How They Hold Onto It: A Strategic Contingency Model of Power." *Organizational Dynamics*, Winter 1977, 2–21.

Sandberg, Thomas. *Work Organization and Autonomous Groups*. Lund: Liber, 1982.

Savage, Charles M. "CIM Management of the Future: Fifth-Generation Management." *Manufacturing Engineering*, January 1989, 59–63.

Schon, Donald. *Beyond the Stable State*. New York: Basic Books, 1971.

Schonberger, Richard J. *Japanese Manufacturing Techniques*. New York: The Free Press, 1982.

Schonberger, Richard J. "Frugal Manufacturing." *Harvard Business Review*, September-October 1987, 95–100.

Schonberger, Richard J. *Building a Chain of Customers*. The Free Press, 1990.

Schuck, Gloria. "Intelligent Technology, Intelligent Workers: A New Pedagogy for the High Tech Work Place." *Organizational Dynamics*, 1985, 66–79.

Schultz-Wild, Rainer. "Transformation Conditions of Future Factory Structures: Technology, Organization, Education and Vocational Training." *Computer-Integrated Manufacturing Systems*, 1(2), 1988, 82–88.

Schultz-Wild, Rainer. "On the Threshold of Computer-Integrated Manufacturing: Diffusion Trends of CIM Technologies in West German Industries." *Computer Integrated Manufacturing Systems*, 2(4), 1989, 240–248.

Schultz-Wild, Rainer. "Process-related Skills: Future Factory Structures and Training." In M. Warner, W. Wobbe, and P. Brodner, Eds., *New Technology and Manufacturing Management: Strategic Choices for Flexible Production Systems*. Chichester, England: Wiley, 1990, pp. 87–99.

Schumacher, E. *Small Is Beautiful*. London: Blond Ltd., 1973.

Scott, Peter B. "Towards Optimized Robotic Assembly." *Omega*, 12(3), 1984, 283–290.

Scott, William R. "Technology and Structure: An Organizational-Level Perspective." In P. S. Goodman, L. S. Sproull, and Associates, Eds., *Technology and Organizations*. San Francisco: Jossey-Bass, 1990, pp. 109–143.

Senker, Peter and Mark Beesley. "Computer Aided Production and Inventory Control Systems: Training Needs for Successful Implementation." Occasional Paper 13. Watford, England: Engineering Industry Training Board, 1985.

Servan-Schreiber, Jean Jacques. *The American Challenge*. New York: Atheneum, 1968.

Sethi, Andrea Krasa, and Suresh Pal Sethi. "Flexibility in Manufacturing: A Survey." *The International Journal of Flexible Manufacturing Systems*, 2, 1990, 289–328.

Shaiken, Harley. *Work Transformed: Automation and Labor in the Computer Age*. New York: Holt, Rinehart and Winston, 1984.

Sheridan, John H. "World-Class Manufacturing: More Than Just 'Playing With the Big Boys.'" *Industry Week*, July 2, 1990, 36–38, 40–46.

Sherwood, John J. "Creating Work Cultures with Competitive Advantage." *Organizational Dynamics*, Winter 1988, 5–27.

Silverman, David. *The Theory of Organizations: A Sociological Framework*. New York: Basic Books, 1981.

Skeats, Arthur E. "Employment Security: Is It Free? *National Productivity Review*, Autumn 1987, 307–313.

Skinner, Wickham. *Manufacturing in the Corporate Strategy*. New York: John Wiley, 1978.

Skinner, Wickham. *Manufacturing: The Formidable Competitive Weapon*. New York: Wiley, 1985.

Skinner, Wickham. "Operations Technology: Blind Spot in Strategic Management." *Interfaces*, 14(1), 1984, 116–125.

Skinner, Wickham. "The Focussed Factory." *Harvard Business Review*, May-June 1974, 113–121.

Skoog, Hans and Ulf Holmquist. "Matching the Equipment to the Job." *Assembly Automation*, November 1983.

Slack, Nigel. "Manufacturing Systems Flexibility: An Assessment Procedure." *Computer-Integrated Manufacturing Systems*, 1(1), 1988, 25–31.

Slack, Nigel. "Flexibility of Manufacturing Systems." *International Journal of Operations and Production Management*, 7(4), 1987, 35–45.

Slack, Nigel. "Flexibility as a Manufacturing Objective." *International Journal of Operations and Production Management*, 3(3), 1983, 4–13.

Sloan, Alfred. *My Years with General Motors*. New York: Doubleday, 1964.

SME Sociotechnical Study Committee. *Challenges and Opportunities for Manufacturing Engineers*. Dearborn, Michigan: Society of Manufacturing Engineers, 1989.

Sorge, Arndt, Gert Hartman, Malcolm Warner, and Ian Nicholas. *Microelectronics and Manpower in Manufacturing*. Aldershot, England: Gower, 1983.

Sorge, Arndt and Wolfgang Streeck. "Industrial Relations and Technical Change: The Case for an Extended Perspective." In R. Hyman and W. Streeck, Eds., *New Technology and Industrial Relations*. Oxford, England: Basil Blackwell, 1988.

Spector, Bert, and Michael Beer. "Sedalia Engine Plant (A)." In M. Beer, B. Spector, P. R. Lawrence, D. Q. Mills, and R. E. Walton, Eds., *Human Resource Management: A General Manager's Perspective*. New York: The Free Press, 1985, pp. 607–640.

Sproull, Lee S. and Paul S. Goodman. "Technology and Organizations: Integration and Opportunities." In Paul S. Goodman, Lee S. Sproull and Associates (Eds.), *Technology and Organizations*, San Francisco: Jossey-Bass, 1990, pp. 254–265.

Stalk, George, Jr. "Time—The New Source of Competitive Advantage." *Harvard Business Review*, July-August 1988, 41–51.

Stalk, George Jr. and Thomas M. Hout. *Competing Against Time*. New York: The Free Press, 1990.

Stjernberg, Torbjorn. *Oragnizational Change and the Quality of Life*. Stockholm: Economic Research Institute, Stockholm School of Economics, 1977.

Stjernberg, Torbjorn and A. Philips. *Long-term Effects of Organizational Developments—QWL Efforts Since the Early '70s in Swedish Companies*. Stockholm: Economic Research Institute, Stockholm School of Economics, 1984.

Stymne, Bengt. "Demands of the Training of Managers in Situations of Profound Technological Change: The Case of Teli.'" Paper presented at the EGOS Colloquium, Berlin, July 11–14, 1989.

Susman, Gerald I. "Design for Manufacturability." Advanced Manufacturing Forum Highlight No. 17, University Park, PA: Pennsylvania State University Press, 1989.

Susman, Gerald I. *Product Life Cycle Management*. University Park, PA: Pennsylvania State University Business School, 1988.

Susman, Gerald I. *Autonomy at Work: A Sociotechnical Analysis of Participative Management*. New York: Praeger Publishers, 1976.

Susman, Gerald I. and James W. Dean. "Strategic Uses of Computer-Integrated Manufacturing in the Emerging Competitive Environment." *Computer-Integrated Manufacturing Systems*, 2(3), 1989, 133–138.

Susman, Gerald I. and Richard B. Chase. "A Sociotechnical Analysis of the Integrated Factory." *Journal of Applied Behavioral Science*, 22(3), 1986, 257–270.

Swamidass, Paul M. *Manufacturing Flexibility*. Operations Management Association, Monograph No. 2, 1988.

Swamidass, Paul M. and William T. Newell. "Manufacturing Strategy, Environmental Uncertainty, and Performance: A Path Analytic Model." *Management Science*, 33(4), 1987, 509–524.

Swedish Work Environment Fund. *Towards a Learning Organization*. Stockholm: Swedish Work Environment Fund, 1988.

Swenson, Peter W. "Fast Track to CIM: Automated Machining." *Manufacturing Engineering*. December 1990, 31–33.

Symon, Gillian. "Human-Centered Computer Integrated Manufacturing." *Computer Integrated Manufacturing Systems*, 3(4), 1990, 223–229.

Talaysum, Adil T., M. Zica Hassan, and Joel D. Goldhar. "Uncertainty Reduction Through Flexible Manufacturing." *IEEE Transactions on Engineering Management*, EM-34(2), 1987, 85–91.

Taylor, Frederick W. *The Principles of Scientific Management*. New York: Harper & Row, 1911.

Taylor, James C. "Long-Term Sociotechnical Systems Change in a Computer Operations Department." *The Journal of Applied Behavioral Science*, 22(3), 1986, 303–313.

Taylor, James C. and Robert A. Asadorian. "The Implementation of Excellence: STS Management." *Industrial Management*, July–August 1985, 5–15.

Thompson, James D. *Organizations in Action*. New York: McGraw-Hill, 1967.

Thorsrud, Einar. "A Strategy for Research and Social Change in Industry: A Report on the

Industrial Democracy Project in Norway." *Social Science Information*, 9(5), 1970, 65–90.

Tombak, Mihkel and Arnaud DeMeyer. "Flexibility and FMS: An Empirical Analysis." *IEEE Transactions on Engineering Management*, EM-35(2), 1988, 101–107.

Tornatzky, Louis D., J. D. Eveland, M. G. Boylan, William A. Hetzner, E. C. Johnson, David Roitman, and J. Schneider. *The Process of Technical Innovation: Reviewing the Literature*. Washington, DC: National Science Foundation, 1983.

Tornatzky, Louis D. and Mitchell Fleischer. *The Processes of Technological Innovation*. Lexington, MA: Lexington Books, 1990.

Tornatzky, Louis D. and K. J. Klein. "Innovation Characteristics and Innovation Adoption-Implementation." *IEEE Transactions on Engineering Management*, EM-29(1), 1982, 28–45.

Tosi, Henry and Lisa Tosi. "What Managers Need to Know About Knowledge-Based Pay." *Organizational Dynamics*, Winter 1986, 52–64.

Trist, Eric L. "The Evolution of Socio-Technical Systems." *Issues in the Quality of Working Life*, Vol. 2. Toronto, Ontario: Ontario Quality of Working Life Centre, 1981.

Trist, Eric L., & Bamforth, K. "The Social and Psychological Consequences of the Long Wall Method of Coal Getting." *Human Relations*, 4, 1951, 3–38.

Trist, Eric, G. W. Higgin, Henry Murray, and A. B. Pollock. *Organizational Choice*. London: Tavistock, 1963.

Trist, Eric and Henry Murray. Eds. *The Social Engagement of Social Science: A Tavistock Anthology. Vol. I: The Socio-Psychological Perspective*. Philadelphia: The University of Pennsylvania Press, 1990.

Trist, Eric L., Gerald I. Susman, and Grant R. Brown. An Experiment in Autonomous Working in an American Underground Coal Mine." *Human Relations*, 30(3), 1977, 201–36.

Ulrich, Karl et al., *A Framework for Including the Value of Time in Design-For-Manufacturing Decision Making*. Cambridge, MA: Sloan School of Management, The MIT Press, 1991.

U.S. Congress. *Making Things Better: Competing in Manufacturing*. Washington, DC: Office of Technology Assessment, OTA-ITE-443, February 1990.

U.S. Congress. *Computerized Manufacturing Automation: Employment, Education, and the Workplace*. Washington, DC: Office of Technology Assessment, OTA-CIT-235, April 1984.

U.S. Department of Commerce. "Current Industrial Reports: Manufacturing Technology 1988." SMT (88)-1, Bureau of Census, May 1989.

U.S. Department of Commerce. "A Competitive Assessment of the U.S. Flexible Manufacturing Systems Industry." Office of Capital Goods and International Construction Sector Group, International Trade Administration, 1985.

Utterback, James and William J. Abernathy. "A Dynamic Model of Process and Product Innovation." *Omega*, 3(6), 1975, 639–656.

Utterback, James and Linsu Kim. "Invasion of a Stable Business by Radical Innovation." In Paul R. Kleindorfer, Ed., *The Management of Productivity and Technology in Manufacturing*. New York: Plenum Press, 1985, 113–151.

Van de Ven, Andrew H. "Central Problems in the Management of Innovation." *Management Science*, 32(5), 1986, 590–607.

Van de Ven, Andrew H. and Gordon Walker. "The Dynamics of Interorganizational Coordination." *Administrative Science Quarterly*, 29, 1984, 598–621.

Venkatesan, Ravi. "Cummins Engine Flexes Its Factory." *Harvard Business Review*, March-April 1990, 120–127.

Von Bertalanffy, Ludwig. *General System Theory*. New York: George Braziller, 1968.

Von Hippel, Eric. *The Sources of Innovation*. Oxford, England: Oxford University Press, 1988.

Voss, Christopher A. "Implementation: A Key Issue in Manufacturing Technology: The Need for a Field of Study." *Research Policy*, 17, 1988, 55–63.

Voss, Christopher A. *The Management of New Manufacturing Technology*. London, England: London Business School, 1984.

Wabalickis, Roger N. "Justification of FMS with the Analytic Hierarchy Process." *Journal of Manufacturing Systems*, 7(3), 1988, 175–182.

Wall, Toby D., Chris W. Clegg, and Nigel J. Kemp. *The Human Side of Advanced Manufacturing Technology*. Chichester, England: Wiley, 1987.

Wall, Toby D., J. Martin Corbett, Robin Martin, Chris W. Clegg, and Paul R. Jackson. "Advanced Manufacturing Technology, Work Design and Performance: A Change Study." *Journal of Applied Psychology*, 75(6), 1990, 691–697.

Wall, Toby D. and Nigel Kemp. "The Nature and Implications of Advanced Manufacturing Technology: Introduction." In T. D. Wall, C. W. Clegg, and N. J. Kemp, Eds., *The Human Side of Manufacturing*. Chichester, England: Wiley, 1987, pp. 1–14.

Wall, Toby D., Nigel J. Kemp, Paul R. Jackson, and Chris W. Clegg. "Outcomes of Autonomous Workgroups: A Long-Term Field Experiment." *Academy of Management Journal*, 29(2), 1986, 280–304.

Walton, Richard E. "From Control to Commitment in the Workplace." *Harvard Business Review*, March-April 1985, 77–84.

Walton, Richard E. "Social Choice in the Development of Advanced Information Technology." In H. Kolodny and H. van Beinum, Eds., *The Quality of Working Life in the 1980s*. New York: Praeger, 1983.

Walton, Richard E. "Establishing and Maintaining High Commitment Work Systems." In J. John Kimberley, and R. H. Miles, Eds., *The Organizational Life Cycle*. San Francisco: Jossey-Bass, 1980.

Walton, Richard E. "How to Counter Alienation in the Plant." *Harvard Business Review*, November-December 1972, 72–81.

Walton, Richard E. and Paul R. Lawrence. *Human Resource Management Trends and Challenges*. Boston: Harvard Business School, 1985.

Weick, Karl E. "Technology as Equivoque: Sensemaking in New Technologies." In P. S. Goodman, L. S. Sproull and Associates, Eds., *Paul S. Goodman, Lee S. Sproull and Associates*, San Francisco: Jossey-Bass, 1990.

Weick, Karl E. *The Social Psychology of Organizing*. Reading, MA: Addison-Wesley, 1969.

Weiner, Elizabeth, Dean Foust, and Dori Jones. "Why Made-in-America Is Back in Style." *Business Week*, November 7, 1988, 116, 117, 120.

Weisbord, Marvin. *Productive Workplaces: Organizing and Managing for Dignity, Meaning and Community*. San Francisco: Jossey-Bass, 1987.

Weiss, Andrew R. and Philip H. Birnbaum. "Technological Infrastructure and the Implementation of Technological Strategies." *Management Sciences*, 35(8), 1989, 1014–1026.

Wemmerlov, Urban and Nancy L. Hyer. "Cellular Manufacturing in the U.S. Industry: A Survey of Users." *International Journal of Production Research*, 27(9), 1989, 1511–1530.

Wemmerlov, Urban and Nancy L. Hyer. "Research Issues in Cellular Manufacturing." *International Journal of Production Research*, 25(3), 1987, 413–431.

Wernerfelt, Birger and Aneel Karnani. "Competitive Strategy Under Uncertainty." *Strategic Management Journal*, 8, 1987, 187–194.

Wharton, T. J. and Edna W. White. "Flexibility and Automation: Patterns of Evolution." *Operations Management Review*, 6(3–4), 1988, 1–8.

Whitney, Daniel E. "Manufacturing by Design." *Harvard Business Review*, July-August 1988, 83–91.

Whitney, Daniel E. "Real Robots Do Need Jigs." *Harvard Business Review*, May-June 1986, 110–116.

Whitney, Daniel E., James L. Nevins, Thomas L. De Fazio, R. E. Gustavson, R. W. Metzinger, J. M. Rourke and D. S. Seltzer. *The Strategic Approach to Product Design*. Cambridge, MA: Charles Stark Draper Laboratory, 1986.

Wildemann, Horst. "Justification on Strategic Planning for New Technologies." *Human Systems Management*, 6(3), 1986, 253–263.

Willenborg, J. A. M. and J. J. Krabbendam. "Industrial Automation Requires Organizational Adaptations." *International Journal of Production Research*, 25(11), 1987, 1683–1691.

Wilkinson, B. *The Shopfloor Politics of New Technology*. London: Heinemann Educational Books, 1983.

Williams, David J. *Manufacturing Systems*. New York: Halsted Press, 1988.

Winch, Graham. *The Implementation of Integrating Innovations: The Case of CAD/CAM*. Coventry, England: Warwick Business School, University of Warwick, 1989.

Winch, Graham. "Organization Design for CAD/CAM." In Graham Winch, Ed., *Information Technology in Manufacturing Processes: Case Studies in Technological Change*. London: Rossendale, 1983.

Wobbe, Warner. "A European View of Advanced Manufacturing in the United States." In M. Warner, W. Wobbe, and P. Brodner, Eds., *New Technology and Manufacturing Management: Strategic Choices for Flexible Production Systems*. Chichester, England: Wiley, 1990, pp. 227–235.

Womack, James P., Daniel T. Jones, and Daniel Roos. *The Machine That Changed the World*. New York: Rawson Associates, 1990.

Woodward, Joan. *Industrial Organization: Theory and Practice*. London: Oxford University Press, 1965.

Zaltman, Gerald, Robert B. Duncan, and Jonny Holbek. *Innovations and Organizations*. New York: Wiley, 1973.

Zeitz, Gerald. "Interorganizational Dialectics." *Administrative Science Quarterly*, 25, 1980, 72–88.

Zelenovic, Dragutin M. "Flexibility—A Condition for Effective Production Systems." *International Journal of Production Research*, 20(3), 1982, 319–337.

Zuboff, Shoshana. *In the Age of the Smart Machine: The Future of Work and Power*. New York: Basic Books, 1988.

Zuboff, Shoshana. "Automate/Informate: The Two Faces of Intelligent Technology." *Organizational Dynamics*, 1985, 5–18.

Zuboff, Shoshana. "Technologies That Informate: Implications for Human Resource Management in the Computerized Industrial Workplace." In R. E. Walton and P. R. Lawrence, Eds., *Human Resource Management: Trends and Challenges*. Boston, MA: Harvard Business School Press, 1986.

AUTHOR INDEX

Abegglen, J.C., 29, 30, 31
Abernathy, W.J., 2, 13, 15, 23, 24, 44, 93, 95, 99, 100, 101, 102, 103, 104, 107, 137, 138, 231, 237, 239, 306
Ackoff, R.L., 145
Adler, P., 130, 159, 160, 171, 180, 189, 193, 194, 196, 199, 211, 213, 215, 220, 221
Agervald, O., 178, 179
Aguren, S., 117, 162, 171, 172, 176, 177, 178, 179, 181, 183, 214, 324
Aldrich, H., 137
Andersson, S., 116, 130
Argote, L., 293
Argyris, C., 108
Arnopoulos, S., 154, 158, 172, 223
Asadorian, R.A., 147, 153
Ashby, W.R., 125, 135
Ayres, R.V., 15, 103, 238, 239
Azzone, G., 56

Baily, M.N., 14, 16, 239, 240
Baldridge, J.V., 292
Bamforth, K., 143, 144, 149, 171
Barley, S.R., 120, 122
Beatty, C.A., 52, 274, 293, 319
Bechler, B.A., 174
Beckman, S.L., 293
Beer, M., 125, 140, 148, 172, 222, 223, 224, 314
Beesley, M., 25

Bennett, R.E., 51, 73, 238, 240, 241, 250, 281, 324
Benson, J.K., 295
Berger, P.L., 146
Berliner, C., 50, 58, 65, 293
Bertele, U., 56
Bessant, J., 122, 126, 129, 140, 192, 195, 197, 209, 210, 212, 241, 265, 266, 272, 273, 275, 304
Betcherman, G., 172, 173, 316, 324
Birnbaum, P.H., 302
Bjorn-Anderson, N., 270
Blumberg, M., 112, 180, 182, 192, 193, 194, 203, 204, 210, 211, 213, 250, 265
Blundell, W.R.C., 112
Boaden, R.J., 48, 63
Boddy, D., 2, 46, 62, 242, 250, 254, 262
Boer, H., 233, 234, 250, 254, 255, 260, 276, 277, 280, 287, 324
Bolwijn, P.T., 1, 28, 33, 35, 44, 45, 79, 130, 138
Bomba, S.J., 126, 140, 180
Bower, J., 242, 243
Boylan, M.G., 229
Bradbury, R.J., 194
Braun, E., 273
Braverman, H., 192
Bredbacka, C., 177
Bright, J.C., 83, 89, 101, 199
Brimson, J.A., 50, 58, 65, 293
Brooks, H., 232, 235, 238, 281, 293, 325

Brown, J.D., 238
Browne, J., 8, 40, 304
Buchannan, D.A., 2, 46, 62, 242, 250, 254, 262
Buckley, W., 145
Buitendam, A., 130
Bullock, R.J., 222
Burbridge, J.L., 174
Burnes, B., 200, 202
Burnham, R.A., 292
Bushe, G.R., 158
Butera, F., 181
Buzacott, J.A., 78, 81

Cagley, J.W., 253
Campbell, A., 124
Canada, J.R., 53, 58, 60
Canter, R.R., 152, 209
Cardozo, R.N., 253
Carlsson, M., 106
Carnall, C.A., 172
Carpenter, A.J., 234, 235
Carter, E.E., 242, 243
Carter, J.R., 41
Cavestro, W., 192, 218, 219, 220
Chakrabarti, A.K., 14, 16, 239, 240
Chandler, A.D.J., 134
Chase, R.B., 116, 122, 149, 193, 213
Cherns, A.B., 144, 145, 152
Child, J., 22, 121, 256, 257
Choate, P., 15, 93, 238, 239
Clark, K.B., 2, 13, 24, 99, 103, 107, 235, 251, 258
Clarkson, M., 154
Clegg, C.W., 151, 192, 193, 326
Cohen, S.S., 15, 50
Cole, R.E., 171
Collins, P.D., 4, 8, 232, 256
Cooley, M., 192, 194, 205
Corbett, J.M., 158, 189, 197, 206, 207, 326
Corlett, E.N., 194
Cross, K.F., 77, 97, 98, 309
Cummings, T., 112, 150, 194, 210, 213
Cunningham, M.T., 253
Curington, W.P., 155, 222
Cyert, R.M., 243, 263

d'Iribarne, A., 124
Daft, R.L., 142, 291, 292
Damanpour, F., 257, 265
Darrow, W.P., 12
Davis, L.E., 113, 122, 144, 145, 147, 150, 151, 152, 156, 157, 171, 187, 188, 189, 194, 199, 209
Davis, S.M., 99, 117, 131, 135, 136

De Meyer, A., 22, 24, 38
De Fazio, T.L., 10
Dean, J.W., 21, 24, 129, 130, 131, 132, 138, 149, 172, 198, 233, 243, 244, 288, 320, 324
Denis, H., 118, 121, 124, 125, 196, 209, 313
Dertouzos, M.L., 13, 14, 17, 238, 239, 288
Deschamps, I., 276
Dewar, R.D., 122
Dickson, K.E., 265, 266
Dooner, M., 40, 49
Downs, G.W., 290
Dresner, B., 148, 219
Dubois, D., 40
Duffy, W., 156
Duncan, R.B., 121, 232, 235, 268
During, W.E., 233, 234, 250, 254, 255, 260, 280, 287, 324
Dyer, D., 108
Dyer, J.S., 60

Edgren, J., 117, 170, 171, 172, 179, 181, 183, 214, 324
Emery, F.E., 123, 143, 144, 145, 147, 149, 154, 212, 213
Emery, M., 154
Engelstad, P.H., 164
Epple, D., 293
Ettlie, J.E., 70, 76, 78, 131, 180, 220, 250, 265, 295, 324
Eurojobs, 181
Evan, W.M., 257, 265
Evanisko, M.J., 292
Eveland, J.D., 229

Farley, J.V., 70, 241
Faux, V., 223
Fenner, D.B., 188, 192, 290
Ferdows, K., 24
Fidler, L.A., 276
Fine, C.H., 53
Fitter, M., 200, 202
Fleischer, M., 285
Fligstein, N., 121, 134
Flykt, S., 183
Forslin, J., 116, 130
Foster, G., 76, 237, 256, 263
Foust, D., 139

Galbraith, J., 130, 135
Ganter, H.-D., 256
Gardell, B., 164
Gerwin, D., 4, 15, 36, 38, 47, 52, 61, 67, 78, 99, 104, 107, 116, 121, 122, 123, 180, 182,

192, 193, 203, 204, 211, 213, 231, 233, 235, 236, 237, 242, 244, 245, 250, 253, 255, 256, 260, 262, 263, 264, 265, 266, 268, 273, 274, 279, 280, 289, 302, 316, 317, 318, 319, 320, 325
Giles, H., 217
Gilkey, R.W., 224
Glick, W., 211
Godin, J., 172, 324
Goes, J.B., 229, 292
Gold, B., 25, 50, 53, 240
Goldhar, J.D., 32, 33
Goodman, P.S., 122, 188, 192, 290, 321
Gordon, J.R.M., 52, 274, 293, 319
Gould, L., 43, 46, 68
Graham, M.B.W., 43, 68, 70, 180, 233, 234, 235, 251, 253, 254, 297, 298, 302, 325
Greenhalgh, L., 224
Greenwood, R., 122
Greiner, L., 223
Griffith, G.L., 188, 192, 290
Grondahl, P. 184
Groover, M.P., 11, 46
Grunt, M., 241, 272, 275, 304
Gulowson, J.A., 149, 150, 171
Gunn, T.G., 4, 5, 7, 18, 25, 79, 103
Gupta, N., 155, 222
Gustavsen, B., 164
Gustavsson, S.-O., 80
Gwynne, S.C., 140
Gyllenhammar, P.G., 162, 178, 185

Haas, A., 164
Hackman, J.R., 171, 172, 174, 210, 211, 212
Hage, J., 4, 122, 232, 233, 235, 236, 248, 256, 260, 266, 268, 271, 272, 318, 319
Hall, R.H., 17, 119, 121
Hall, R.W., 43
Halpern, N., 155, 158, 172, 196, 201, 216, 325
Hamblin, D.J., 53, 61
Hammonds, K.H., 130
Hancock, P.A., 174
Hanifin, L.E., 8
Hansen, J.A., 190, 192, 194, 195, 196, 199, 206, 210, 212, 214
Hansson, R., 162, 177, 178
Harhen, J., 8, 304
Harris, L., 14
Hartman, G., 204
Hassan, M.Z., 32
Havn, E., 159
Hayes, R.H., 2, 13, 21, 23, 29, 30, 31, 44, 92, 95, 102, 137, 138, 235, 237, 239, 251, 258, 308

Haywood, B., 122, 126, 129, 140, 197
Hazelhurst, R.J., 194
Hedberg, B., 277
Hendricks, J.A., 51
Herbst, P.G., 149
Hetzner, W.A., 229
Hicks, D.A., 13, 99, 104, 106, 107, 232, 280, 325
Hickson, D.J., 77, 122, 123, 269, 319
Higgin, G.W., 121, 143, 265
Hill, P., 265
Hill, M.M., 50, 52, 276, 277, 311, 324
Hinings, C.R., 122, 269
Hirschhorn, L., 4, 113, 123, 188, 207, 208, 304
Hoerr, J., 181, 183, 220
Hoffman, J., 152, 209
Holbek, J., 232, 235, 268
Holland, J.R., 45, 55
Holmquist, U., 11
Horngren, C.T., 76, 237, 256, 263
Hottenstein, M., 130
Hout, T.M., 23, 32
Howard, R., 159, 160, 196, 199, 220, 221
Howell, R.A., 238, 241, 244, 256, 257, 325
Hull, F., 4, 8, 232, 256
Hundy, B.B., 53, 61
Hutchinson, G.K., 45, 55
Hyer, N.L., 155, 170, 171, 174, 175

Ihregren, K., 177
Industrial Management Roundtable, 130

Jackson, P.R., 151, 326
Jaikumar, R., 2, 16, 70, 76, 80, 82, 92, 93, 129, 175, 253, 255, 290, 308, 321, 325
Jelinek, M., 33, 288, 320
Jenkins, D., 162
Jenkins, G.D., 155, 222
Johnson, E.C., 229
Johnson, J.D., 276
Johnson, H.T., 65, 238
Johnston, R., 118, 138
Jones, D., 139
Jones, D.T., 2, 164, 168
Jonsson, B., 140
Jonsson, S., 290, 321
Jordan, N., 143, 194, 207

Kahn, B., 70
Kaimann, R.A., 174
Kanter, R.M., 108
Kantrow, A.M., 2, 24, 99, 103, 107
Kaplan, R.S., 53, 54, 65, 133, 238
Kaplinsky, R., 83, 126

Kapstein, J., 181, 183
Karasek, R.A., 211
Karlsson. K.G., 162, 177, 178
Karmarkar, U.S., 257
Karnani, A., 32
Katz, H.C., 216
Kearney and Trecker Corporation, 249
Keen, P.G.W., 63
Kekre, S., 33
Keller, M., 16, 18, 54, 266, 318
Kelley, J.E., 160
Kelley, M.R., 232, 236, 238, 281, 293, 325
Kemp, N.J., 151, 192, 193
Keys, D.E., 51
Khandwalla, P.N., 122
Kierser, A., 256
Kiggundu, M., 211
Kimberly, John R., 267, 292, 293
King, R.W., 130
Kirby, P., 153, 158, 172
Kirton, J.A., 62, 63, 311
Klein, J.A., 118, 140, 147, 150, 153, 154, 155, 172, 191, 214, 215, 291, 325
Kochan, T.A., 164, 165, 216, 224
Koelsch, J., 126
Kolodny, H., 116, 117, 118, 121, 124, 125, 135, 136, 144, 145, 148, 157, 163, 169, 170, 196, 209, 210, 219, 313
Krabbendam, J.J., 68
Krabbendam, K., 277, 324
Krafcik, J.F., 164, 166, 167, 180, 325
Kulatilaka, N., 49, 57
Kumar, V., 78
Kumpe, T., 130, 138

Lawler, E.E., 144, 155, 211, 222
Lawrence, P.R., 99, 108, 117, 118, 123, 126, 128, 130, 131, 132, 133, 135, 136, 138, 140, 145, 161, 198, 216
Lederer, P.J., 257
Ledford, G.E., 155, 222
Lee, C.A., 269
Lee, G., 130
Lehmann, D.R., 70
Leonard, R., 49, 50, 242
Leonard-Barton, D., 268, 269, 276, 278, 279, 280, 325
Lester, R.K., 13, 14, 17, 288
Levi, D.J., 254
Levitt, B., 278, 293
Lim, S.H., 26, 81
Lindholm, R., 116, 126, 140, 171, 183, 214
Linger, J.K., 15, 93, 238, 239

Liu, M., 118, 121, 124, 125, 196, 209, 313
Long, R.L., 222, 223
Lorsch, J., 123, 126, 128, 130, 131, 133, 135, 145, 161, 198
Luckmann, T., 146
Lund, R.T., 180, 192, 194, 195, 196, 199, 206, 210, 212, 214
Lynch, R.L., 77, 97, 98, 309

MacCrimmon, K.R., 245
MacDuffie, J.P., 164, 166, 167, 180, 325
Maidique, M.A., 235
Majchrzak, A., 130, 174, 180, 188, 191, 192, 216, 217, 218, 220, 221, 254, 267
Mansell, J., 145
Mansfield, E., 28, 35
Manufacturing Studies Board, 116, 125, 136, 216, 221
March, J.G., 243, 263, 277, 278, 279, 293
Margirier, G., 70, 79
Martin, R., 326
Martin, J.M., 129, 167, 175
Maton, B., 160
McKersie, R.B., 216, 224
Meredith, J.R., 50, 52, 53, 61, 63, 233, 234, 235, 281, 289, 311
Meyer, A.D., 229, 292
Michaels, R.E., 33
Miles, R., 137
Miller, D., 122
Miller, J.G., 24, 35, 76, 93
Mills, D.Q., 140
Mintzberg, H., 122, 267
Mishne, P.D., 132, 172
Mohr, L.B., 122, 237, 290, 302, 316
Molloy, E.S., 150
Moore, B.E., 222
Moore, W.L., 70
More, R.A., 295, 301, 302
Moseley, R., 273
Mumford, E., 147, 159
Munro, H., 232, 255
Munson, F.C., 237, 316
Murray, H., 121, 143, 265

Nag, A., 15, 81
Nagel, R., 11
Nakane, J., 24
Nevins, James L., 10
Newell, W.T., 26, 78, 326
Newton, K., 172, 324
Nicholas, I., 204
Nightingale, D.V., 222, 223
Noble, D., 192

Nohara, H., 168
Noori, H., 5, 232, 255
Norggren, F., 164
Norstedt, J.-P., 116, 176
Nunnally, J.C., 76
Nutt, P.C., 98, 271, 275

Odrey, N.O., 11
Oldham, G.R., 174, 210, 211
Olsen, J.P., 277, 279
Olshavsky, R.W., 33
Oney, M.K., 180, 181
Orne, D.L., 8

Painter, B., 176, 182, 191, 200, 201, 202, 203, 204, 205, 220, 221, 325, 326
Palframan, D., 289
Pasmore, W.A., 147, 159
Patterson, W.P., 188, 193
Pava, C.H.P., 147, 159
Pedersen, P.H., 270
Peganoff, J., 224
Pelz, D.C., 237, 316
Pennings, J.M., 269
Peters, T.J., 235, 248
Pfeffer, J., 32, 234, 267, 269, 295
Pheysey, D.C., 77, 122, 123
Pollock, A.B., 121, 143, 265
Porter, M.E., 21, 101
Poza, E., 153
Primrose, P.L., 49, 50, 242
Pugh, D.S., 77, 122, 123

Qualls, W., 33

Rahimi, M., 174
Rankin, T., 145, 216, 224
Ranney, J., 147
Ranson, S., 122
Rathmill, K., 40
Reich, R.B., 1, 15, 70, 93, 103, 238
Rhenman, E., 125, 135
Rice, A.K., 144, 149, 265
Roberts, E.B., 229, 252
Roberts, K.H., 211
Roethlisberger, F.J., 193
Rohan, T., 171, 182, 183
Roitman, D., 229
Roos, D., 2, 164, 168
Rosenthal, S.R., 43, 68, 70, 73, 180, 234, 235, 241, 248, 251, 253, 254, 266, 297, 298, 302, 325, 326
Rosow, J., 152
Ross, T.L. 22

Roth, A.V., 35, 76, 93
Rothwell, S., 192, 216, 217
Rubenowitz, S., 164
Rudnicki, Edward J., 51
Rumelt, R., 134

Saaty, T.L., 58, 60
Sakurai, M., 34, 240, 241
Salancik, G.R., 269, 295
Salzman, H., 130
Sandberg, T., 162, 163
Savage, C.M., 121, 139
Schneck, R.E., 269
Schneider, J., 229
Schon, D., 312
Schonberger, R.J., 41, 92, 198
Schoonhaven, C.B., 320
Schuck, G., 122
Schultz-Wild, R., 11, 124, 197
Schumacher, E., 255, 265
Scott, P.B., 84
Scott, W.R., 122, 123
Seed, A.H., 238
Senker, P., 25, 192, 195, 209, 210, 212
Servan-Schreiber, J.J., 134
Sethi, A.K., 40, 116
Sethi, S.P., 40, 116
Shani, A.B., 158
Sheridan, J.H., 118
Sherwood, J.J., 156
Shivnan, J., 8, 304
Silverman, D., 122
Skeats, A.E., 224
Skinner, W., 21, 22, 29, 31, 82, 130, 140, 171, 182, 183, 248, 265
Skoog, H., 11
Slack, N., 25, 26, 27, 28, 40, 96
Slem, C.M., 254
Sloan, A., 134
SME Sociotechnical Study Committee, 128
Snow, C., 137
Solow, R.M.S., 13, 14, 17, 288
Sorge, A., 204, 321
Soucy, S.R., 238
Spector, B., 125, 140, 148, 172, 222, 224, 314
Sproull, L.S., 122
Srinivasan, K., 33
Stecke, K., 40
Stalk, G., Jr., 23, 29, 30, 31, 32, 306
Stjernberg, T. 118, 157, 210
Stoll, H.W., 131
Streeck, W., 321
Stymne, B., 118, 205, 121, 124, 125, 196, 209, 313

Sullivan, W.G., 53, 58, 60
Suresh, N.C., 53, 61, 63
Susman, G.I., 21, 24, 116, 122, 129, 130, 131, 132, 138, 149, 150, 171, 172, 193, 198, 213, 214
Swamidass, P.M., 22, 26, 41, 78, 326
Swedish Work Environment Fund, 122
Swenson, P.W., 186
Symon, G., 158, 159, 160

Talaysum, A.T., 32
Tannenbaum, A.S., 164
Tarondeau, J.-C., 15, 38, 47, 67, 78, 99, 104, 107, 192, 231, 235, 236, 245, 253, 255, 256, 317, 318, 325
Taylor, F.W., 161
Taylor, J.C., 122, 147, 151, 153, 187, 188, 189, 194, 199
Thompson, J.D., 111, 118, 123, 125, 128, 133, 145, 161, 199, 231
Thorsrud, E., 144, 147, 149, 212, 265
Thulestedt, B.-M., 116, 130
Tolchinsky, P.D., 159
Tombak, M., 22, 38
Tornatsky, L.D., 229, 230, 262, 278, 285, 291
Tosi, L., 222
Tosi, T., 222
Treece, J., 130
Trist, E.L., 119, 121, 143, 144, 145, 147, 149, 171, 208, 212, 255, 265
Trygg, L., 106

Ulrich, K., 53
U.S. Congress, 5, 7, 13, 14
U.S. Department of Commerce, 5, 11, 232, 237, 250, 295
Utterback, J., 99

Van de Ven, A.H., 291, 296, 298
Van Beinum, H., 144
Venkatesan, R., 18, 117, 127
Von Bertalanffy, L., 145
Von Hippel, E., 23, 74, 237
Voss, C.A., 262, 287

Wabalickis, R.N., 60
Walker, G., 296, 298
Wall, T.D., 150, 151, 153, 172, 192, 193, 197, 208, 209, 216, 326
Walton, R.E., 140, 144, 145, 146, 147, 172, 216, 316
Warner, M., 124, 204
Waterman, R.H., 235, 248
Wehrung, D.A., 245
Weick, K.E., 121, 122, 135, 189, 190, 194, 196, 197, 199
Weiner, E., 138
Weir, M., 147, 159
Weisbord, M., 154
Weiss, A.R., 302
Weiss, M., 11
Wemmerlov, U., 155, 170, 171, 174, 175
Wernerfelt, B., 32
Wharton, T.J., 22
Wheelwright, S.C., 2, 21, 29, 30, 31, 99, 102, 235, 251, 258, 308
Whetten, D.A., 137
White, E.M., 22
White, J.G., 253
Whitney, D.E., 10, 23, 38, 39, 67, 82, 130
Wildemann, H., 58, 59
Wilkinson, B., 192, 194, 209, 218
Willenborg, J.A.M., 68
Williams, D.J., 11
Winch, G., 244, 280, 320
Womack, J.P., 2, 164, 166, 168, 180, 313
Woodward, J., 123, 161, 231

Young, A., 254

Zager, R., 152
Zaltman, G., 232, 235, 268
Zeitz, G., 295, 298
Zelenovic, D.M., 22
Zimmerman, J.L., 257
Zuboff, S., 189, 195, 196, 197, 199
Zysman, J. 15, 50

SUBJECT INDEX

Abernathy-Utterback model, 99
Activity analysis, 299
  customer-vendor relations, 301
Adaptation, 94–95
Adoption decision, confidence in task force, 243–245
Adoption process, 235, 246
  hypotheses, 245–247
Advanced manufacturing technology:
  alternative renewal process, 104
  benefits, intangible, 50
  costs, intangible, 51
  description of, 4–8
  emerging roles of, 13–18
  environments, 167
  glossary, 4
  individual level, 186
  integrative characteristics, 190
  introduction, 1–4
  investment per employee, 315
  justification, 49
  strategy, 306
  utilization, 11–13
AGV, 176
Approaches:
  integrative, 16–17, 197, 207, 305
  technological, 16, 22, 304
Assembly, batch and mass, 10–11
Assembly cells, 170, 175–178
Assembly plants, 101

Attentional commitment, 189
Automated technologies, embedded nature, 194
Autonomy, 214–215
  criteria of, 150

Boundaries:
  between operators and engineers, 129
  de-layering, 128
  de-massing, 128
  flattening out, 128
  organizational, collapse of, 128, 141, 201
  vertical, 129
Boundary management, 152
Buffer, technical core, 128

CAM environment, 186
CAM infrastructure, 248
CAM innovation, 316
  customer-vendor relationship, 295
  user's viewpoint, 295
CAM technology:
  adoption task force bias, 243
  and changeover flexibility, 43
  divisibility as a basis for implementation, 266
  flexibility, 311
  with mix flexibility, 43
  power implications, 270
Cell teams, 175
Cellular manufacturing, 155, 175
Centralization-decentralization, 134

Ceteris paribus characteristics, 291, 294
Champion, innovation, 233–235
Change in global flexibility, 105
Changing values, 154
Choice in the control of work, 208–209
CIM, 47
CNC operators, 153
Codetermination agreement, 164
Committees, cross-plant, 120
Compatibility, 154
Competitiveness, 13–14
Complementary relationship, automated manufacturing systems, 207
Computer-aided manufacturing, 8–9
Computer-based technologies, skill requirements, 201
Conceptual framework, 94
  innovation, 231
  manufacturing flexibility, 94
Conceptual model, the, 22–26, 41
Concurrent engineering, 130
Conflict:
  power, 268
  and resistance to change, 267
  role, 268
  role changes, 268
Contingency organization:
  models, 123
  national societal factors, 124
  theory, 119, 161
Continuous events, reliability, 190
Coordination:
  computer integrated, 191
  integrating departments, 128
  mechanisms, 130, 135
  organizational, 125, 191
  teams, 128
Corporate culture, 113
Cross-training, 127, 160, 174
Crown Cork and Seal, 150
Cummins Engine Company, 148, 150, 154, 219

Decentralized decision making, 201
Dematurity theory, 104, 106
Democratic participation, 164
Design committee, 156, 172
Design methodology, 147
Design principles, 152, 315
Design process, 152, 156, 157
  comprehensive, integrated organization, 31
  continuous, 140
  participative, 140
Design teams, 154

Design-to-manufacture, 135
  boundary, 112, 129
  interface, 129–132
Deskilling, 192
  criticism of new technology, 192
  curvilinear model, 193
  labor displacement, 192
  or reskilling, 192–194
Diagonal slice, 157
Differentiation, 126, 128
Diffusion:
  of information, 293
  information brokers, 293
  synergistic, 292
Discrepancy analysis, 66–72, 308
Discretion, 188

Engineered organizational design, 162–164
Equipment and process technology, characteristics, 100

Fabrication, batch and mass, 9–10
Factories-within-factories, 140, 171
Factory of the future, 210
FAS, 171
Feedback loops:
  phases I–IV, 96–98
  relationships, 92
Flexibility, 24, 40, 310
  bank, 23, 25
  designing CAM for, 42
  gaps, 97
  hierarchy, 26
  manufacturing, 22
  in practice, 81
  range aspects, 27–28
  strategic implications, 21–22
  time aspects, 27–28
Flexibility, changeover, 33–36, 93, 112, 117, 181
  and CAM technology, 43
  changes in, 86
  range aspects, 33
  scale measurement, 80
  time aspects, 34
Flexibility, changes, 83–91
  changes in, 90
Flexibility, market
  changes in, 90
Flexibility, material, 38, 47, 94, 118
  changes in, 86
  scale measurement, 81
Flexibility, measurement, 73–81
  actual performance, 69, 75
  construct validity, 76

SUBJECT INDEX 391

content validity, 75
reliability, 76
required performance, 66, 74
scales, 77–81
Flexibility, mix, 28–33, 74, 93, 116, 181
  changes in, 86
  scale measurement, 78
  with CAM technology, 43
Flexibility, modification:
  changes in, 90
  scale measurement, 36, 44, 81, 93, 117
Flexibility, rerouting:
  changes in, 86
  range, aspects, 37
  scale measurement, 37, 46, 81, 94, 117, 181
  time aspects, 37
Flexibility, volume:
  changes in, 86
  scale measurement, 37, 81, 93, 117, 136, 182
Flexibility cycle, 322
Flexible assembly systems, 170, 181–182
Flexible factory, 95–99
Flexible focused factories, 170, 182–184
Flexible manufacturing cells, 174
Flexible manufacturing systems, 170, 180–181
FMS, 167, 171, 180
Focused factories, 29, 163, 182, 308
Fragile/lean, 166
Functional forms, 133
Functional organization, 134

Gain sharing systems, 223
GE Canada, 114, 314–315
General Motors, 165
  BOC Division, 17
  CPC Division, 17
  Hamtramck plant, 16
  Saturn, 274
General Motors Corporation of Canada, 176, 182, 190
Group technology, 173–175. *See also* Manufacturing cells

High commitment work systems, 144
Horizontal task, 150
Human resource management, 215
  practices, 165
  support, 215–224
Human-centered design, 158

Implementation:
  as a control process, 287, 318
  hypotheses, 283
  interpersonal cooperation, 268

learning, 290
learning transfers across firms, 293
process model, 282
removing discrepancies, 277
Implementing innovation:
  perceived managerial support, 276
  verbal communication, 276
Information processing, 135
Information processing capacity, 136
Infrastructure:
  external institutional constraints, 256
  sophistication, 255–256
Infrastructure design:
  long-run issues, 255–257
  short-run issues, 252–255
Infrastructure knowledge, 249–250
  transfer of, 250–252
Innovation:
  administrative, 257
  awareness, 235
  conceptual framework, 231
  definition, 232
  expectations, 262
  originality level, 237
  performance, 263
  performance gap, 236
  and size, 232
  stages, 232
  technical, 257
Innovation and infrastructure, 266
  compatability, 264
Innovation champion, 233–235
  characteristics, 234
Innovation process, 285
Innovation research:
  comparative approaches, 291
  implications, 290–294
  process-oriented approaches, 291
  time-dependent relationships, 292
Innovation task force, participation, 254
Integrate:
  between spheres, 126
  within spheres, 126
Integration, requirements, 126
Integration and coordination, 190–192
Integration hypothesis, 166
Integrative approach, 305
Integrative approaches, 16–17, 118, 197, 207
Integrator, 131
Intensification of labor, 161
Interdependence, 125, 128
  and coordination, 124–128
  pooled, 125
  reciprocal, 125, 128

Interdependence (*continued*)
    sequential, 125
    technological, 112
Interdependencies, 112
    internal, 113
Internal diffusion, 280
International Motor Vehicle Program, 164–168
Interorganizational relations:
    control, 296
    innovation, 294–299
    uncertainty, 296
Islands of automation, 127, 142

Job characteristics, 210
Job design:
    models, 210–212
    and new skills, 209–210
Job rotation, 127, 178
Joint optimization, 145, 146
Justification:
    AMT, 49
    and flexibility, 48
    capital, 48
    concept, 48
    evaluation, 62
    experimental AMT, 63, 288
Justification methods:
    modified discounted cash flow, 54–55
    multiattribute approaches, 58–60
    simulation, 55–57
    strategic approach, 61

Lateral relationships, 141
Lead users, 74
Learning strategy:
    adaption to new technology, 278
    benign technological adjustments, 278
    malign technological adjustments, 276
    reduce expectations, 279
Level of analysis, 112, 120

Machining cell, 126
Mail study, 87–90
Manufacturing, focused, 29
Manufacturing cells, 117, 167, 170, 173, 175. *See also* Group technology
Manufacturing flexibility, 168
Manufacturing sign-off, 131
Manufacturing technology, integration of, 191
Material handling, 178–180
Minimal critical specifications, 155
Multiskilled workers, 154
Multiskilling, 127, 154, 160, 212
Mutual adjustment, 128

Net present value, 54–57
New paradigm of work, 143, 161, 314
New technology:
    abstract events, 189
    continuous events, 189
Nonroutine activities, 128
NUMMI, 165

Open system, 145
Operator:
    control, 208
    knowledge and experience, 187
    tasks, 187
Operators, arousal state, 190
Organizational arrangements, flexible, 118
Organizational choice, 144–146
Organizational conflict, 318
Organizational forms, 132, 133, 137
    divisional, 134
    functional, 133
    integrative, 139–140
    network organizations, 137–139
    project management/matrix, 134–137
Organizational lag, 265, 318
Organizational philosophy, 153
Organizational structure, 121. *See also* Structure
    alternative, 119
Organizational values, 153
Other discrepancies, 71–72

Parallel arrangements, 179
Parallelization, 116, 175
Parallel line, 175
Pay for knowledge system, 155
Philosophies, new work, 113
Pilot study, 82–87
Plants-within-plants, 31, 140, 171, 183
Pratt and Whitney Canada, 172, 316
Preparation, 248
Preparation process (hypotheses), 259
Primary work system, 144, 147–149
Principle of incompletion, 155
Problem identification:
    interorganizational, 238
    within organizations, 238
Process manager, 287, 320
Product-focused forms, 136, 168–184, 185
    hierarchy, 169–171
Product-focused units, 116
Product innovation, 100
Production, flow line, 45
Product-process design department, 132
Product-process matrix, 102

SUBJECT INDEX   393

Product shops, 170, 184
Product variety, 31

Quality circles, 171
Quality of working life, 144
   redefinition, 23

Redesign suggestions, 320
Reducing conflict:
   evolution, 272
   isolation, 273
   methods, 271–276
   participation, 273
   persuasion, 272
   revolution, 271
Resistance to change, 318
Reskilling, 192, 193
Reward systems, 222–224
Robust/buffered, 166
Role changes in AMT environments, 200–206
Roles:
   computer programming, 201
   computer specialists, 204
   draftspersons, 205
   electricians, 201
   first-level management team, 203
   maintenance workers, 187, 201–202
   operators, 187, 200–201, 208
   predictive maintenance, 202
   process/shift team coordinator, 203
   professional engineers and managers, 205–206
   shift supervisor/shift superintendent, 203
   supervisors, 202–204
   technicians and computer specialists, 204–205
   unit/zone manager or coordinator, 203
Routine operations, 188
Routinization, 280

Saab, 176
Scientific management, 161, 163
Selection, 216
Self-regulation, *see* Work groups
Short lead times, 31
Short-term orientation, 238–240
   direct impacts, 240–241
   indirect impacts, 242–243
Simultaneous engineering, 35
Skill-based pay system, 155, 222–223
Skills:
   attentiveness, 199
   cognitive understanding, 194
   conceptual, 194
   conceptual thinking, 196
   individual responsibility, 199

   intellective, 196
   motor, 194
   nature of, 194–200
   programming, 187
   statistical inference, 198
   understanding of process phenomena, 197–198
   verbal communication, 198
   visualization, 195–196
Small factory, 183
Sociotechnical principles, 152
Sociotechnical systems, 141
   critiques, 159
   design, 145–147
   design principles, 152–156
   design processes, 156–159
   methodology, 35
   theory, 142–145
Specialist control, 208
Speed of response, 299
Steering committee, 156, 172
Stiga AB, 179
Strategic considerations, 306
Strategic management, 317
Strategy, organization and innovation, 321
Structure:
   action perspective, 122
   configuration of design parameters, 122
   determinant, 123
   organizational, 120–122
   resistance, 124
   and structuring, 120, 122
   systems view, 121
   and technology, 122–124
Study teams, 156
Supervisory roles, 152
Support congruence, 156
Support teams, 173
System designers, AMT, 206

Teams, *see also* Work Groups
   cross-functional, 131
   self-regulating, 117
   structures, 116
   work, 212–214
Technical complexity, 266, 297
   and infrastructure, 289, 291
Technological approach, 16, 22, 304
Technological change, 313
Technologies, computer-based manufacturing, 189
Technology:
   automating, 197
   deterministic, 113, 130, 142, 164, 188
   flexible, 118

Technology (*continued*)
  group, 116
  informating, 197
  stochastic, 113, 188
  structure, 120, 122
  and workers, 207–208
Technology and infrastructure, 276
Technology-centered design, 189
Technology-structure relationship, 120, 133
Texas Instruments' OST System, 288
Time orientation, strategic management, 288
Total productive maintenance, 118
Toyota, 165
Training:
  automated manufacturing technologies, 216–222
  content programs, 219
  program design, 221–223
  skills, 217
  sources, 218–219
  time, 220–221
  for whom, 219–220
Trust and confidence, 138

Uncertainty, 111, 307
  effects on innovation, 263
  breakdowns, 188
  and change, 286
  environmental, 128
  financial, 231
  implications, 303
  innovation capabilities, 267
  and interdependence, 111–114
  market, 23
  new equipment generated, 231
  process, 113, 117–120
  product, 116–117
  product and process, 116
  product-market, 111
  reduction, 23, 94–95

  social, 231
  structural adaption to, 114–116
  technical, 231
  unpredictable demands, 115
Uncertainty, strategy and flexibility, 27–28
Uncertainty avoidance, 317
Union-management relationships and AMT, 224
User-involvement, 158

Value-added chains, 68
Value chain, 138
Values, changing, 113
Variances:
  analysis, 152
  control, 21
  key, 152
  nonroutine, 189
  system induced, 206
  unpredictable and unanticipated, 206
Vertical slice, 158
Vertical task, 115, 150
Visualization, data-based reasoning, 195
Volvo:
  Kalmar, 177
  Tuve, 183
  Uddevalla, 140, 183

Westinghouse Canada, 174
Wildemann's approach, 59–60
Work and technology, changing relationship, 187
Work design, 206–215
Work groups, 149–150, 171–173. *See also* Teams
  autonomous, 151
  interdisciplinary design, 158
  self-directed, 150
  self-maintaining organizational units, 150
  self-managing, 150
  self-regulating, 120, 150, 172
  semi-autonomous, 149–150, 171